LONDON MATHEMATICAL SOCIETY LECTURE NOTE SERIES

Managing Editor: Professor N.J. Hitchin, Mathematical Institute,
University of Oxford, 24–29 St. Giles, Oxford OX1 3LB, United K

The titles below are available from booksellers, or from Cambridge Universit

161 Lectures on block theory, BURKHARD KÜLSHAMMER
163 Topics in varieties of group representations, S.M. VOVSI
164 Quasi-symmetric designs, M.S. SHRIKANDE & S.S. SANE
166 Surveys in combinatorics, 1991, A.D. KEEDWELL (ed)
168 Representations of algebras, H. TACHIKAWA & S. BRENNER (eds)
169 Boolean function complexity, M.S. PATERSON (ed)
170 Manifolds with singularities and the Adams-Novikov spectral sequence, B. BOTVINNK
171 Squares, A.R. RAJWADE
172 Algebraic varieties, GEORGE R. KEMPF
173 Discrete groups and geometry, W.J. HARVEY & C. MACLACHLAN (eds)
174 Lectures on mechanics, J.E. MARSDEN
175 Adams memorial symposium on algebraic topology 1, N. RAY & G. WALKER (eds)
176 Adams memorial symposium on algebraic topology 2, N. RAY & G. WALKER (eds)
177 Applications of categories in computer science, M. FOURMAN, P. JOHNSTONE & A. PITTS (eds)
178 Lower K-and L-theory, A. RANICKI
179 Complex projective geometry, G. ELLINGSRUD *et al*
180 Lectures on ergodic theory and Pesin theory on compact manifolds, M. POLLICOTT
181 Geometric group theory I, G.A. NIBLO & M.A. ROLLER (eds)
182 Geometric Group Theory II, G.A. NIBLO & M.A. ROLLER (eds)
183 Shintani Zeta Functions, A. YUKIE
184 Arithmetical Functions, W. SCHWARZ & J. SPILKER
185 Representations of solvable groups. O. MANZ & T.R. WOLF
186 Complexity: knots, colotigs and counting, D.J.A. WELSH
187 Surveys in combinator. 1993 K. WALKER (ed)
188 Local'analysis for the odd order theorem, H. BENDER & G. GLAUBERMAN
189 Locally presentable and accessible categories, J. ADAMEK & J. ROSICKY
190 Polynomial inwariants of finite groups, D.J. BENSON
191 Finite geometry and combinatorics, F. DE CLERCK *et al*
192 Symplectic geometry, D. SALAMON (ed)
194 Independent random variables and rearrangment invariant spaces, M. BRAVERMAN
195 Arithmetic of blowup algebras, WOLMER VASCONCELOS
196 Microlocal analysis for differential operators, A. GRIGIS & J. SJÖSTRAND
197 Two-dimensional homotopy and combinatorial group theory, C. HOG-ANGELONI *et al*
198 The algebraic characterization of geometric 4-manifolds, J.A. HILLMAN
199 Invariant Potential theory in rhe unit ball of C^n, MANFRED STOLL
200 The Grothendieck theory of dessins d'enfant, L. SCHNEPS (ed)
201 Singularities, JEAN-PAUL BRASSELET (ed)
202 The technique of pseudodifferential operators, H.O. CORDES
203 Hochschild cohomology of von Neumann algebras, A. SINCLAIR & R. SMITH
204 Combinatorial and geometric group theory, A.J. DUNCAN, N.D. GILBERT & J. HOWEB (eds)
205 Ergodic theory and its connections with harmonic analysis, K. PETERSEN & I. SALAMA (eds)
207 Groups of Lie type and their geometries, W.M. KANTOR & L. DI MARTINO (eds)
208 Vector bundles in algebraic geometry, N.J. HITCHIN, P. NEWSTEAD & W.M. OXBURY (eds)
209 Arithmetic of diagonal hypersutrfaces over finite fields, Q. GOUVÉA & N. U
210 Hilbert C*-modules, E.C. LANCE
211 Groups 93 Galway / St Andrews I, CM. CAMPBELL *et al* (eds)
212 Groups 93 Galway / St Andrews II, C.M. CAMPBELL *et al* (eds)
214 Generalised Euler-Jacobi inversion formula and asymptotics beyond all orders, V. KOWALENKO *el al*
215 Number theory 1992–93, S. DAVID (ed)
216 Stochastic partial differential equations, A. ETHERIDGE (ed)
217 Quadratic form wirh applications to algebraic geometry and topology, A. PFISTER
218 Surveys in-combinatorics, 1995, PETER ROWLINSON (cd)
220 Algebraic set theory, A. JOYAL & I. MOERDIJK
221 Harmonic approximation, S.J. GARDINER
222 Advances in linear logic, J.-Y. GIRARD. Y. LAFONT & L. REGNIER (eds)
223 Analytic semigroups and semilinear initial boundary value problems, KAZUAKI TAIRA
224 Comrutabilitv, enumerability, unsolvability, S.B. COOPER, T.A. SLAMAN & S.S. WAINER (eds)
225 A mathematical introduction to string theory, S. ALBEVERIO *et al*
226 Novikov conjectures, index theorems and rigidity I. S. FERRY, A. RANICKI & 1. ROSENBERG (eds)
227 Novikov conjectures, index theorems and rigidity II, S. FERRY, A. RANICK1 & J. ROSENBERG (eds)
228 Ergodic theory of Z_d actions, M. POLLICOTT & K. SCHMIDT (eds)
229 Ergodicity for infinite dimensional systems, G. DA PRATO & J. ZABCZYK
230 Prolegomena to a middlebrow arithmetic of curves of genus 2, J.W.S. CASSELS & E.V. FLYNN
231 Semigmup theory and its applications. K.H. HOFMANN & M.W. MISLOVE (eds)
232 The descriptive set theory of Polish group actions, H. BECKER & A.S. KECHRIS
233 Finite fields and applications, S. COHEN & H. NIDERREITER (eds)
234 Introduction to subfactors, V. JONES & V.S. SUNDER
235 Number theory 1993–94. S. DAVID (ed)
236 The James forest, H. FETTER & B. GAMBOA DE BUEN

237 Sieve methods, exponential sums, and their applications in number theory, G.R.H. GREAVES *et al*
238 Representation theory and algebraic geometry, A. MARTSINKOVSKY & G. TODOROV (eds)
240 Stable groups, FRANK 0. WAGNER
241 Surveys in combinatorics, 1997, R.A. BAILEY (ed)
242 Geometric Galois actions I, L. SCHNEPS & P. LOCHAK (eds)
243 Geometric Galois actions II, L. SCHNEPS & P. LOCHAK (eds)
244 Model theory of groups and automrphism groups, D. EVANS (ed)
245 Geometry, combinatorial designs and related structures, J.W.P HIRSCHFELD *et al*
246 p-Automorphisms of Finite-groups, E. KHUKHRO
247 Analytic number theory, Y. MOTOHASHI (ed)
248 Tame topology and o-minimal structures, LOU VAN DEN DRIES
249 The atlas of finite gmups: ten years on, ROBERT CURTIS & ROBERT WILSON (eds)
250 Characters and blocks of finite groups, G. NAVARRO
251 Groner bases and applications, B. BUCHBERGER & E WINKLER (eds)
252 Geometry and cohomology in group theory, P. KROPHOLLER, G. NIBLO, R. STÖHR (eds)
253 The q-Schur algebra, S. DONKIN
254 Galois representations in arithmetic algebraic geometry, A.J. SCHOLL & R.L. TAYLOR (eds)
255 Symmetries and integrability of difference equations, P.A. CLARKSON & F.W. NIJHOFF (eds)
256 Aspects of Galois theory, HELMUT VOLKLBIN *et al*
257 An introduction to nancommutative differential geometty and its physical applications 2ed, J. MADORE
258 Sets and proofs, S.B. COOPER & J. TRUSS (eds)
259 Models and computability, S.B. COOPER & J. TRUSS (eds)
260 Groups St Andrews 1997 in Bath, I, CM. CAMPBELL *et al*
261 Groups St Andrews 1997 in Bath, II, C.M. CAMPBELL *et al*
262 Analysis and logic, C.W. HENSON, J. IOVINO, A.S. KECHRIS & E. ODELL
263 Singularity theoy, BILL BRUCE & DAVID MOND (eds)
264 New trends in algebraic geometry, K. HULEK, F. CATANESE, C. PETERS & M. REID (eds)
265 Elliptic curves in cryptography, I. BLAKE, G. SEROUSSI & N. SMART
267 Surveys in combinatorics, 1999, J.D. LAMB & D.A. PREECE (eds)
268 Spectral asymptatics in the semi-classical limit, M. DIMASSI & J. SJÖSTRAND
269 Ergodic theory and topological dynamics, M.B. BEKKA & M. MAYER
270 Analysis on Lie groups, N.T. VAROPOULOS & S. MUSTAPHA
271 Singular perturbations of differential operators, S. ALBEVERIO & P. KURASOV
272 Character theory for the odd order theorem, T. PETERFALVI
273 Spectral theory geometry, E.B. DAVIES & Y. SAFAROV (eds)
274 The Mandlebrot set, theme and variations, TAN LEI (ed)
275 Descriptive set theory and dynamical systems, M. FOREMAN *et al*
276 Singularities of plane curves, E. CASAS-ALVERO
277 Computatianal and geometric aspects of modern algebra, M.D. ATKINSON *et al*
278 Glbal attractors in abstract parabolic problems, J.W. CHOLEWA & T. DLOTKO
279 Topics in symbolic dynamics and applications, F. BLANCHARD, A. MAASS & A. NOGUEIRA (eds)
280 Chamctw and automorphism groups of compact Riemann surfaces, THOMAS BREUER
281 Explicit birational geometry of 3-folds, ALESSIO CORTI & MILES REID (eds)
282 Auslander-Buchweitz approximations of equivariant modules, M. HASHIMOTO
283 Nonlinear elasticity, Y. FU & R.W. OGDEN (eds)
284 Foundations of computational mathematics, R. DEVORE, A. ISERLES & E. SÜLI (eds)
285 Rational points on curves over finite fields, H. NIEDERREITER & C. XING
286 Clifford algebras and spinors 2ed, p. LOUNESTO
287 Topics on Riemann surfaces and Fuchsian groups, E. BUJALANCE, A.F. COSTA & E. MARTINEZ (eds)
288 Surveys in cambinatorics, 2001, J. HIRSCHFELD (ed)
289 Aspects of Sobolev-type inequalities, L. SALOFF-COSTE
290 Quantum groups and Lie theory, A. PRESSLEY (ed)
291 Tits buildings and the model theory of groups, K. TENT (ed)
292 A quantum groups primer, S. MAJID.
293 Second order partial diffetcntial equations in Hilbett spaces, G. DA PRATO & I. ZABCZYK
294 Imwduction to the theory of operator spaces, G. PISIER
295 Geometry and integrability, LIONEL MASON & YAVUZ NUTKU (eds)
296 Lectures on invariant theory, IGOR DOLGACHEV
297 The homotopy category of simply connected 4-manifolds, H.-J. BAUES
299 Kleinian groups and hyperbolic 3-manifolds, Y. KOMORI, V. MARKOVIC, & C. SERIES (eds)
300 Introduction to Möbius differential geometry, UDO HERTRICH-JEROMIN
301 Stable modules and the D(2)-problem, F.E.A. JOHNSON
302 Diicrete and continuous nonlinear Schrödinger systems, M.J. ABLOWITZ, B. PRINARI, & A.D. TRUBATCH
303 Number theory and algebraic geometry, MILES REID & ALEXEI SKOROBOOATOV (eds)
304 Groups St Andrews 2001 in Oxford Vol. I, COLIN CAMPBELL, EDMUND ROBERTSON
305 Groups St Andrews 2001 in Oxford Vol. 2, C.M. CAMPBELL, E.F. ROBERTSON & G.C. SMITH (eds)
307 Surveys in combinatorics 2003, C.D. WENSLEY (ed)
308 Topology, Geometry and Quantum Field Theory, U. TILLMAN (ed)
309 Corings and comodules, TOMASZ BRZEZINSKI & ROBERT WISBAUER
310 Topics in dynamics and ergodic theory, SERGEY BEZUGLYI & SERGIY KOLYADA (eds)
311 Groups, T.W. MÜLLER
312 Foundations of computational mathematics, Minneapolis 2002, FELIPE CUCKER *et al* (eds)
313 Transcendental Aspects of Algebraic Cycles, S. MÜLLER-STACH & C. PETERS (eds)

London Mathematical Society Lecture Note Series. 324

Surveys in Combinatorics 2005

Edited by

B. S. WEBB
The Open University

CAMBRIDGE
UNIVERSITY PRESS

CAMBRIDGE UNIVERSITY PRESS
Cambridge, New York, Melbourne, Madrid, Cape Town, Singapore, São Paulo

Cambridge University Press
The Edinburgh Building, Cambridge CB2 2RU, UK

www.cambridge.org
Information on this title: www.cambridge.org/9780521615232

First published 2005

Printed in the United Kingdom at the University Press, Cambridge

A catalogue record for this book is available from the British Library

Library of Congress Cataloguing in Publication data

ISBN-13 978-0-521-61523-2 paperback
ISBN-10 0-521-61523-2 paperback

Contents

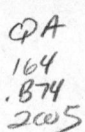

Preface

The Twentieth British Combinatorial Conference was organised jointly by the University of Durham and the Open University. It was held at Durham in July 2005. The British Combinatorial Committee had invited nine distinguished combinatorial mathematicians to give survey lectures in areas of their expertise, and this volume contains the survey articles on which these lectures were based.

In compiling this volume I am indebted to the authors for preparing their articles so accurately and in such a timely manner, and to the referees for their prompt replies and their attention to detail while commenting on the articles. I would also like to thank Roger Astley at Cambridge University Press, and Mike Grannell at the Open University for their advice and help. Finally, without the previous efforts of editors of earlier *Surveys*, my job would have infinitely more difficult!

The British Combinatorial Committee gratefully acknowledges the financial support provided by the London Mathematical Society, the Institute of Combinatorics and its Applications, and from the EPSRC.

Bridget S. Webb
The Open University
b.s.webb@open.ac.uk

February 2005

Finite field models in additive combinatorics

Ben Green

Abstract

The study of many problems in additive combinatorics, such as Szemerédi's theorem on arithmetic progressions, is made easier by first studying models for the problem in $\mathbb{F}_p^n$, for some fixed small prime p. We give a number of examples of finite field models of this type, which allows us to introduce some of the central ideas in additive combinatorics relatively cleanly. We also give an indication of how the intuition gained from the study of finite field models can be helpful for addressing the original questions.

1 Introduction

This article is concerned with a variety of problems in additive and combinatorial number theory. The following two examples will convey the general flavour:

Problem 1.1 (3-term APs) *What is $r_3(N)$, the cardinality of the largest subset of $\{1,\dots,N\}$ containing no three distinct elements $x, x+d, x+2d$ in arithmetic progression?*

Problem 1.2 (Sets with small doubling) *If $A \subseteq \mathbb{Z}$, write $A+A$ for the set of all sums $a+a'$, $a,a' \in A$. What can be said about the structure of A if A is nearly closed under addition in the sense that $|A+A| \leqslant K|A|$?*

What, then, is the "general flavour"? Of course, both of these problems are of an additive combinatorial flavour. Furthermore, they may both be asked in a general abelian group. Regarding Problem 1.1, we may define the quantity $r_3(G)$ for any finite abelian group G. And Problem 1.2 makes sense in any abelian group.

The ability to generalise to an arbitrary G will be a common feature of many of the questions we discuss. An important observation is that not all abelian groups were created equal. It turns out that both Problems 1.1 and 1.2 are both considerably easier in groups other than those in which they were originally asked ($\mathbb{Z}/N\mathbb{Z}$ for Problem 1.1[1] and $\mathbb{Z}$ for Problem 1.2). Indeed, Meshulam [41] observed that Problem 1.1 is naturally addressed in $\mathbb{F}_3^n$, whereas Ruzsa [47] saw that Problem 1.2 is particularly pleasant in $\mathbb{F}_2^\infty$. Here, $\mathbb{F}_p$ denotes the finite field with p elements, and $\mathbb{F}_p^\infty$ is our notation for a vector space of countable dimension over $\mathbb{F}_p$.

Roughly speaking, the reason that finite field models are nice to work with is that one has the tools of linear algebra, including such notions as *subspace* and *linear independence*, which are unavailable in general abelian groups.

Historically, questions such as Problems 1.1 and 1.2 were investigated in their original settings, and it was observed only later that analogous arguments worked in the finite field setting and in fact looked rather simpler. More recently, there has been a trend in the opposite direction. This has been fuelled by an idea of Bourgain [10] which, suitably interpreted, can be viewed as a way of converting

[1] In many questions, the difference between $\mathbb{Z}/N\mathbb{Z}$ and $\{1,\dots,N\}$ is purely technical.

arguments in the finite field setting to arguments which work for an arbitrary group G by using a kind of "approximate linear algebra". The author [26] produced a result about sets of integers with few solutions to $x + y = z$ which would have been very difficult to attain without first considering a finite field model, and more work of this sort is in progress. It is an interesting feature of many problems that progress for the groups G which are "of interest", such as $\mathbb{Z}/N\mathbb{Z}$, is scarcely simpler than for general abelian G.

The format of this article is as follows. After setting up a little notation and a few definitions, we will discuss a number of finite field problems of "Szemerédi type", that is to say along the lines of Problem 1.1. We will strive for a uniform treatment of three such problems: 3-term APs (§4), right-angled triangles (§5) and 4-term APs (§6). We will discuss a fourth problem in §7, which concerns solutions to $x + y = z$ and is in a somewhat similar spirit.

After these four sections we will, in §8, sketch an argument of Bourgain, which is currently being developed by the author and others, including T.Tao and I. Shkredov, into a machine for converting arguments in the finite field setting into arguments that work in any finite abelian group G. This is often of some interest when $G = \mathbb{Z}/N\mathbb{Z}$, because in that case it is often possible to infer results concerning the integers.

After that there follow three further sections of a somewhat miscellaneous nature dealing with finite field analogues of problems in additive number theory.

Since this is a survey article we have not gone into a great deal of technical detail. There are, however, two areas we discuss which are not well covered in the literature. Thus on the author's webpage one may find two supplementary documents [29, 30]. The first of these gives details of the finite field version of Shkredov's argument, which is outlined in §5. The second supplies proofs for the result of Ruzsa discussed in §10.

Our scope in this article is a little limited, in that our main interest is in additive combinatorial problems which can be usefully studied in $\mathbb{F}_p^n$ for fixed p, regarding n as a variable parameter. Secondly, I have unashamedly prioritised areas in which I have personally worked. There are most assuredly other areas of mathematics where finite field models have proved invaluable, such as the study of the Kakeya and restriction phenomena. We do not touch upon these matters here, referring the reader instead to the article [42] as well as in the surveys [37, 58, 59].

2 Notation and Basic Definitions

Let p be a prime (p will be either 2, 3 or 5). Write $\mathbb{F}_p$ for the finite field with p elements, which may be identified with $\mathbb{Z}/p\mathbb{Z}$, and for an integer $n \geqslant 1$ write $\mathbb{F}_p^n$ for a vector space of dimension n over $\mathbb{F}_p$. This will be understood to have been given to us with a fixed basis $(e_1, \dots, e_n)$, relative to which we will occasionally write a given $x \in \mathbb{F}_p^n$ as a coordinate vector $(x_1, \dots, x_n)$. We will always write $N = p^n$ for the cardinality of the space $\mathbb{F}_p^n$.

Once we have a basis the *Fourier transform* of a function $f : \mathbb{F}_p^n \to \mathbb{C}$ can be written down in a concrete form. A complete set of characters $\gamma : \mathbb{F}_p^n \to S^1$ is given by the maps γ_ξ, defined by

$$\gamma_\xi(x) = \gamma_{\xi_1,\dots,\xi_n}(x) = \omega^{\xi_1 x_1 + \cdots + \xi_n x_n} = \omega^{\xi^T x},$$

where $\xi \in \mathbb{F}_p^n$ and $\omega = e^{2\pi i/p}$. Thus, for any $\xi \in \mathbb{F}_p^n$, we define

$$\widehat{f}(\xi) := \sum_x f(x)\gamma_\xi(x) = \sum_x f(x)\omega^{\xi^T x}.$$

We may also write this as $f^\wedge(\xi)$ on occasion. The basic facts concerning the Fourier transform are summarised in the following lemma.

Lemma 2.1 (The Fourier Transform) *Let $f, g : \mathbb{F}_p^n \to \mathbb{C}$ be two functions. Then*

1. $\widehat{f}(0) = \sum_x f(x)$;

2. (Plancherel) $\sum_x f(x)\overline{g(x)} = N^{-1}\sum_\xi \widehat{f}(\xi)\overline{\widehat{g}(\xi)}$;

3. (Inversion) $f(x) = N^{-1}\sum_\xi \widehat{f}(\xi)\omega^{-\xi^T x}$;

4. (Convolution) *Write* $(f * g)(x) = \sum_y f(y)g(x-y)$. *Then* $(f*g)^\wedge(\xi) = \widehat{f}(\xi)\widehat{g}(\xi)$.

Very often, we will be concerned with functions f which are the characteristic functions of sets $A \subseteq \mathbb{F}_p^n$. It is very convenient to abuse notation and write $A(x)$ for such a function. Thus $A(x) = 1$ if $x \in A$, and $A(x) = 0$ otherwise. This notation is by now reasonably widespread in the literature, as are alternative notations such as χ_A or $\mathbf{1}_A$.

It will be very convenient to use the language of conditional expectation. Suppose that x is a variable or set of variables, and that f is a real-valued function of x. Then we write

$$\mathbb{E}(f(x)|x \in B) := |B|^{-1}\sum_{x\in B} f(x)$$

for the average of $f(x)$ over all $x \in B$.

3 Uniformity

A notion which will feature repeatedly in this article is that of *uniformity*, also referred to in various related guises as regularity, pseudorandomness or quasirandomness.

Definition 3.1 *Let $A \subseteq \mathbb{F}_p^n$ be a set, and let $\eta \in (0,1)$ be a parameter. We will say that A is η-uniform if*

$$\sup_{\xi \neq 0} |\widehat{A}(\xi)| \leqslant \eta N.$$

Observe that if A is η-uniform then it is also η'-uniform for all $\eta' \geqslant \eta$.

The basic philosophy behind this definition is as follows. A truly random set A (generated, say, by including each $x \in \mathbb{F}_2^n$ in A independently at random with probability 1/2) will be η-uniform with very high probability. In fact, using a large deviation estimate such as Chernoff's bound (see [5] for example) one can show that this is true even for $\eta = N^{-1/2+\epsilon}$. A truly random set will have many other properties almost surely. Remarkably, many of these are consequences of A being η-uniform. This phenomenon was investigated in the context of graphs by Thomason [62, 63] and by Chung, Graham and Wilson [14]. Chung and Graham [13] later defined quasi-randomness for subsets of $\mathbb{Z}/N\mathbb{Z}$. Quasirandomness has been most thoroughly explored in the context of graphs, for which the reader should consult the excellent survey articles [38, 39]. The notions of uniformity in $\mathbb{Z}/N\mathbb{Z}$ and in $\mathbb{F}_p^n$ differ in little more than notation.

As an example of uniformity/quasirandomness at work, and to get comfortable with the notation, let us prove that uniformity is more-or-less equivalent to a combinatorial condition involving $M(A)$, the number additive quadruples in A (solutions to $a_1 + a_2 = a_3 + a_4$, $a_i \in A$).

Lemma 3.2 *Let $A \subseteq \mathbb{F}_p^n$ have cardinality αN.*

1. *Suppose that A is η-uniform. Then $M(A) \leqslant (\alpha^4 + \eta^2\alpha)N^3$.*
2. *Suppose that $M(A) \leqslant (\alpha^4 + \epsilon)N^3$. Then A is $\epsilon^{1/4}$-uniform.*

Remark An easy application of the Cauchy-Schwarz inequality confirms that $M(A) \geqslant \alpha^4 N^3$, so this lemma concerns sets with close to the minimum number of additive quadruples.

Proof The proof of this Lemma rests on the identity

$$M(A) = N^{-1}\sum_{\xi}|\widehat{A}(\xi)|^4,$$

which may be proved by observing that $M(A) = \sum_x (A * A)(x)^2$ and using Lemma 2.1 (2) and (4). To prove (1), assume that A is η-uniform, so that $|\widehat{A}(\xi)| \leqslant \eta N$ for all $\xi \neq 0$. Then we have

$$\begin{aligned} NM(A) &= |\widehat{A}(0)|^4 + \sum_{\xi\neq 0}|\widehat{A}(\xi)|^4 \ \leqslant\ |A|^4 + \sup_{\xi\neq 0}|\widehat{A}(\xi)|^2\sum_{\xi}|\widehat{A}(\xi)|^2 \\ &= \alpha^4 N^4 + \sup_{\xi\neq 0}|\widehat{A}(\xi)|^2\cdot\alpha N^2 \ \leqslant\ (\alpha^4 + \eta^2\alpha)N^4, \end{aligned}$$

as required. To prove (2), assume that $M(A) \leqslant (\alpha^4 + \epsilon)N^3$. Then for any $\xi \neq 0$ one has

$$|\widehat{A}(\xi)|^4 \ \leqslant\ \sum_{\xi}|\widehat{A}(\xi)|^4 - |\widehat{A}(0)|^4 \ =\ NM(A) - |A|^4 \ \leqslant\ \epsilon N^3,$$

which is what we wanted to prove. □

We observe that if $A \subseteq \mathbb{F}_p^n$, and if $H + g$ is a coset of some subspace $H \leqslant \mathbb{F}_p^n$, then there is a natural notion of what it means for A to be η-regular relative to $H + g$. Indeed we may define a set $A_H^{+g} \subseteq H$ by setting

$$A_H^{+g}(x) \ =\ A(x+g)$$

for $x \in H$. Since H is a subgroup, it is isomorphic to $\mathbb{F}_p^m$ for some $m \leqslant n$ and it makes sense to talk about the Fourier transform on H. We say that A is η-uniform on $H+g$ if A_H^{+g} is η-uniform, considered as a subset of H.

The key reason for uniformity being so important to us in the present survey is that it allows us to count solutions to certain linear equations in sets which are sufficiently uniform. Lemma 3.2 was of course a rather special example of this (the linear equation being $a_1+a_2=a_3+a_4$). The next proposition illustrates this further.

Proposition 3.3 *Let p be a prime and suppose that $A \subseteq \mathbb{F}_p^n$. Suppose that $\lambda_1, \ldots, \lambda_k$, $k \geqslant 3$, are non-zero integers coprime to p. Let $H \leqslant \mathbb{F}_p^n$ be a subspace, and let $g_1, \ldots, g_k \in \mathbb{F}_p^n$ satisfy $\sum \lambda_i g_i = 0$. Suppose that the density of A on $H+g_i$ is α_i, and that A is η-uniform. Then M, the number of solutions to $\sum \lambda_i a_i = 0$ with $a_i \in H + g_i$ for $i = 1, \ldots, k$, satisfies*

$$|M - \alpha_1 \ldots \alpha_k |H|^{k-1}| \leqslant \eta^{k-1}(\alpha_1 \ldots \alpha_k)^{1/k}|H|^{k-1}.$$

Proof With the notation introduced above we can write

$$M = \sum_{\substack{h \in H \\ \sum \lambda_i h_i = 0}} A_H^{+x_1}(h_1) \ldots A_H^{+x_k}(h_k).$$

This can be written in terms of the Fourier transform on H as

$$M = |H|^{-1} \sum_{\xi} \widehat{A_H^{+x_1}}(\lambda_1 \xi) \ldots \widehat{A_H^{+x_k}}(\lambda_k \xi).$$

Separating off the term $\xi = 0$ and bounding the other term using Hölder's inequality, we get

$$\begin{aligned}
|M - \alpha_1 \ldots \alpha_k |H|^{k-1}| &\leqslant |H|^{-1} \sum_{\xi \neq 0} |\widehat{A_H^{+x_1}}(\lambda_1 \xi) \ldots \widehat{A_H^{+x_k}}(\lambda_k \xi)| \\
&\leqslant \sup_{\xi \neq 0} |\prod_j \widehat{A_H^{+x_j}}(\lambda_j \xi)|^{1-2/k} \cdot \prod_j (\sum_{\xi} |\widehat{A_H^{+x_j}}(\lambda_j \xi)|^2)^{1/k} \\
&\leqslant \eta^{k-2}(\alpha_1 \ldots \alpha_k)^{1/k} |H|^{k-1}.
\end{aligned}$$

This concludes the proof. □

Of particular importance to us will be two cases of the above with $k = 3$: $(\lambda_1, \lambda_2, \lambda_3) = (1, 1, -2)$, which corresponds to arithmetic progressions of length 3, and $(\lambda_1, \lambda_2, \lambda_3) = (1, 1, -1)$, corresponding to what are known as *Schur triples* (solutions to $x + y = z$).

A particularly nice feature of finite fields is that the notion of a set $A \subseteq \mathbb{F}_p^n$ being uniform is closely related to that set being well-distributed in cosets of codimension one hyperplanes. We will use this principle several times in the sequel, so let us state and prove a quantitative version of it now.

Lemma 3.4 *Suppose that $A \subseteq \mathbb{F}_p^n$ is a set of size αN ($N = p^n$) and that A is not η-uniform, so that there is $\xi \neq 0$ with $|\widehat{A}(\xi)| > \eta N$. Let $H = \langle \xi \rangle^{\perp}$, and write $h(x) = H(x)/|H|$. Then*

1. $\mathbb{E}(A * h(x)^2) \geqslant \alpha^2 + \eta^2$;

2. $\sup_x A * h(x) \geqslant \alpha + \frac{\eta^2}{\alpha}$;

3. $\sup_x A * h(x) \geqslant \alpha + \eta/2$.

Remark $A * h(x)$ is the density of A on the coset $H + x$.

Proof To prove (1), observe that

$$\begin{aligned} N\sum_x A * h(x)^2 &= \sum_\gamma |\widehat{A}(\gamma)|^2|\widehat{h}(\gamma)|^2 \\ &\geqslant |\widehat{A}(0)|^2|\widehat{h}(0)|^2 + |\widehat{A}(\xi)|^2|\widehat{h}(\xi)|^2 \\ &\geqslant (\alpha^2 + \eta^2)N^2. \end{aligned}$$

Statement (2) is a simple corollary of this:

$$\alpha N \sup_x A * h(x) = \left(\sum_x A * h(x)\right) \sup_x A * h(x) \geqslant \sum_x A * h(x)^2.$$

Statement (3) is proved by working directly with the definition of $\widehat{A}(\xi)$. It leads to somewhat better qualitative bounds than (2).

Let, then, $H + x_j$, $j = 0, 1, \dots, p-1$, be a complete set of cosets of H. Then

$$\widehat{A}(\xi) = \sum_j |A \cap H_j|\omega^j = \sum_j a_j\omega^j,$$

where $a_j = |A \cap H_j| - \alpha|H|$. Thus $\sum_j |a_j| \geqslant \eta N$. Observe, however, that $\sum a_j = 0$; it follows that $\sum_j |a_j| + a_j \geqslant \eta N$, and whence from the pigeonhole principle that $|a_j| + a_j \geqslant \eta N/p$ for some j. For such a j, we have $a_j \geqslant \eta N/2p$. □

4 Roth's Theorem and the iteration method

Let us begin by recalling Problem 1.1.

Problem 1.1 *What is the cardinality of the largest subset of* $\{1, \dots, N\}$ *containing no three distinct elements* $x, x+d, x+2d$ *in arithmetic progression?*

This question was first raised by Erdős and Turán in 1936 [16], and was addressed by Klaus Roth [46]. Define $r_3(N)$ to be the answer to Problem 1.1. Roth proved that $r_3(N) \ll N/\log\log N$, a bound which was improved to $N(\log N)^{-c}$ independently by Heath-Brown [36] and Szemerédi [57], and then to $r_3(N) \ll N(\log\log N/\log N)^{1/2}$ by Bourgain [10]. We are still a long way from a complete understanding of $r_3(N)$; the best known lower bound is Behrend's [6] 1946 example showing that $r_3(N) \gg Ne^{-c\sqrt{\log N}}$.

It is natural to define $r_3(G)$ for any group G with no 2-torsion (though see [40]). A particularly appealing case, which fits with the discussion of this article, is $G = \mathbb{F}_3^n$. In this case it turns out that the four proofs [10, 36, 46, 57] can all be adapted to give the following result.

Theorem 4.1 *We have* $r_3(\mathbb{F}_3^n) \ll N/\log N$ $(= O(3^n/n))$.

In fact, all four proofs look the same in the finite field setting. Roth's proof was adapted to the finite field setting by Meshulam [41] and the argument we give to prove Theorem 4.1 is the same as his.

There are two key ingredients. The first is a special case of Proposition 3.3, asserting that if A is sufficiently uniform then we can count solutions to the equation $a_1 + a_2 = 2a_3$ (that is, arithmetic progressions of length three).

Lemma 4.2 *Suppose that* $A \subseteq \mathbb{F}_3^n$ *has cardinality* αN, *and that* A *is* η*-uniform. Then there are at least* $(\alpha^3 - \eta\alpha)N^2$ *solutions to the equation* $a_1 + a_2 = 2a_3$ *with* $a_i \in A$. *In particular if* $\eta = \alpha^2/2$ *and* $N > 2/\alpha^2$ *then* A *contains a 3-term AP* $(x, x+d, x+2d)$ *with* $d \neq 0$.

Proof The first part is just a matter of setting $H = \mathbb{F}_3^n$ and $(\lambda_1, \lambda_2, \lambda_3) = (1, 1, -2)$ in Proposition 3.3. To verify the second statement, one must simply check that if $\eta = \alpha^2/2$ and $N > 2/\alpha^2$ then $(\alpha^3 - \eta\alpha)N^2$ is greater than αN, the number of "trivial" 3-term APs (x, x, x) in A. □

The second key ingredient is Lemma 3.4 (3), which asserts that if A is not η-uniform then it has increased density on some coset of a hyperplane. In combination with Lemma 4.2 this leads naturally to an *iterative* method for proving Theorem 4.1.

Proof of Theorem 4.1 Set $A_0 = A$, $H_0 = \mathbb{F}_3^n$, $\alpha_0 = \alpha$. For each $i = 0, 1, \dots$ we perform the following algorithm:

- If A_i is $\alpha_i^2/2$-uniform then STOP.
- Otherwise by Lemma 3.4 find a hyperplane $H_{i+1} \leqslant H_i$ and an $x \in H_i$ such that $|A_i \cap (H_{i+1} + x)| > (\alpha_i + \alpha_i^2/4)|H_{i+1}|$. Now set $A_{i+1} = (A_i - x) \cap H_{i+1}$ and set $\alpha_{i+1} = |A_i \cap (H_{i+1} + x)|/|H_{i+1}|$.

Note that if A_i contains a 3-term AP then so does A.

The algorithm cannot be repeated forever, since the sequence $(\alpha_i)_{i=1}^\infty$ satisfies $\alpha_0 = \alpha$ and $\alpha_{i+1} \geqslant \alpha_i + \alpha_i^2/4$ then $\alpha_i > 1$ for $i > 50/\alpha$. Thus we reach a STOP at step K of the algorithm, for some $K < 50/\alpha$. At this stage, A_K is $\alpha_K^2/2$-uniform. If in addition $|H_K| > 2/\alpha_K^2$ then, by Lemma 4.2, A_K contains a 3-term AP. Since $|H_K| = 3^{-K}N > 3^{-50/\alpha}N$ and $\alpha_K > \alpha$, we see that the original set A contains a 3-term AP if $\alpha > C/\log N$ for some C. □

We call the above an *iteration argument* for obvious reasons. We will encounter several such arguments in this survey, so let us take the opportunity to look at the important features of it.

Our concern was with certain *configurations* **Config**, which in this section were the three-term arithmetic progressions $(x, x+d, x+2d)$, $d \neq 0$.

A key feature of the argument was a collection **Struct** of *structures*, which in this case was the collection of all cosets of subspaces of $\mathbb{F}_3^n$. There was also some measure of the *complexity* $\omega(S)$ of a given structure $S \in$ **Struct**, this being the codimension of the subspace. For a given set $A \subseteq \mathbb{F}_3^n$ and for any $S \in$ **Struct** there was a notion of the *density* $\delta_S(A)$ of A relative to S. Finally, there was a *norm* $\|\cdot\|_S$ on functions $f : S \to [-1,1]$, for any $S \in \mathbf{Struct}$ (in the example above, this was the L^∞ norm of the Fourier transform of f, regarded as a function on $\dot{S}$). This we used to define a notion of *uniformity* relative to some $S \in \mathbf{Struct}$; a set was η-uniform if $\|A - \delta_S(A)\|_S \leqslant \eta$.

The "iteration step" of Roth's argument can be presented in the following way. Let $S \in \mathbf{Struct}$, and let $A \subseteq \mathbb{F}_3^n$ be a set with $\delta_S(A) = \alpha$. Then one of the following three alternatives holds:

1. (generalised von Neumann theorem[2]) $\|A - \delta_S(A)\|_S \leqslant \alpha^2/2$, in which case A contains some $c \in \mathbf{Config}$;

2. (density increment) $\|A - \delta_S(A)\|_S > \alpha^2/2$, in which case we may find $S' \in \mathbf{Struct}$, $\omega(S') \leqslant \omega(S) + 1$, such that $\delta_{S'}(A) \geqslant \delta_S(A) + \alpha^2/4$;

3. (endpoint) $|S| < 2/\alpha^2$.

Several subsequent arguments will have the same general form, with different notions of **Struct**, **Config**, ω and $\|\cdot\|_S$. The choice of **Struct** and, perhaps more importantly, of the norm $\|\cdot\|_S$ is vitally important. $\|\cdot\|_S$ must be "strong" enough for us to be able to prove a von Neumann theorem, yet "weak" enough that one may obtain a density increment.

To conclude this section, let use return to the question of estimating $r_3(\mathbb{F}_3^n)$, which I regard as a very interesting one. It seems to dramatically expose our lack of understanding of 3-term arithmetic progressions. There does not seem to be an analogue of Behrend's example in the finite field setting (Behrend's construction makes important use of convexity in $\mathbb{R}^n$). The best known lower bounds on $r_3(\mathbb{F}_3^n)$ come from design theory, where a set in $\mathbb{F}_3^n$ with no 3-term AP is known as a *cap*. Write $f(n)$ for the cardinality of the largest cap in $\mathbb{F}_3^n$. In [15] one finds the estimate

$$\mu(3) := \limsup_{n\to\infty} \frac{\log_3(f(n))}{n} \geqslant 0.724851,$$

which seems to be the best known. In that paper it is stated as an interesting research problem to determine if $\mu(3) = 1$. I believe that this is not so.

Conjecture 4.3 $\mu(3) < 1$. *That is, there is an absolute constant* $\delta > 0$ *such that* $r_3(\mathbb{F}_3^n) \leqslant (3-\delta)^n$.

I would expect any methods used to make progress on this conjecture to assist with the Problem 1.1. At present the best known bound is that given in Theorem 4.1.

[2]This term is one that Tao and I are trying to popularize to emphasise the connection with results in ergodic theory such as [19, Lemma 3.1]. Such results tend to be established using several applications of the Cauchy-Schwarz inequality – see for example [33, §5]. The phrase "key lemma" was used for a related concept in the theory of graph regularity in the excellent survey [39]: now the more descriptive term "counting lemma" is popular (cf. [23, 26, 43]).

5 Right-angled triangles—an argument of Shkredov

In this section we write $V_n = \mathbb{F}_2^n$, and $N = |V_n| = 2^n$.

We are concerned with a sort of two-dimensional generalisation of Problem 1.1:

Problem 5.1 *What is $r_\angle(N)$, the cardinality of the largest subset of $\{1,\dots,N\} \times \{1,\dots,N\}$ containing no* corner $((x,y),(x+d,y),(x,y+d))$, $d \neq 0$*?*

Ajtai and Szemerédi [2] proved that $r_\angle(N) = o(N)$, and various subsequent authors [54, 64] have obtained explicit bounds of the shape $r_\angle(N) \ll N/(\log_* N)^c$. Here $\log_* N$ is the number of times one must take the logarithm of N in order to produce a number less than 2.

Very recently Shkredov [53] produced the first "sensible" bound

$$r_\angle(N) \ll N/(\log\log\log N)^c.$$

In this section we give the finite field version of his argument, in which the details are greatly simplified.

Let G be an abelian group of size N, and consider the collection of corners in $G \times G$, by which we mean triples $((x,y),(x+d,y),(x,y+d))$, $d \neq 0$. Write $r_\angle(G)$ for the cardinality of the largest set $A \subseteq G \times G$ which does not contain any corner.

Theorem 5.2 (Shkredov) *We have $r_\angle(\mathbb{F}_2^n) \ll N^2/(\log\log N)^{1/25}$.*

It is natural to try and use the iteration method, in the form outlined in the previous section. The most naïve attempt at doing this would involve taking **Struct** to be the set of cosets of products $H \times H$, where $H \leqslant \mathbb{F}_2^n$ is a subspace, and the definition of uniformity to be much the same as before. The notion of having no large Fourier coefficients makes perfect sense in $H \times H$. Unfortunately, however, this notion of uniformity is not subtle enough to give good control on the number of corners, essentially because it does not "see" the coordinate structure of $H \times H$. The following example is instructive:

Example 5.3 Let B be a random (and hence highly uniform) subset of V_n with cardinality βN, and let $A \subseteq \mathbb{F}_2^n \times \mathbb{F}_2^n$ be the set $B \times B$. Then A is also highly uniform. The density of A is $\alpha = \beta^2$. A corner in A corresponds to a quadruple of points $(x, x+d, y, y+d) \in B^4$, and we know from Lemma 3.2 that there are roughly $\beta^4 N^3 = \alpha^2 N^3$ such configurations. If A were truly random, however, it would have more like $\alpha^3 N^3$ corners.

The next idea, then, might be to define a somewhat finer notion of uniformity which respects the coordinate structure somewhat more. Using Proposition 3.2 as a guide, we might define $A \subseteq \mathbb{F}_2^n \times \mathbb{F}_2^n$ to be *rectilinearly η-uniform* if the number of configurations $((x,y),(x+d,y),(x,y+e),(x+e,y+e))$ in A^4 is at most $(\alpha^4+\eta)N^4$. Such a notion does, as we will see, give some control on the number of corners in A. Unfortunately passing to a new structure $S \in$ **Struct** on which the density increases is now problematic.

To see why, consider again example (5.3). It is easy to see that A fails to be rectilinearly uniform, but there is no product set $(H+x) \times (H'+x')$, H, H' large subspaces of $\mathbb{F}_2^n$, on which the density of A increases markedly.

Note, however, that in this example there *is* at least some structure on which the density of A increases, and that is the product set $B \times B$ (of course, the density of A on this set is one). This behaviour is more-or-less typical: if a set $A \subseteq \mathbb{F}_2^n \times \mathbb{F}_2^n$ has substantially more than $\alpha^4 N^4$ rectangles then it has increased density on some product $B_1 \times B_2$. This can be proved graph-theoretically by associating to A the bipartite graph Γ_A with vertex sets two copies of $\mathbb{F}_2^n$, an edge xy being deemed to lie in Γ_A precisely if $(x, y) \in A$. A rectangle in A then corresponds to a copy of C_4 in Γ_A, and we are reduced to showing that if Γ_A has substantially more than $\alpha^4 N^4$ copies of C_4 then there are large vertex sets B_1, B_2 such that the edge density of Γ_A restricted to $B_1 \cup B_2$ is much greater than α. Shkredov in effect provides a spectral proof of this statement, which is in the spirit of [18]. A purely combinatorial proof is more traditional, and somewhat simpler – the details may be found in [29].

The discussion of the last paragraph might suggest that we should enlarge **Struct** to include all translates of products $B \times B$. This turns out to be too much of a compromise – one cannot establish a useful generalised von Neumann theorem.

The above discussions motivate Shkredov's main advance, which is an appropriate definition of **Struct**. The definition depends on the global density α of A, a feature which has no analogue in other applications of the iterative method discussed in this paper.

Definition 5.4 *Let $\alpha > 0$. Define* $\mathbf{Struct}_\alpha$ *to consist of all translates of product sets $S = E_1 \times E_2$, where E_1, E_2 are subsets of some $H \leqslant \mathbb{F}_2^n$, $|E_i| = \beta_i|H|$ and E_i is a $(2^{-36}\beta_1^{12}\beta_2^{12}\alpha^{36})$-uniform subset of H for $i = 1, 2$.*

Definition 5.5 *Suppose that $S = E_1 \times E_2$ is a product set, and that $f : S \to [-1, 1]$ is a function. Then we define the* rectangle norm *of f, $\|f\|_S$ by*

$$\|f\|_S^4 = \mathbb{E}\big(f(x, y)f(x', y)f(x, y')f(x', y')|x, x' \in E_1, y, y' \in E_2\big).$$

It is not totally obvious that $\|\cdot\|_S$ is a norm, but this is in fact the case. Let us now look at how the argument fits together, starting with a generalised von Neumann theorem.

Proposition 5.6 (Generalised von Neumann) *Let $S \in$* $\mathbf{Struct}_\alpha$*, so that $\mathcal{S} = E_1 \times E_2$ be a product set, where $E_1, E_2 \subseteq H$, $|E_i| = \beta_i|H|$ and E_i is $(2^{-36}\beta_1^{12}\beta_2^{12}\alpha^{36})$-uniform for $i = 1, 2$. Let $A \subseteq S$ be a set with $\delta_S(A) \geqslant \alpha$. Suppose that and that $\|A - \delta_S(A)\|_\square^4 \leqslant 2^{-8}\alpha^{12}$. Then A has at least $\alpha^3\beta_1^2\beta_2^2N^3/2$ corners.*

The proof of this statement involves a number of applications of Cauchy-Schwarz.

To complement the generalised von Neumann theorem, we must establish a density increment result. The following can be obtained by simple graph theory (or alternatively by spectral methods, as done in [53]).

Proposition 5.7 (Density increment on a product set) *Let $S = E_1 \times E_2$ be a product set, and suppose that $A \subseteq S$ has $\delta_S(A) = \alpha$ and $\|A - \alpha\|_S^4 \geqslant \eta$. Then there are sets $F_i \subseteq E_i$ with $|F_i| \geqslant 2^{-8}\eta|E_i|$ such that the density of A on $S' = F_1 \times F_2$ satisfies $\delta_{S'}(A) \geqslant \alpha + 2^{-14}\eta^2$.*

Remark There is no need to assume that the sets E_1, E_2 are uniform in this proposition.

Proposition 5.7 has a significant deficiency, which means that it cannot be used in combination with Proposition 5.6 to provide an iterative proof of Theorem 5.2. This is that the sets F_1, F_2 which it outputs need not be uniform, and so it is quite possible that $S' \notin \mathbf{Struct}_\alpha$. The following further result is required.

Proposition 5.8 (Uniformising a product set) *Let $\alpha, \tau, \sigma \in (0,1)$ be parameters, and let $S' = F_1 \times F_2$ be a product set in $W \times W$ with $|F_i| = \delta_i N$. Suppose that $A \subseteq S'$ is a set with $\delta_{S'}(A) = \alpha + \tau$, and that*

$$|W| \geqslant \exp(16\sigma^{-2}\delta^{-1}\tau^{-1}). \tag{5.1}$$

Then there is a subspace $W' \subseteq W$, $\dim W' \geqslant \dim W - 8\sigma^{-2}\delta^{-1}\tau^{-1}$ and $t_1, t_2 \in W$ such that if $E_1' = (F_1 - t_1) \cap W'$, $E_2' = (F_2 - t_2) \cap W'$ and $S'' = E_1' \times E_2'$ then

1. *$|S''| \geqslant \delta_1\delta_2\tau|W'|^2/2$;*
2. *E_1', E_2' are 2σ-uniform as subsets of W';*
3. *$\delta_{S''}(A - (t_1, t_2)) \geqslant \alpha + \tau/8$.*

The proof of this theorem also proceeds by a version of the iterative method, and in this sense Skhredov's argument is a sort of double iteration method. The most important content of the proposition is that if $S \subseteq W \times W$ then we may pass to a translate of $W' \times W'$ on which S looks uniform, where $W' \leqslant W$ is a subspace of somewhat large codimension. If this really was our only aim, then we could proceed as follows. Either S is already uniform, or else S has a large Fourier coefficient ξ. In the latter case, S has increased density on some translate of $\xi^\perp$, by Lemma 3.4 (2). $\xi^\perp$ obviously contains a set of the form $W' \times W'$, with W' having codimension at most two. Now simply iterate the argument.

The one further issue is that we also need to keep control of the density of A, which sits inside S. To achieve this it is necessary to *partition* $W \times W$ into pieces which are translates of products $W' \times W'$, such that S is uniform on almost all of them. By a simple pigeonhole argument there must be some piece on which S is uniform, and on which the relative density of A is still quite large. Note that the subspaces W' need not be the same for each piece; this is important from the point of view of obtaining bounds, or else one runs into examples such as that in §9 of [26].

To get this decomposition into pieces one uses the iterative argument with one small modification. At the jth stage of the iteration we will have a collection $\mathcal{C}_j$ of pieces, each being a translate of some product $W' \times W'$. If $c \in \mathcal{C}_j$, write $\delta(c)$ for the relative density of S on the piece c. Our previous proposal was to ensure that $\sup_{c \in \mathcal{C}_j} \delta(c)$ increases at each step of the iteration, this idea having served us well in the past. What one does instead is to increase the L^2 average $\mathbb{E}(\delta(c)^2 | c \in \mathcal{C}_j)$. This can be accomplished by using Lemma 3.4 (1).

Propositions 5.7 and 5.8 together give the requisite density increment result to go with the generalised von Neumann theorem of Proposition 5.6. Thus we can employ an iteration argument. Working out the bounds gives Theorem 5.2.

6 Progressions of Length Four

In this section we give another example of the iterative method at work.

Problem 6.1 *Estimate $r_4(N)$, the cardinality of the largest subset of $\{1, \dots, N\}$ containing no four distinct elements $x, x+d, x+2d, x+3d$ in arithmetic progression?*

This question was, like Problem 1.1, raised by Erdős and Turán in 1936. Szemerédi [56] was the first to show that $r_4(N) = o(N)$. It was not until as recently as 1998 that the first "sensible" upper bound, $r_4(N) \ll N(\log\log N)^{-c}$, was provided by Gowers [21]. Gowers' argument was iterative, like the arguments of §4 and 5.

Of course, one can define $r_4(G)$ for any abelian group G of size N. Recently, T. Tao and the author [34] studied the case $G = \mathbb{F}_5^n$, starting from Gowers' work. Certain features of [21] become rather simpler in this setting, and we were able to run the iterative method quite efficiently, obtaining the following theorem.

Theorem 6.2 (G.–Tao) $r_4(\mathbb{F}_5^n) \ll N(\log N)^{-c}$ *for some* $c > 0$.

Write **Config** for the collection of all four-term progressions in $\mathbb{F}_5^n$. Any hope of proving a generalised von Neumann theorem with the same uniformity norm that we used in §4 is dashed by the following example:

Example 6.3 (Gowers; Furstenberg-Weiss) There is a set $A \subseteq \mathbb{F}_5^n$ with density $1/5$, which is highly uniform, but which does not contain roughly $5^{-4}N^2$ four-term arithmetic progressions.

Proof Let $A = \{x \in \mathbb{F}_5^n : x^T x = 0\}$. Then A certainly has density approximately $1/5$. To see that A is highly uniform, write

$$\widehat{A}(\xi) = \frac{1}{5} \sum_{\lambda \in \mathbb{F}_5} \sum_{x \in \mathbb{F}_5^n} \omega^{\lambda x^T x - \xi^T x} = \frac{1}{5} \sum_{\lambda} \prod_{j=1}^{n} \omega^{\lambda x_j^2 - \xi_j x_j}.$$

If $\lambda \neq 0$ then each term in the product has magnitude $\sqrt{5}$, giving a total contribution of $5^{n/2}$; if $\lambda = 0$ then, provided $\xi \neq 0$, at least one term in the product vanishes. It follows that $\sup_{\xi \neq 0} |\widehat{A}(\xi)| \leqslant 5^{n/2} = \sqrt{N}$.

However, A has roughly $5^{-3}N^2$ progressions of length four. Indeed, since A is so highly uniform we know from Proposition 3.3 that it contains roughly this many progressions of length *three*. However if x, $x+d$ and $x+2d$ all lie in A then $x+3d \in A$ automatically, in view of the easily verified identity

$$x^T x - 3(x+d)^T(x+d) + 3(x+2d)^T(x+2d) - (x+3d)^T(x+3d) = 0. \qquad \square$$

Remark Gowers has shown us an example of a subset of $\mathbb{Z}/N\mathbb{Z}$ which is uniform and has density α, but has many *fewer* than $\alpha^4 N^2$ four-term arithmetic progressions.

Similar examples can be constructed using any quadratic form $q(x) = x^T M x + r^T x + b$ in place of $x^T x$. Remarkably, there are essentially no other examples. We shall formalise this statement in what follows.

Definition 6.4 (Gowers norm) *Let $f : \mathbb{F}_5^n \to [-1,1]$ be a function. Then the Gowers U^3-norm of f, $\|f\|_{U^3}$, is defined by*

$$\begin{aligned}\|f\|_{U^3}^8 &= \mathbb{E}\big(f(x)f(x+a)f(x+b)f(x+c)f(x+a+b)f(x+a+c)\\ &\qquad f(x+b+c)f(x+a+b+c)|x,a,b,c\big). \end{aligned} \tag{6.1}$$

Again, it is not completely obvious that $\|\cdot\|_{U^3}$ is a norm, but this is not to hard to show. The following result is due to Gowers [21]. As with the other generalised von Neumann theorems we have mentioned, the proof involves several applications of the Cauchy-Schwarz inequality.

Theorem 6.5 (Generalised Von Neumann theorem) *Suppose that $A \subseteq \mathbb{F}_5^n$ has density α, and that $\|A - \alpha\|_{U^3} \leqslant \delta^4/100$. Then A has at least $\alpha^4 N^2/2$ four-term arithmetic progressions.*

The next theorem is proved in [34] by adding a single new idea, the so-called "symmetry argument", to the ideas of Gowers [21]. This theorem clarifies the sense in which the "quadratic" examples of Furstenberg and Weiss are in a sense the only ones:

Theorem 6.6 (Gowers; G.–Tao) *Suppose that $\|A-\alpha\|_{U^3} \geqslant \delta$. Then A has quadratic bias, meaning that there is some quadratic form $q(x) = x^T Mx + r^T x + b$ such that A has density at least $\alpha + C(\delta)$ on the zero set $S = \{x : q(x) = 0\}$.*

The reader may note that these two theorems do not, in their present incarnations, dovetail together to give an iteration argument because there is no natural definition of **Struct**. Roughly speaking, Gowers took **Struct** to be the collection of translates of subspaces of $\mathbb{F}_5^n$. It is possible to deduce from the conclusion of Theorem 6.6 that A has increased density on some $S \in$ **Struct**, but unfortunately the codimension $\omega(S)$ might be exceedingly large (perhaps $n - n^{1/100}$). This does not, then, lead to a very efficient iterative argument.

In [34] a much less appetising approach is forced to work, which leads to superior bounds. Roughly, this involves taking **Struct** to be the collection of all *quadratic submanifolds*, our name for an intersection

$$S = \bigcap_{j=1}^{k} \{x : q_j(x) = 0\},$$

where $q_j(x) = x^T M_j x + r_j^T x + b_j$ are quadratic forms. The "roughly" is quite important. We must in fact assume that S is "generic", meaning that the matrices M_j are not too linearly dependent. In practise this means that they satisfy a *rank condition* such as $\mathrm{rk}(\lambda_1 M_1 + \cdots + \lambda_k M_k) \geqslant 10k$ for all possible choices of scalars $\lambda_j \in \mathbb{F}_5$. We also allow our quadratic forms to be defined only on a subspace $W \leqslant \mathbb{F}_5^n$ of not-too-large codimension. This is because of the very useful observation that an arbitrary quadratic submanifold can be made generic after passing to an appropriate subspace W.

Generalising the notion of Gowers U^3-norm to such a setting is straightforward; in fact the definition is the same except that the expectation in (6.1) is taken over S. Proving an analogue of Theorem 6.5 is substantially more involved, but it is possible and reads as follows.

Theorem 6.7 *Let $S \in$* **Struct**. *That is to say, S is a generic quadratic submanifold in some $W \leqslant \mathbb{F}_5^n$, this being the zero set of some k quadratic forms $q_1, \dots, q_k$ on W. Then S has approximately $|S|^3/|W|$ four-term arithmetic progressions.*

1. (Generalised von Neumann theorem) *Suppose that $A \subseteq S$ has density α, and that $\|A-\alpha\|_{U^3(S)} \leqslant \alpha^{20}$. Then A has at least $\alpha^4|S|^3/2|W|$ four-term arithmetic progressions.*

2. (Gowers-type inverse theorem) *Suppose that $\|A - \alpha\|_{U^3(S)} \geqslant \delta$. Then A has quadratic bias, meaning that there is some quadratic form $q_{k+1}(x) = x^T M_{k+1}x + r_{k+1}^T x + b_{k+1}$ such that A has density at least $\alpha + C(\delta)$ on the set $S \cap \{x : q_{k+1}(x) = 0\}$.*

A key feature of the theorem is that the density increment $C(\delta)$ is independent of the number of quadratic forms k. The proof of the theorem is long and somewhat difficult and occupies the bulk of [34].

Theorem 6.7 of course allows one to set up an iteration scheme. If $A \subseteq \mathbb{F}_5^n$ is a set with density α which contains no four-term progressions, then one may find a sequence

$$\mathbb{F}_5^n = S_0 \supseteq S_1 \supseteq S_2 \supseteq \dots$$

of generic quadratic manifolds, defined on subspaces

$$\mathbb{F}_5^n = W_0 \geqslant W_1 \geqslant W_2 \geqslant \dots$$

such that the density of A on S_j is at least $\alpha + jC(\alpha^{20})$. This leads to a contradiction after $C(\alpha^{20})^{-1}$ iterations.

Unfortunately, this still leads to a bound of the shape $r_4(\mathbb{F}_5^n) \ll N(\log\log N)^{-c}$, since we have only been able to establish Theorem 6.7 with a function $C(\delta)$ which behaves like $\exp(-\delta^{-B})$, and this results in a very large number of iterations. We conjecture that a better bound holds, but we cannot prove this even in the less general context of Theorem 6.6. I regard this as one of the key open questions in this area of arithmetic combinatorics.

Conjecture 6.8 (Polynomial Gowers Inverse Conjecture) *Let $f : \mathbb{F}_5^n \to [-1, 1]$ be a function with $\mathbb{E}f = 0$. Suppose that $\|f\|_{U^3} \geqslant \delta$. Then there is a quadratic form q on $\mathbb{F}_5^n$ such that*

$$|\mathbb{E}f(x)\omega^{q(x)}| \gg \delta^C,$$

for some absolute constant C.

We do know this with δ^C replaced by a function of exponential type. An affirmative answer to the PGI conjecture would be implied by an affirmative answer to the Polynomial Freiman-Ruzsa conjecture (PFR), which is discussed in some detail in §10.

Fortunately, for the purposes of obtaining a bound on $r_4(\mathbb{F}_5^n)$ one can get by with a weaker conclusion in Theorems 6.5 and 6.6. In Theorem 6.5, one can obtain a "polynomial" density increment, leading to a much shorter iterative process, by passing to a set of the form $\{x : q(x) = 0\} \cap (W + t)$, where $W \subseteq \mathbb{F}_5^n$ is a subspace. One can allow the codimension of W to be a power of α^{-1}, which is just as well since this is the best bound we have.

7 Szemerédi Regularity in Groups

The object of this section is to state some results and open problems from [26]. The results are slightly different from those in the previous section in the problem addressed is not quite of "Szemerédi type". However what we discuss here is certainly in a similar spirit, being concerned with solutions of linear equations in sets of integers, and can furthermore be interpreted as an application of the iteration method.

We will be somewhat brief: more details can of course be found in the paper [26] itself, which is written from a viewpoint rather similar to that of the present survey.

Szemerédi's regularity lemma is a famous result in graph theory. It can be regarded as structure theorem for all graphs, in the sense that it shows that one can decompose a completely arbitrary graph into a bounded number of pieces, almost all of which are pseudorandom. There are many excellent articles on this topic – see for example [39].

One consequence of Szemerédi's regularity lemma is the following interesting result.[3]

Theorem 7.1 *Let Γ be a graph on N vertices, and suppose that one must remove δN^2 edges from G in order to destroy all triangles in Γ. Then Γ has at least $C_1(\delta)N^3$ triangles, for some $C(\delta) > 0$.*

Put another way, if a graph is almost triangle-free (i.e. contains few triangles) then it can be made truly triangle-free by the removal of a small number of edges.

Our investigations in [26] were motivated by an "arithmetic" question related to the above theorem.

Theorem 7.2 (See [26]) *Let G be an abelian group of size N, and suppose that $A \subseteq G$ is a set. Suppose that one must remove δN elements from A in order to create a sum-free set (that is, a set with no solutions to $x + y = z$). Then A has at least $C_2(\delta)N^2$ Schur triples (triples (x, y, z) for which $x + y = z$).*

This result may be regarded as a structure theorem for sets which are almost sum-free; they can be made truly sum-free by the removal of a few elements.

This theorem is deduced from a result which we call a Szemerédi-type regularity lemma for abelian groups. This result is a perfect example for the present survey, since in the context of a general abelian group it requires substantial preparation to even state the result. When $G = \mathbb{F}_2^n$, however, things are much easier.

Theorem 7.3 (Regularity lemma for $\mathbb{F}_2^n$) *Let $A \subseteq \mathbb{F}_2^n$ be a set, and let $\epsilon > 0$ be a parameter. Then there is a subspace $H \subseteq \mathbb{F}_2^n$ with codimension at most $M(\epsilon)$, and such that A is ϵ-uniform on at least a proportion $1 - \epsilon$ of the cosets of H.*

Let us say, for the rest of this section, that A is ϵ-regular relative to H if it satisfies the conclusion of this theorem.

[3] We have not attributed this result, as it is not clear to us where it was first stated. A slightly weaker result was obtained by Ruzsa and Szemerédi in 1976 [52]. The result is also well-known in the literature concerning "property testing": see, for example, [3].

Let us sketch the deduction of Theorem 7.2 from Theorem 7.3. Suppose that $A \subseteq \mathbb{F}_2^n$ is a set with the property that one must remove at least δN elements from A to leave a set which is sum-free. Apply Theorem 7.3 with $\epsilon = (\delta/10)^3$, giving a subspace H of codimension at most $M(\epsilon)$ such that A is ϵ-uniform for a proportion at least $1-\epsilon$ of the cosets of H. For each coset $H+x$, we ask two questions:

- Is A ϵ-uniform on $H+x$?
- Is the density of A on $H+x$ at least $(2\epsilon)^{1/3}$?

If the answer to either of these questions is *no* then we simply remove all of $A \cap (H+x)$ from A. Let the set remaining after we have asked the above questions for all cosets $H+x$ be called A'. It is easy to see that

$$|A'| > |A| - 10\epsilon^{1/3}N = |A| - \delta N.$$

We claim that A' is sum-free. Indeed, were it not there would be x_1, x_2, x_3 with $x_1 + x_2 = x_3$, such that A is ϵ-uniform and has density $\alpha_i \geqslant (2\epsilon)^{1/3}$ on each $H + x_i$. By Proposition 3.3 this means that K, the number of solutions to $a_1 + a_2 = a_3$ with $a_i \in A \cap (H + x_i)$, satisfies

$$|K - \alpha_1\alpha_2\alpha_3|H|^2| \leqslant \epsilon|H|^2,$$

which means that $K \geqslant \epsilon|H|^2/2$. Thus certainly the number of Schur triples in A is certainly at least $\epsilon|H|^2/2$, which is at least $\epsilon 2^{-2M(\epsilon)-1}N^2$.

The proof of Theorem 7.3 is very much in the spirit of the iterative method. One again takes **Struct** to be the collection of all subspaces $H \leqslant \mathbb{F}_2^n$, but here there is no **Config**. Let $A \subseteq \mathbb{F}_2^n$ be a set, and let $H \in$ **Struct**. We define the L^2-density of A with respect to H by

$$\delta_H(A) = \mathbb{E}\Big(\frac{|A \cap (H+x)|^2}{|H|^2} \mid x \in \mathbb{F}_2^n\Big).$$

In [26] this is called the *index*, and is written $\mathrm{ind}(A; H)$.

The key to the proof is the following lemma (Lemma 2.2 of [26]), which can be proved by elaborating somewhat on the proof of Lemma 3.4 (1).

Lemma 7.4 *Let $\epsilon \in (0, \frac{1}{2})$ and suppose that $H \leqslant \mathbb{F}_2^n$ is a subgroup which is not ϵ-regular for A. Then there is a subgroup $H' \leqslant H$ such that* $\mathrm{codim}(H') \leqslant 2^{\mathrm{codim}(H)}$ *and $\delta_{H'}(A) \geqslant \delta_H(A) + \epsilon^3$.*

Theorem 7.3 is simply a matter of applying Lemma 7.4 iteratively. Since $\delta_H(A) \leqslant 1$ for any H, the number of iterations is no more than $\lceil 1/\epsilon^3 \rceil$.

An unfortunate feature of Theorem 7.3 and its proof is that $M(\epsilon)$ grows like a tower of twos of height ϵ^{-3}. This is because each application of Lemma 7.4 results in an exponentiation of the codimension of H. By adapting a brilliant construction of Gowers [20], which shows that Szemerédi's regularity lemma for graphs must have tower type bounds, we were able to show that $M(\epsilon)$ *must* be at least as bad as a tower of twos of height about $\log(1/\epsilon)$.

We were not able to produce a similar example in the setting of Theorem 7.2.

Problem 7.5 *Find a "reasonable" bound for $C_2(\delta)$, the quantity appearing in Theorem 7.2, or prove that no such bound exists.*

In fact for $G = \mathbb{F}_2^n$ I am not able to exclude the possibility that $C_2(\delta)$ can be a polynomial in δ. This need not be the case for $G = \mathbb{Z}/N\mathbb{Z}$, due to the Behrend example of a large subset of $\{1, \dots, N\}$ containing no 3-term AP. See [26] for a further discussion.

The corresponding for graphs (relating to Theorem 7.1) is also wide open, though again it is known that $C_1(\delta)$ cannot be taken to be polynomial in δ.

8 From Finite Fields to $\{1, \dots, N\}$

We have now seen several examples concerning additive combinatorics in finite fields. However, for many of the problems we have considered it is an analogue in $\{1, \dots, N\}$ or (more-or-less equivalently) in $\mathbb{Z}/N\mathbb{Z}$ which is actually of interest.

In recent years the passage from finite fields to the integers, at least for problems concerning configurations of the type we have been discussing in the last four sections, has started to form into something resembling a theory. This is thanks to the work of Bourgain [10] on finding good bounds for $r_3(N)$.

Bourgain's ideas are developed in detail in his original paper, of course, and have also been discussed in [26] and [60]. In this section we restrict ourselves to a few remarks which illustrate the important points.

Consider the problem of finding a bound for $r_3(G)$ using the iteration method, where G is an abelian group with order N and no 2-torsion. It is not hard to see (essentially by changing the letter ξ to γ in Proposition 3.3) that if $A \subseteq G$ has density α and substantially fewer than $\alpha^3 N^2$ 3-term APs then A has a non-trivial large Fourier coefficient, that is to say

$$\widehat{A}(\gamma) := \sum_x A(x)\gamma(x)$$

has magnitude a large fraction of N for some non-trivial character $\gamma \in G^*$.

It is not immediately clear how to use this information. We can no longer assert that A has increased density on a subspace, because in a general group G there is no such thing as a subspace. What we can show, rather painlessly, is that A has increased density on a translate of a *Bohr set*, that is to say a set of the form $x + B(\{\gamma\}, \epsilon)$, where

$$B(\{\gamma\}, \epsilon) := \{x \in G : |1 - \gamma(x)| \leqslant \epsilon\}.$$

The reader who has followed the various iterative arguments in the last four sections might now suggest that we define **Struct** to be the collection of all Bohr sets $B(\Gamma, \epsilon)$, where $\Gamma = \{\gamma_1, \dots, \gamma_d\}$ is a set of characters and

$$B(\Gamma, \epsilon) := \{x \in G : |1 - \gamma_j(x)| \leqslant \epsilon \qquad \text{for all } j = 1, \dots, d\}.$$

Note that in $\mathbb{F}_3^n$ a Bohr set is the same thing as a subspace when $\epsilon < 1/4$. Such a strategy is clearly not going to be without its difficulties. If $S \in$ **Struct**, it looks as though we are going to have to make some sense of what it means to do Fourier analysis on S. Since $B(\Gamma, \epsilon)$ is not a group, this will certainly not be a trivial matter.

In fact, $B = B(\Gamma, \epsilon)$ is quite a long way from being a group. The homomorphism $(\gamma_1, \ldots, \gamma_d) : G \to \mathbb{T}^d$ carries B into a small d-dimensional box D. If one picks x, x' at random in D, the chance that $x+x' \in D$ is just 2^{-d}. Hence one expects that typically $|B+B| \approx 2^d|B|$, which compares unfavourably with the result $|H+H| = |H|$ which holds if $H \leqslant G$ is a genuine subspace.

We will not, in this survey, go into the details of what we mean by Fourier analysis on B, nor how the large doubling constant of B is unpleasant in this context. We hope the reader will believe us when we say that reducing the doubling constant is a very helpful thing to do.

Bourgain's advance is to consider B not by itself, but together with another Bohr set $B' := B(\Gamma, \epsilon')$, where ϵ' is much smaller than ϵ . Then if $x \in B$ and $x' \in B'$ we have $x + x' \in B(\Gamma, \epsilon + \epsilon')$, a set which ought not to be much larger than B. Thus $|B + B'| \approx |B|$, and we may think of the pair (B, B') as behaving like an approximate group. Roughly speaking, it turns out to indeed be possible to run an iterative argument in which **Struct** is the collection of all such pairs (B, B').

There are a number of further technicalities to be overcome. One interesting one is that our assertion that $B(\Gamma, \epsilon + \epsilon')$ is not much larger than $B(\Gamma, \epsilon)$ is not true in general. Suppose, for example, that $G = \mathbb{F}_5^n$, that the characters in Γ are linearly independent and that $\epsilon < 2\sin(\pi/5), \epsilon + \epsilon' > 2\sin(\pi/5)$. Then $|B(\Gamma, \epsilon)| = 5^{n-d}$, whilst $|B(\Gamma, \epsilon + \epsilon')| = 3^{-d}5^n$. Bourgain circumvents this difficulty by using an averaging argument to show that for a typical ϵ the size of $B(\Gamma, \epsilon)$ is roughly invariant under small perturbations of ϵ. Tao [60] observed that one could also replace Bohr sets by *smoothed* Bohr sets, and then such difficulties go away. I implemented this idea slightly differently in [26], defining the a smoothed Bohr "set" by

$$\widetilde{B}(\Gamma, \epsilon)(x) := \int_0^\infty B(K, t)(x) \frac{e^{-t/\epsilon}}{\epsilon}\, dt.$$

We conclude this section by giving an up-to-date summary of the extent to which the problems of the last four sections have been given Bourgain's treatment. Of course, in the original paper [10] the question of $r_3(G)$ was treated (actually, Bourgain only treats $r_3(N)$ but it is clear that his methods work in an arbitrary G). In [26] the results of §7 are all fully generalised to any finite abelian G, and in particular Theorem 7.2 is proved in this general setting. As regards adapting the methods of §5 to obtain a bound of the form $r_{\angle}(G) \ll N(\log\log N)^{-c}$, this ought to be possible (Shkredov, work in progress). Finally there is the issue of transferring the arguments of §6 to obtain a bound of the form $r_4(G) \ll N(\log N)^{-c}$. In particular one would like this for $G = \mathbb{Z}/N\mathbb{Z}$, which would imply that $r_4(N) \ll N(\log N)^{-c}$. Since the argument for $r_4(\mathbb{F}_5^n)$ is already rather difficult, one should not expect this to be at all straightforward. Even describing the correct generalisation of the notion of quadratic form to an arbitrary G is not straightforward [35].

9 Progressions in Sumsets

As promised, we now move onto questions of a somewhat more miscellaneous nature. This section concerns the following problem.

Problem 9.1 *Let $A \subseteq \{1, \ldots, N\}$ be a set of size $N/10$ (say). Must $A + A$ contain a long arithmetic progression?*

Bourgain [9] proved that the answer is "yes"; $A + A$ must contain a surprisingly long arithmetic progression. If $L(N, \alpha)$ is the smallest l for which there is a set $A \subseteq \{1, \ldots, N\}$ of cardinality αN such that $A + A$ does not contain a progression of length l, then Bourgain showed that $L(N, 1/10) \gg \exp(c(\log N)^{1/3})$. In [25] this was improved to $L(N, 1/10) \gg \exp(c(\log N)^{1/2})$. An example of Ruzsa [50] shows that $L(N, 1/10) \ll \exp(c_\epsilon(\log N)^{2/3+\epsilon})$.

It seems as though the natural finite field analogue of Problem 9.1 involves replacing "arithmetic progression" by "coset of a subspace".

Problem 9.2 *Write $D(n, \alpha)$ for the smallest d for which there is $A \subseteq \mathbb{F}_2^n$ of density α such that $A + A$ does not contain a coset of a subspace of dimension d. Estimate $D(n, \alpha)$.*

The techniques of [25] adapt to this situation in a straightforward manner, and one obtains the following.

Theorem 9.3 *Suppose that $\alpha \geqslant n^{-1/4}$. Then $D(n, \alpha) \geqslant \alpha^2 n/80$.*

A detailed proof of this fact may be found in [28]. In keeping with the philosophy of this survey, some of the details are rather cleaner than in the original argument [25] which applied to subsets of $\{1, \ldots, N\}$.

A more dramatic difference between the finite field case and the original setting of Problem 9.1 can be observed when one tries to adapt Ruzsa's construction to the finite field setting.

Theorem 9.4 (Ruzsa's niveau sets in $\mathbb{F}_2^n$) $D(n, 1/4) \leqslant n - \sqrt{n}$.

Proof Let A be the set of all vectors $x \in \mathbb{F}_2^n$ with at least $n/2 + \sqrt{n}/2$ ones with respect to the standard basis. By the central limit theorem the number of ones in a random vector $(x_1, \ldots, x_n)$ is roughly normally distributed with mean $n/2$ and standard deviation $\sqrt{n}/2$, and so for large n we have $|A| \geqslant 2^{n-2}$. Now any vector $x \in A + A$ must have at least $\sqrt{n}$ zeros. Using this fact, we shall prove that $A + A$ meets all translates of all $(n - \lfloor\sqrt{n}\rfloor)$-dimensional subspaces. Indeed, write $d = \lfloor\sqrt{n}\rfloor$ and suppose that U is a translate of some subspace of dimension $n = d$. U can be written as

$$U = \{a_0 + \lambda_1 a_1 + \cdots + \lambda_{n-d} a_{n-d} : \lambda_i \in \mathbb{F}_2\},$$

where the a_i are linearly independent. Write a_i in component form as $(a_i^{(j)})_{j=1}^n$. The column rank of the matrix (a_{ij}) is $n - d$, and hence so is the row rank. Without loss of generality, suppose that the first $n - d$ rows $(a_1^{(j)}, \ldots, a_{n-d}^{(j)})$, $j = 1, \ldots, n - d$, are linearly independent. Then we can solve the $n - d$ equations

$$a_0^{(j)} + \lambda_1 a_1^{(j)} + \cdots + \lambda_{n-d} a_{n-d}^{(j)} = 1$$

for the λ_i, giving a vector in U with no more than d zeros. □

Problem 9.5 *Narrow the gap between Theorems 9.3 and 9.4.*

My suspicion is that the upper bound of Theorem 9.4 is closer to the truth.

I cannot resist mentioning two problems which were raised at the AIM conference on additive combinatorics. The first is due to Croot:

Problem 9.6 *Fix $\theta \in (0,1)$. What is*

$$l(\theta) = \limsup_{N\to\infty} \min_{A \subseteq [N], |A| = N^{1-\theta}} (\textit{length of the longest progression in } A + A)?$$

In words, we are interesting in finding subsets $A \subseteq [N]$ with density $N^{-\theta}$ such that $A + A$ contains no long arithmetic progression. Croot states that the bounds $2/\theta - 1 \leqslant l(\theta) \ll 2^{1/\theta}$ are known. The upper bound comes by considering a multi-dimensional progression of dimension about $\theta \log_2 N$: it would be interesting to see whether a construction related to niveau sets gives anything better.

The second question is due to Katznelson:[4]

Problem 9.7 *What is the supremum of the measures of open subsets A of the torus $\mathbb{T}^d$ for which $A - A$ does not contain a 1-dimensional subgroup? In particular, is it 2^{-d}?*

10 Freiman's Theorem

A great deal of the material in this section was communicated to me in person by Imre Ruzsa, and is reproduced here and in the supplementary document [30] (which contains proofs) with his kind permission. The reader will also wish to consult Ruzsa's own survey article [51], as well as the material from the AIM conference on Additive Combinatorics [1].

This section concerns Problem 1.2 of the introduction. Let $A \subseteq \mathbb{F}_2^\infty$ have doubling at most K, meaning that we have the inequality $|A + A| \leqslant K|A|$. What can be said about the structure of A?

It is hard to think of any examples of sets A with this property other than cosets of subspaces, and large subsets of them. In fact, these are the only such examples as was shown by Imre Ruzsa [47]. The best known bounds for a result of this type are due to Ruzsa and the author [31]:

Theorem 10.1 (Freiman's theorem in $\mathbb{F}_2^\infty$) *Let $A \subseteq \mathbb{F}_2^\infty$ be a finite set with $|A+A| \leqslant K|A|$. Then A is contained within a coset of some subgroup $H \leqslant \mathbb{F}_2^\infty$ with $|H| \leqslant K^2 2^{2K^2-2}|A|$.*

A version of this result, with somewhat weaker bounds, will be a consequence of Proposition 10.2 below (which is also due to Imre Ruzsa).

Theorem 10.1 gives, in a weak sense, a complete description of sets with small doubling. We showed that if $|A + A| \leqslant K|A|$ then A is contained in a coset of a subspace of size at most $K^2 2^{2K^2-2}|A|$; conversely, if A has this property then it is clear that $|A + A| \leqslant K^2 2^{2K^2-2}|A|$. It would be of great interest to have a structure theorem which does not result in exponential losses in K of this sort. Perhaps one can even arrange things so that one has a result of the form

$$\text{doubling constant } K \Longrightarrow \text{structure} \Longrightarrow \text{doubling constant } K',$$

where K' is *polynomial* in K.

[4] Since the conference, Bourgain has shown that the answer is in fact $\frac{1}{2}$.

It is easy to see that such a structure theorem would have to take a form somewhat different from Theorem 10.1. Indeed if one takes $A \subseteq \mathbb{F}_2^\infty$ to be a subspace H together with K points $x_1, \dots, x_K$ such that $\mathrm{Span}(x_1, \dots, x_K) \cap H = \{0\}$ then it is clear that $|A + A| \leqslant K|A|$, but that the smallest coset-of-a-subspace containing A has size roughly $2^K|A|$.

Ruzsa [47] reports that Katalin Marton has suggested that one should be looking for a covering of A by a small number $C_1(K)$ of cosets of some rather smaller subspace of size $C_2(K)|A|$. I agree with this, and it is to some extent believable that $C_1(K)$ and $C_2(K)$ can be polynomial in K. Ruzsa was probably the first to actually dare to conjecture this, and he certainly states such a conjecture explicitly in [51]. Such matters are also touched upon (in the $\mathbb{Z}$-setting) in [11, 24].

Imre Ruzsa indicated to me a large part of the following proposition giving a number of statements equivalent to such a structure theorem. The proof may be found in [30].

Proposition 10.2 (Ruzsa) *The following five statements are equivalent.*

1. *If $A \subseteq \mathbb{F}_2^\infty$ has $|A+A| \leqslant K|A|$, then there is $A' \subseteq A$, $|A'| \geqslant |A|/C_1(K)$, which is contained in a coset of some subspace of size at most $C_2(K)|A|$.*

2. *If $A \subseteq \mathbb{F}_2^\infty$ has $|A + A| \leqslant K|A|$, then A may be covered by at most $C_3(K)$ cosets of some subspace of size at most $C_4(K)|A|$.*

3. *If $A \subseteq \mathbb{F}_2^\infty$ has $|A+A| \leqslant K|A|$, and if additionally there is a set B, $|B| \leqslant K$, such that $A + B = A + A$, then A may be covered by at most $C_5(K)$ cosets of some subspace of size at most $C_6(K)|A|$.*

4. *Suppose that $f : \mathbb{F}_2^m \to \mathbb{F}_2^\infty$ is a function with the property that $|\{f(x)+f(y) - f(x+y) : x, y \in \mathbb{F}_2^m\}| \leqslant K$. Then f may be written as $g+h$, where g is linear and $|Im(h)| \leqslant C_7(K)$.*

5. *Suppose that $f : \mathbb{F}_2^m \to \mathbb{F}_2^\infty$ is a function with the property that for at least $2^{3m}/K$ of the quadruples $(x_1, x_2, x_3, x_4) \in \mathbb{F}_2^m$ with $x_1 + x_2 = x_3 + x_4$ we have $f(x_1) + f(x_2) = f(x_3) + f(x_4)$. Then there is an affine linear function $g : \mathbb{F}_2^m \to \mathbb{F}_2^\infty$ such that $f(x) = g(x)$ for at least $2^m/C_8(K)$ values of x.*

Furthermore if $C_i(K)$ is bounded by a polynomial in K for all $i \in I$, where I is any of the sets $\{1, 2\}, \{3, 4\}, \{5, 6\}, \{7\}, \{8\}$ then in fact $C_i(K)$ is bounded by a polynomial in K for all i.

Remark Statement (4) is perhaps the most elegant and natural one here. Observe also that (4) is rather easy with the bound $C_7(K) = 2^K$. Thus Proposition 10.2 implies a weak version of Theorem 10.1. It is the possibility of polynomial bounds for $C_i(K)$ that is the most interesting feature of this proposition. Let us call this the PFR conjecture:

Conjecture 10.3 (Polynomial Freiman-Ruzsa conjecture for $\mathbb{F}_2^n$) *The function $C_7(K)$ (and hence all of the other functions $C_i(K)$, $i = 1, \dots, 8$), can be taken to be polynomial in K.*

The following question has implications for PFR.

Question 10.4 *Let $A \subseteq \mathbb{F}_2^n$ be a set of density α. Then $2A - 2A$ contains a subspace with codimension $f(\alpha)$. What is the behaviour of $f(\alpha)$?*

Using a Fourier-analytic technique of Bogolyubov [8] one may show that $f(\alpha) \ll \alpha^{-2}$, and a refinement of this technique due to Chang [12] allows one to improve this to $f(\alpha) \ll \alpha^{-1}\log(1/\alpha)$. We have not been able to rule out the possibility that $f(\alpha) \ll \log(1/\alpha)$, which if true would imply PFR.

The proof of Proposition 10.2 uses an important result known as Plünnecke's inequality [45], a new proof of which was found by Ruzsa [49]. This states that if A is a subset of any abelian group G, and if $|A + A| \leqslant K|A|$, then we have the inequality $|sA - tA| \leqslant K^{s+t}|A|$ for any positive integers s, t. The reader may observe that (1) of Proposition 10.2 implies a much stronger bound for some large subset $A' \subseteq A$, for large s, t, at least if there is a good bound on $C_2(K)$. We may call such an A' *subplünnecke*. Nets Katz asked me to formulate a principle to the effect that A being subplünnecke implies that A is very economically contained in some coset of a subspace. The following result is my best effort so far in this direction:

Proposition 10.5 *Let $A \subseteq \mathbb{F}_2^\infty$, and suppose that there is a constant B such that $|tA| \leqslant t^B|A|$ for all $t \geqslant B\log B$. Then A is contained in a union of $2^{CB\log B}$ cosets of some subspace having size at most $2^{CB\log B}|A|$.*

The hope, of course, is that one might be able to show that if $|A + A| \leqslant K|A|$ then A has a large subset A' which is subplünnecke in the sense of Proposition 10.5, for some reasonably small B (ideally, $B = O(\log K/\log\log K)$, which would imply PFR).

For me the most important reason for wanting to understand the PFR conjecture is the implications it would have for our understanding of quadratic Fourier coefficients. In particular, PFR in $\mathbb{F}_5^n$ (the formulation is obvious) would imply a positive solution to the PGI Conjecture (Conjecture 6.8).

Proposition 10.6 *Suppose that* PFR *is true in $\mathbb{F}_5^n$. Then* PGI *is true. That is, let $f : \mathbb{F}_5^n \to [-1, 1]$ be a function with $\mathbb{E}f = 0$, and suppose that $\|f\|_{U^3} \geqslant \delta$. Then there is a quadratic form q on $\mathbb{F}_5^n$ such that*

$$|\mathbb{E}f(x)\omega^{q(x)}| \gg \delta^C,$$

for some absolute constant C.

For more concerning this see [35].

In my opinion it would be very interesting to determine whether PGI has any implications for PFR. It is just plausible that this represents the most natural way to attack PFR, though at the moment we have little idea how to carry out such a programme.

The results of this section may be discussed in the context of general abelian groups G. However, the issues are of a rather different nature to those discussed in §8. Freiman's original work concerned subsets of $\mathbb{Z}$, and was quite geometric in feel. See [7, 17, 27] for a further discussion. Ruzsa's proof [48] has proved much more adaptable, and recently Ruzsa and the author [32] were able to obtain a structure theorem for sets with small doubling which is valid in any abelian group.

Theorem 10.7 (G. – Ruzsa) *Let G be an abelian group, and suppose that $A \subseteq G$ has $|A + A| \leqslant K|A|$. The A is contained in a set of the form $H + P$, where H is a subgroup, P is a generalised arithmetic progression, the dimension of P is $\leqslant C_9(K)$ and $|H||P| \leqslant C_{10}(K)$.*

Remark A generalised arithmetic progression of dimension d is a set of the form

$$\{a_0 + \lambda_1 a_1 + \cdots + \lambda_d a_d : 0 \leqslant \lambda_i \leqslant L_i \qquad \text{for } i = 1, \ldots, d.\}$$

We obtain the bounds $C_9(K) \ll K^C$ and $C_{10}(K) \ll e^{K^C}$, for some absolute constant C.

References

[1] Problems from the AIM conference on Additive Combinatorics (Palo Alto, 2004), online notes, available at `http://www.aimath.org`, 2004.

[2] M. Ajtai and E. Szemerédi, Sets of lattice points that form no squares, *Stud. Sci. Math. Hungar.* **9** (1974), 9–11.

[3] N. Alon, Testing subgraphs in large graphs, in *Random structures and algorithms (Poznan, 2001) Random Struct. Algorithms* **21** (2002), 359–370.

[4] N. Alon, R.A. Duke, H. Lefmann, V. Rödl and R. Yuster, The algorithmic aspects of the regularity lemma, *J. Algorithms* **16** (1994), 80–109.

[5] N. Alon and J. Spencer, *The probabilistic method*, Wiley, New York (2000).

[6] F.A. Behrend, On sets of integers which contain no three elements in arithmetic progression, *Proc. Nat. Acad. Sci* **23** (1946), 331–332.

[7] Y. Bilu, Structure of sets with small sumset, in *Structure Theory of Set Addition Astérisque*, 258, (1999), pp. 77–108.

[8] N.N. Bogolyubov, Sur quelques propriétés arithmétiques des presquepériodes, *Ann. Chaire Math. Phys. Kiev* **4** (1939), 185–194.

[9] J. Bourgain, On arithmetic progressions in sums of sets of integers, in *A Tribute to Paul Erdős* (eds. A. Baker, B. Bollobás & A. Thomason), CUP, Cambridge (1990), pp. 105–109.

[10] J. Bourgain, On triples in arithmetic progression, *Geom. Funct. Anal.* **9** (1999), 968–984.

[11] J. Bourgain, On the dimension of Kakeya sets and related maximal inequalities, *Geom. Funct. Anal.* **9** (1999), 256–282.

[12] M.-C. Chang, A polynomial bound in Freiman's theorem, *Duke Math. J.* **113** (2002), 399–419.

[13] F.R.K. Chung and R.L. Graham, Quasi-random subsets of $\mathbb{Z}_n$, *J. Combin. Theory Ser. A* **61** (1992), 64–86.

[14] F.R.K. Chung, R.L. Graham and R.M. Wilson, Quasi-random graphs, *Combinatorica* **9** (1989), 345–362.

[15] Y. Edel, Extensions of generalized product caps, *Designs, Codes and Cryptography* **31** (2004), 5–14.

[16] P. Erdős and P. Turán, On some sequences of integers, *J. London Math. Soc.* **11** (1936), 261–264.

[17] G.R. Freiman, *Foundations of a Structural Theory of Set Addition*, *Translations of Mathematical Monographs*, 37, Amer. Math. Soc., Providence, RI, USA (1973).

[18] A. Frieze and R. Kannan, A simple algorithm for constructing Szemerédi's regularity partition, *Electron. J. Combin.* **6** (1999), Research Paper 17, 7pp (electronic).

[19] H. Furstenberg, Y. Katznelson and D. Ornstein, The ergodic theoretical proof of Szemerédi's theorem, *Bull. Amer. Math. Soc* **7** (1982), 527–552.

[20] W.T. Gowers, Lower bounds of tower type for Szemerédi's uniformity lemma, *Geom. Funct. Anal.* **7** (1997), 322–337.

[21] W.T. Gowers, A new proof of Szemerédi's theorem for progressions of length four, *Geom. Funct. Anal.* **8** (1998), 529–551.

[22] W.T. Gowers, A new proof of Szemerédi's theorem, *Geom. Funct. Anal.* **11** (2001), 465–588.

[23] W.T. Gowers, Hypergraph regularity and the multidimensional Szemerédi theorem, preprint, 2004.

[24] W.T. Gowers, Rough structure and classification, in *GAFA 2000, Special Volume* (1999), pp. 79–117.

[25] B.J. Green, Arithmetic progressions in sumsets, *Geom. Funct. Anal.* **12** (2002), 584–597.

[26] B.J. Green, A Szemerédi-type regularity lemma in abelian groups, *GAFA*, in press. Available at `http://www.arxiv.org/`, 2003.

[27] B.J. Green, Edinburgh lecture notes on Freiman's theorem, notes, available on the author's webpage `http://www.dpmms.cam.ac.uk/~bjg23`, 2001.

[28] B.J. Green, Restriction and Kakeya Phenomena, notes from a course given in Part III of the Mathematical Tripos, Cambridge University, available on the author's webpage `http://www.dpmms.cam.ac.uk/~bjg23/`, 2002.

[29] B.J. Green, An argument of Shkredov in the finite field setting, expository note, available on the author's webpage `http://www.dpmms.cam.ac.uk/~bjg23/`, 2004.

[30] B.J. Green, The Polynomial Freiman-Ruzsa Conjecture: some notes, expository note, available on the author's webpage `http://www.dpmms.cam.ac.uk/~bjg23/`, 2004.

[31] B.J. Green and I.Z. Ruzsa, Sets with small sumset and rectification, *Bull. London Math. Soc.*, in press. Available at `http://www.arxiv.org/`, 2003.

[32] B. J. Green and I.Z. Ruzsa, Freiman's theorem in an arbitrary abelian group, in preparation, 2004.

[33] B.J. Green and T.C. Tao, The primes contain arbitrarily long arithmetic progressions, preprint, available at `http://www.arxiv.org`, 2004.

[34] B.J. Green and T.C. Tao, New bounds for Szemerédi's theorem for progressions of length 4 in finite field geometries, in preparation, 2004.

[35] B.J. Green and T.C. Tao, An inverse theorem for the Gowers U3-norm, manuscript, 72pp, 2004.

[36] D.R. Heath-Brown, Integer sets containing no arithmetic progressions, *J. London Math. Soc.* **35** (1987), 385–394.

[37] N.H. Katz and T.C. Tao, Recent progress on the Kakeya conjecture, in *Proceedings of the 6th International Conference on Harmonic Analysis and Partial Differential Equations (El Escorial, 2000)* Publ. Mat. 2002, Vol. Extra, pp. 161–179.

[38] M. Krivelevich and B. Sudakov, Pseudo-random graphs, preprint, available at `http://www.math.princeton.edu/~bsudakov`

[39] J. Komlós and M. Simonovits, Szemerédi's regularity lemma and its applications in graph theory, in *Combinatorics, Paul Erdős is eighty, Vol. 2 (Keszthely, 1993)* (eds. D. Miklós, V.T. Sós and T. Szőnyi), *Bolyai Soc. Math. Stud.*, 2, János Bolyai Math. Soc., Budapest (1996), pp. 295–352.

[40] V. Lev, Progression-free sets in finite abelian groups, *J. Number Theory* **104** (2004), 162–169.

[41] R. Meshulam, On subsets of finite abelian groups with no 3-term arithmetic progressions, *J. Combin. Theory Ser. A* **71** (1995), 168–172.

[42] G. Mockenhaupt and T.C. Tao, Restriction and Kakeya phenomena in finite fields, *Duke Math. J.* **121** (2004), 35–74.

[43] B. Nagle, V. Rödl and M. Schacht, The counting lemma for k-uniform hypergraphs, preprint, 2004.

[44] M.B. Nathanson, *Additive number theory. Inverse problems and the geometry of sumsets*, *Graduate Texts in Math.*, 307, Springer-Verlag, New York (1996).

[45] H. Plünnecke, Eigenschaften und Abschätzungen von Wirkingsfunktionen, BMwF-GMD-22 Gesellschaft für Mathematik und Datenverarbeitung, Bonn 1969

[46] K.F. Roth, On certain sets of integers, *J. London Math. Soc.* **28** (1953), 104–109.

[47] I.Z. Ruzsa, Structure of sets with small sumset, in *Structure Theory of Set Addition Astérisque*, 258, (1999), 323–326.

[48] I.Z. Ruzsa, Generalized arithmetical progressions and sumsets, *Acta Math. Hungar.* **65** (1994), 379–388.

[49] I.Z. Ruzsa, An application of graph theory to additive number theory, *Scientia, Ser. A* **3** (1989), 97–109.

[50] I.Z. Ruzsa, Arithmetic progressions in sumsets, *Acta Arith* **60** (1991), 191–202.

[51] I.Z. Ruzsa, Sumsets, submitted to proceedings of the European Congress of Mathematicians, 2004.

[52] I.Z. Ruzsa and E. Szemerédi, Triple systems with no six points carrying three triangles, in *Combinatorics (Proc. Fifth Hungarian Colloq., Keszthely, 1976), Vol II Colloquia Mathematica Societatis János Bolyai*, 18, North-Holland, Amsterdam (1978), Vol II, pp. 939–945.

[53] I.D. Shkredov, On a problem of Gowers, preprint, available at `http://www.arxiv.org/`, 2004.

[54] J. Solymosi, Note on a generalization of Roth's theorem, in *Discrete and computational geometry Algorithms Combin.*, 25, Springer, Berlin (2003), pp. 825–827.

[55] E. Szemerédi, Regular partitions of graphs, in *Problèmes combinatoires et théorie des graphes (Colloq. Internat. CNRS, Univ. Orsay, Orsay, 1976) Colloq. Internat. CNRS*, 260, CNRS, Paris (1978), pp. 399–401.

[56] E. Szemerédi, On sets of integers containing no four elements in arithmetic progression, *Acta Math. Acad. Sci. Hungar.* **20** (1969), 89–104.

[57] E. Szemerédi, Integer sets containing no arithmetic progressions, *Acta Math. Hungar.* **56** (1990), 155–158.

[58] T.C. Tao, *Recent progress on the restriction phenomenon* in *Fourier Analysis and Convexity (Milan 2001)* Birkhäuser, Basel (2004),

[59] T.C. Tao, From rotating needles to stability of waves: emerging connections between combinatorics, analysis, and PDE, *Notices Amer. Math. Soc.* **48** (2001), 294–303.

[60] T.C. Tao, Lecture notes 5 from Math 254A, unpublished, available at `http://www.math.ucla.edu/~tao/254a.1.03w/notes5.dvi`, 2001.

[61] T.C. Tao and V. Vu, Additive Combinatorics, book in preparation, 2004.

[62] A. Thomason, Pseudorandom graphs, in *Random graphs '85 (Poznán, 1985) North-Holland Math. Stud.*, 144, North-Holland, Amsterdam (1987), pp. 307–331.

[63] A. Thomason, Random graphs, strongly regular graphs and pseudorandom graphs, in *Surveys in combinatorics 1987 (New Cross, 1987) London Math. Soc. Lecture Note Ser.*, 123, CUP, Cambridge (1987), pp. 173–195.

[64] V.H. Vu, On a question of Gowers, *Ann. Comb.* **6** (2002), 229–233.

School of Mathematics
University of Bristol
University Walk
Bristol BS8 1TW
England
b.j.green@bristol.ac.uk

The subgroup structure of finite classical groups in terms of geometric configurations

Oliver H. King

Abstract

L.E. Dickson's approach to the subgroups of $PSL_2(q)$ (the Linear Fractional Group) gives rise to a description of subgroups as fixing one of: a real point; a pair of real points; a pair of imaginary points; a sub-line; and so on. H.H. Mitchell took a similar approach in describing subgroups of $PSL_3(q)$ and $PSp_4(q)$ (for odd q). In the 1980s, Aschbacher gave a description of subgroups of classical groups as either lying in one of eight classes or being almost simple; the eight classes can largely be described geometrically. The remaining subgroups have not yet been completely determined but a certain amount of geometric structure can be identified. This paper gives a survey of progress towards a geometric description of subgroups of the classical groups.

1 Introduction

There are four classes of Classical Group. Perhaps it is most straightforward to name a significant group within each class and to then describe the various related groups in the class. Four significant groups, then, are $GL_n(q)$, $Sp_n(q)$, $O_n(q)$ and $U_n(q)$. We begin by describing these groups in some detail. This paper is a survey of progress towards a geometric description of the subgroup structure of the classical groups. We shall describe Aschbacher's Theorem in some detail, even though Aschbacher's approach is clearly not geometrical, for Aschbacher's Theorem demonstrates very largely the structure that one should expect to find. It is unrealistic to expect to furnish a proof of Aschbacher's Theorem that is entirely geometric, but it is a reasonable hope that nearly all the maximal subgroups of classical groups can be described geometrically and that maximality can be proven using geometric means. Maximal here means nothing more than maximal relative to inclusion.

The general linear group, $GL_n(q)$, consists of the invertible linear transformations of an n-dimensional vector space $V = V(n,q)$ over the finite field $GF(q)$, where $q = p^e$ with p prime. Above this sits $\Gamma L_n(q)$, the group of all invertible semi-linear transformations of V. Below $GL_n(q)$ sits $SL_n(q)$, the set of all linear transformations of V having determinant 1; $SL_n(q)$ is a characteristic subgroup of $\Gamma L_n(q)$. The centre of $\Gamma L_n(q)$ is the group Z of non-zero scalar linear transformations; this also the centre of $GL_n(q)$. The quotients $P\Gamma L_n(q) = \Gamma L_n(q)/Z$, $PGL_n(q) = GL_n(q)/Z$ and $PSL_n(q) = Z.SL_n(q)/Z$ act naturally on the projective space $PG(n-1,q)$. In this context, $P\Gamma L_n(q)$ may be viewed as the set of all collineations of $PG(n-1,q)$. For some people, this is the natural geometric group. For others, $PGL_n(q)$ is the natural geometric group because its elements can be represented as matrices. From another perspective, $PSL_n(q)$ is sometimes regarded as the most important group in the class because it is usually simple.

The symplectic group, $Sp_n(q)$, is the group of all elements of $GL_n(q)$ preserving a non-degenerate alternating form; the non-degeneracy leads to n being even. There are various ways to describe $Sp_n(q)$ in terms of matrices. One is to consider the matrix $J_S = \begin{bmatrix} 0 & I \\ -I & 0 \end{bmatrix}$. Then $Sp_n(q)$ consists of the matrices A such that $AJ_SA^T = J_S$. Above $Sp_n(q)$ sit $\Gamma Sp_n(q)$ consisting of the invertible semi-linear transformations that preserve the alternating form up to a scalar, and $GSp_n(q) = \Gamma Sp_n(q) \cap GL_n(q)$. A feature of $Sp_n(q)$ is that its elements already have determinant 1 so it is a subgroup of $SL_n(q)$. Each group has a corresponding projective group, namely its image in $P\Gamma L_n(q)$. In geometric terms, $P\Gamma Sp_n(q)$ preserves a symplectic polarity. The geometric objects are the absolute subspaces with respect to this polarity, although $P\Gamma Sp_n(q)$ may be viewed as the group of collineations preserving the set of absolute lines. The group $PSp_n(q)$ is usually simple.

The unitary group, $U_n(q)$, is defined when q is a square and is the group of all elements of $GL_n(q)$ preserving a non-degenerate Hermitian form. In matrix terms, $U_n(q)$ consists of the matrices A such that $A\bar{A}^T = I$, where $\bar{A}$ is obtained from A by raising each entry to the power $\sqrt{q}$. Above $U_n(q)$ sit $\Gamma U_n(q)$ consisting of the invertible semi-linear transformations that preserve the Hermitian form up to a scalar, and $GU_n(q) = \Gamma U_n(q) \cap GL_n(q)$. Below $U_n(q)$ sits $SU_n(q) = U_n(q) \cap SL_n(q)$. Each group has a corresponding projective group, namely its image in $P\Gamma L_n(q)$. In geometric terms, $P\Gamma U_n(q)$ preserves an Hermitian polarity or, put another way, is the group of an Hermitian surface $\mathcal{H}(n-1,q)$ (the set of absolute points). The geometric objects are the subspaces lying on $\mathcal{H}(n-1,q)$. The group $PSU_n(q)$ is usually simple. There are two schools of thought regarding notation for the unitary group: for geometers, the Hermitian surface lies in $PG(n-1,q)$ and so $PSU_n(q)$ is the appropriate terminology; for group theorists, the motivation might come from Chevalley groups or algebraic groups, and then $PSU_n(\sqrt{q})$ or something similar is deemed appropriate. Here we follow the geometric school.

The orthogonal group, $O_n(q)$, is the most difficult to describe. It is the group of all elements of $GL_n(q)$ preserving a non-singular quadratic form Q whose associated symmetric bilinear form B is non-degenerate. The description depends to some extent on the parity of q. Suppose first that q is odd. Then $B(x,y) = Q(x+y) - Q(x) - Q(y)$ and $Q(x) = B(x,x)/2$, so preservation of Q and preservation of B are equivalent; in matrix terms, $O_n(q)$ consists either of the matrices A such that $AA^T = I$ or of the matrices A such that $AJ_OA^T = J_O$, where J_O is the diagonal matrix with entries $1, 1, \ldots, 1, \lambda$ for some non-square λ. These amount to the same thing if n is odd, but when n is even the two possibilities give rise to different types of geometric structure (depending on n and q) and the groups are sometimes written as $O_n^\epsilon(q)$ with $\epsilon = \pm 1$ (we return to this shortly). Above $O_n(q)$ sit $\Gamma O_n(q)$ consisting of the invertible semi-linear transformations that preserve the quadratic form up to a scalar, and the general orthogonal group $GO_n(q) = \Gamma O_n(q) \cap GL_n(q)$. Below $O_n(q)$ sit the special orthogonal group $SO_n(q) = O_n(q) \cap SL_n(q)$ and the commutator subgroup $\Omega_n(q)$ of $O_n(q)$ that has index 2 in $SO_n(q)$. Each group has a corresponding projective group, namely its image in $P\Gamma L_n(q)$. If q is even, then life is more complicated. The bilinear form B is now an alternating form, and Q cannot be recovered from B; an immediate consequence is that n must be even if B is to be non-degenerate. Any element of $GL_n(q)$ preserving Q also preserves B, and this leads to the inclusion $O_n(q) \leq Sp_n(q)$. It is possible, but not straightforward, to determine a matrix criterion for elements of $O_n(q)$, but such a criterion is not used in practice. As with the case where q is odd, there are two canonical forms for a quadratic form, leading to two types of orthogonal group. The fact that $O_n(q) \leq Sp_n(q)$ means that all the elements of $O_n(q)$ have determinant 1. Here the special orthogonal group is defined as the kernel of the *Dickson invariant*; in practical terms $SO_n(q)$ is the subgroup of $O_n(q)$ consisting of the elements having fixed space with even dimension (c.f., [25]); further, $SO_n(q) = \Omega_n(q)$ with just one exception. In geometric terms, the points where Q takes the value 0 are termed *singular* and the set of such points is a quadric. When q is odd, these are precisely the set of absolute points of the polarity corresponding to B. In any case $P\Gamma O_n(q)$ preserves the quadric. The geometric objects are the subspaces of $PG(n-1,q)$ lying on the quadric. For n odd, the quadric is termed parabolic. For n even, one type is termed *hyperbolic* (when there are $(n/2)-1$-dimensional subspaces of $PG(n-1,q)$ lying on the quadric, and we write $O_n^+(q)$) and the other *elliptic* (in which case there are $(n/2)-2$-dimensional subspaces but no $(n/2)-1$-dimensional subspaces lying on the quadric, and we write $O_n^-(q)$). The group $P\Omega_n(q)$ is usually simple for $n \geq 3$. However, there is one notable exception when $n = 4$ and the quadric is hyperbolic; in this case $P\Omega_4^+(q)$ is isomorphic to $PSL_2(q) \times PSL_2(q)$.

The approach we shall take to describing the subgroups structure is to consider the smallest groups in each class of classical groups and to consider the projective groups in most cases. Thus we shall be looking at the subgroup structure of $PSL_n(q)$, $PSp_n(q)$, $P\Omega_n(q)$ and $PSU_n(q)$. The reason for this is that these are usually the cases where results are hardest to prove, with other groups following relatively easily. There will be occasions when it is more convenient to describe subgroups of $SL_n(q)$, $Sp_n(q)$, $\Omega_n(q)$ and $SU_n(q)$, but this does not affect the maximality.

We shall begin, in the next section, by describing the structure of low-dimensional groups. Then we shall describe Aschbacher's Theorem in some detail, with a geometric interpretation, and relate it to the low-dimensional results. After that we shall discuss current progress towards a geometric description of the subgroups of the classical groups in general.

2 Low dimensional groups

The first geometrical description of a subgroup structure appeared more than one hundred years ago. The subgroup structure of $PSL_2(q)$ was determined by E.H. Moore [44] and A. Wiman [51], although Dickson's treatment based on these two papers is often quoted as a source. Not too many years later Mitchell determined the maximal subgroups of $PSL_3(q)$ and then $PSp_4(q)$, both for q odd, and Hartley determined the maximal subgroups of $PSL_3(q)$ for q even. In the process, both Mitchell and Hartley determined the maximal subgroups of $PSU_3(q)$.

2.1 $n = 2$

L.E. Dickson's book ([12]) on linear groups investigates finite classical groups in considerable detail. The terminology appears a little strange nowadays: $GL_n(q)$ and $SL_n(q)$ are written $GLH(n,p^e)$ and $SLH(n,p^e)$, the General and Special Linear Homogeneous groups, and $PSL_n(q)$ is written $LF(n,p^e)$, the Linear Fractional Group. In Chapter 12, Dickson gives a complete determination of the subgroups of $LF(2,p^e)$, based on the work of Moore and Wiman. Mitchell gives another treatment in [42].

Theorem 2.1 ([12]) *The subgroups of $PSL_2(q)$ are as follows:*

(a) *a single class of $q+1$ conjugate abelian groups of order q;*

(b) *a single class of $q(q+1)/2$ conjugate cyclic groups of order d for each divisor d of $q-1$ for q even and $(q-1)/2$ for q odd;*

(c) *a single class of $q(q-1)/2$ conjugate cyclic groups of order d for each divisor d of $q+1$ for q even and $(q+1)/2$ for q odd;*

(d) *for q odd, a single class of $q(q^2-1)/(4d)$ dihedral groups of order $2d$ for each divisor d of $(q-1)/2$ with $(q-1)/(2d)$ odd;*

(e) *for q odd, two classes each of $q(q^2-1)/(8d)$ dihedral groups of order $2d$ for each divisor $d>2$ of $(q-1)/2$ with $(q-1)/(2d)$ even;*

(f) *for q even, a single class of $q(q^2-1)/(2d)$ dihedral groups of order $2d$ for each divisor d of $q-1$;*

(g) *for q odd, a single class of $q(q^2-1)/(4d)$ dihedral groups of order $2d$ for each divisor d of $(q+1)/2$ with $(q+1)/(2d)$ odd;*

(h) *for q odd, two classes each of $q(q^2-1)/(8d)$ dihedral groups of order $2d$ for each divisor $d>2$ of $(q+1)/2$ with $(q+1)/(2d)$;*

(i) *for q even, a single class of $q(q^2-1)/(2d)$ dihedral groups of order $2d$ for each divisor d of $q+1$;*

(j) *a single class of $q(q^2-1)/24$ conjugate four-groups when $q \equiv \pm 3 \pmod 8$;*

(k) *two classes each of $q(q^2-1)/48$ conjugate four-groups when $q \equiv \pm 1 \pmod 8$;*

(l) *a number of classes of conjugate abelian groups of order q_0 for each divisor q_0 of q;*

(m) *a number of classes of conjugate groups of order $q_0 d$ for each divisor q_0 of q and for certain d depending on q_0, all lying inside a group of order $q(q-1)/2$ for q odd and $q(q-1)$ for q even;*

(m) *two classes each of $[q(q^2-1)]/[2q_0(q_0^2-1)]$ groups $PSL_2(q_0)$, where q is an even power of q_0, for q odd;*

(o) *a single class of $[q(q^2-1)]/[q_0(q_0^2-1)]$ groups $PSL_2(q_0)$, where q is an odd power of q_0, for q odd;*

(p) *a single class of $[q(q^2-1)]/[q_0(q_0^2-1)]$ groups $PSL_2(q_0)$, where q is a power of q_0, for q even;*

(q) *two classes each of $[q(q^2-1)]/[2q_0(q_0^2-1)]$ groups $PGL_2(q_0)$, where q is an even power of q_0, for q odd;*

(r) *two classes each of $q(q^2-1)/48$ conjugate S_4 when $q \equiv \pm 1 \pmod 8$;*

(s) *two classes each of* $q(q^2-1)/48$ *conjugate* A_4 *when* $q \equiv \pm 1 \pmod 8$*;*

(t) *a single class of* $q(q^2-1)/24$ *conjugate* A_4 *when* $q \equiv \pm 3 \pmod 8$*;*

(u) *a single class of* $q(q^2-1)/12$ *conjugate* A_4 *when* q *is an even power of 2;*

(v) *two classes each of* $q(q^2-1)/120$ *conjugate* A_5 *when* $q \equiv \pm 1 \pmod{10}$*;*

After close study of the inclusions here, we can write down the following corollary. We exclude the cases $q = 2, 3$ where $PSL_2(q)$ is not simple.

Corollary 2.2 *The maximal subgroups of* $PSL_2(q)$ *are as follows:*

(a) *dihedral groups of order* $q-1$ *for* $q \geq 13$ *odd and* $2(q-1)$ *for* q *even: each stabilizes a pair of points (a hyperbolic quadric);*

(b) *dihedral groups of order* $q+1$ *for* $q \neq 7, 9$ *odd and* $2(q+1)$ *for* q *even: each stabilizes a pair of imaginary points (i.e., points of* $PG(1, q^2)$ *that don't lie in* $PG(1, q)$*, conceivably thought of as an elliptic quadric);*

(c) *a group of order* $q(q-1)/2$ *for* q *odd and* $q(q-1)$ *for* q *even: each stabilizes a point;*

(d) $PSL_2(q_0)$*, where* q *is an odd prime power of* q_0*, for* q *odd, or a prime power of* q_0*, for* q *even: each stabilizes a sub-line;*

(e) $PGL_2(q_0)$*, where* $q = q_0^2$*, for* q *odd: each stabilizes a sub-line;*

(f) S_4 *when* $q \equiv \pm 1 \pmod 8$*, with either* q *prime, or* $q = p^2$ *and* $3 < p \equiv \pm 3 \pmod 8$*;*

(g) A_4 *when* $q \equiv \pm 3 \pmod 8$*, with* $q > 3$ *prime;*

(h) A_5 *when* $q \equiv \pm 1 \pmod{10}$*, with either* q *prime, or* $q = p^2$ *and* $p \equiv \pm 3 \pmod{10}$*.*

A second corollary is the enumeration of maximal subgroups of $PGL_2(q)$ for q odd, because this group lies inside $PSL_2(q^2)$. We do not need to consider q even because in that case $PGL_2(q) = PSL_2(q)$.

Corollary 2.3 *For* $q > 3$ *odd, the maximal subgroups of* $PGL_2(q)$ *are as follows:*

(a) *dihedral groups of order* $2(q-1)$ *for* $q > 5$ *(the stabilizer of a pair of points);*

(b) *dihedral groups of order* $2(q+1)$ *(the stabilizer of a pair of imaginary points);*

(c) *a group of order* $q(q-1)$ *(the stabilizer of a point);*

(d) $PSL_2(q)$*;*

(e) $PGL_2(q_0)$*, where* q *is a prime power of* q_0 *(the stabilizer of a sub-line);*

(f) S_4 *when* $q \equiv \pm 3 \pmod 8$*, with* $q > 3$ *prime;*

2.2 $n = 3$

The collineation groups of the finite projective plane were determined by Mitchell in [42] for q is odd and later by Hartley in [26] for q is even. Mitchell's treatment is perhaps more geometrical than Dickson's in the previous section because there is more room. He defines five canonical transformations: fixing a triangle, fixing two points and two lines, fixing a line and a point on the line, fixing all points on a line and all lines through a point off the line (i.e., an homology), and fixing all points on a line and all lines through a point on the line (i.e., an elation). He then considers in detail how such transformations may lie in a group, alone or in combination. His theorem picks out the maximal subgroups and goes a long way towards identifying all other subgroups. Hartley (a student of Mitchell) employed a similar approach for even q, but phrases his theorem slightly differently. Mitchell and Hartley use the term *hyperorthogonal group* for $PSU_3(q)$ and write it as $HO(3, q)$.

Theorem 2.4 (Mitchell) *Suppose that q is odd. The following is a list of subgroups of $PSL_3(q)$ (with μ the* hcf *of* 3 *and $q - 1$). A subgroup of $PSL_3(q)$ either fixes a point, a line or a triangle (so is a subgroup of (a), (b) or (c) below), or is one of the groups in (d)-(k):*

(a) *the stabilizer of a point, having order $q^3(q+1)(q-1)^2/\mu$;*

(b) *the stabilizer of a line, having order $q^3(q+1)(q-1)^2/\mu$;*

(c) *the stabilizer of a triangle, having order $6(q-1)^2/\mu$;*

(d) *the stabilizer of an imaginary triangle (i.e., a triangle with co-ordinates in $GF(q^3)$), having order $3(q^2+q+1)/\mu$;*

(e) *the stabilizer of a conic, having order $q(q^2-1)$;*

(f) *$PSL_3(q_0)$, where q is a power of q_0;*

(g) *$PGL_3(q_0)$, where q is a power of q_0^3 and* 3 *divides $q_0 - 1$;*

(h) *$PSU_3(q_0^2)$, where q is a power of q_0^2;*

(i) *$PU_3(q_0^2)$, where q is a power of q_0^6 and* 3 *divides $q_0 + 1$;*

(j) *the Hessian groups of orders* 216 *(where* 9 *divides $q - 1$),* 72 *and* 36 *(where* 3 *divides $q - 1$);*

(k) *groups of order* 168 *(when -7 is a square in $GF(q)$),* 360 *(when* 5 *is a square in $GF(q)$ and there is a non-trivial cube root of unity),* 720 *(when q is an even power of* 5*), and* 2520 *(when q is an even power of* 5*).*

The groups of orders 168, 360, 720 and 2520 are isomorphic to $PSL_3(2)$, A_6, $A_6.2$ and A_7 respectively, each is almost simple. The groups of orders 216 and 72 are of symplectic type and are isomorphic to $PU_3(4)$ and $PSU_3(4)$ respectively, with the group of order 36 a subgroup of the latter. The stabilizer of an imaginary triangle is the normalizer of a Singer cyclic subgroup.

Theorem 2.5 (Hartley) *When q is even, the maximal subgroups of $PSL_3(q)$ are given in the following list, where μ is the* hcf *of* 3 *and $q-1$ and is thus* 1 *precisely when q is non-square:*

(a) *the stabilizer of a point, a line, a triangle or an imaginary triangle, as for q odd (and with the same orders);*

(b) $PSL_3(q_0)$*, where q is a prime power of q_0;*

(c) $PGL_3(q_0)$*, when $q = q_0^3$ and q_0 is square;*

(d) $PSU_3(q)$*, where q is square;*

(e) $PU_3(q_0^2)$*, where $q = q_0^6$ and q_0 is non-square;*

(f) *groups of order* 360 *(when $q = 4$).*

The group of order 360 is isomorphic to A_6.

Having broadly identified all the subgroups of $PSL_3(q)$ for q odd, Mitchell is able pick out the subgroups that lie in $PSU_3(q)$.

Theorem 2.6 (Mitchell) *Suppose that q_0 is odd and $q = q_0^2$. The following is a list of subgroups of $PSU_3(q)$ (with ν the* hcf *of* 3 *and q_0+1). A subgroup of $PSU_3(q)$ either fixes a point and a line, or a triangle (so is a subgroup of (a), (b) or (c) below), or is one of the groups in (d)-(i):*

(a) *the stabilizer of the centre and axis of a group of elations (i.e., the stabilizer of a point on $\mathcal{H}(2,q)$ together with its polar line), having order $q_0^3(q_0+1)(q_0-1)/\nu$;*

(b) *the stabilizer of the centre and axis of an homology (i.e., a point not on $\mathcal{H}(2,q)$ together with its polar line), having order $q_0(q_0+1)^2(q_0-1)/\nu$;*

(c) *the stabilizer of a triangle, having order $6(q_0-1)^2/\nu$;*

(d) *the stabilizer of an imaginary triangle, having order $3(q_0^2-q+1)/\nu$;*

(e) *the stabilizer of a conic, having order $q_0(q_0^2-1)$;*

(f) $PSU_3(q_1)$*, where q is an odd power of q_1;*

(g) $PU_3(q_1)$*, where q is an odd power of q_1^3 and* 3 *divides $\sqrt{q_1}+1$;*

(h) *the Hessian groups of orders* 216 *(where* 9 *divides q_0+1),* 72 *and* 36 *(where* 3 *divides q_0+1);*

(i) *groups of order* 168 *(when -7 is a non-square in $GF(q_0)$),* 360 *(when* 5 *is a square in $GF(q_0)$ but there is not a non-trivial cube root of unity),* 720 *(when q_0 is an odd power of* 5*), and* 2520 *(when q_0 is an odd power of* 5*).*

As with $PSL_3(q)$, Hartley only lists the maximal subgroups of $PSU_3(q)$.

Theorem 2.7 (Hartley) *Suppose that q_0 is even and $q = q_0^2$. The maximal subgroups of $PSU_3(q)$ are given in the following list (where ν is the* hcf *of* 3 *and q_0+1):*

(a) *the stabilizer of a point and a line, a triangle or an imaginary triangle, as for q odd (and with the same orders);*

(b) $PSU_3(q_1)$, *where q is an odd prime power of q_1;*

(c) $PU_3(q_1)$, *where $q = q_1^3$ and $\sqrt{q_1}$ is non-square;*

(d) *groups of order* 36 *(when $q = 4$).*

2.3 $n = 4$

In considering $n = 4$, Mitchell's approach to $PSL_3(q)$ proved difficult to extend to $PSL_4(q)$. However he did manage to deal with the slightly smaller group $PSp_4(q)$, identifying all the maximal subgroups. In Dickson's notation, this is the group $A(4, p^e)$.

Theorem 2.8 (Mitchell) *Assume that $p > 2$. The maximal subgroups of $PSp_4(q)$ are as follows:*

(a) $PSp_4(q_0)$, *where q is an odd prime power of q_0;*

(b) $PGSp_4(q_0)$, *where $q = q_0^2$;*

(c) *the stabilizer of a point and a plane, having index $q^3 + q^2 + q + 1$;*

(d) *the stabilizer of a parabolic congruence, having index $q^3 + q^2 + q + 1$;*

(e) *the stabilizer of a hyperbolic congruence, having index $q^2(q^2 + 1)/2$;*

(f) *the stabilizer of a elliptic congruence, having index $q^2(q^2 - 1)/2$;*

(g) *the stabilizer of a quadric, having index $q^3(q^2 + 1)(q + 1)/2$ (for $q > 3$);*

(h) *the stabilizer of a quadric, having index $q^3(q^2 + 1)(q - 1)/2$ (for $q > 3$);*

(i) *the stabilizer of a twisted cubic, having index $q^3(q^4 - 1)$ (for $p > 3, q > 7$);*

(j) *groups of orders* 1920 *(for q prime and $\equiv \pm 1 \pmod 8$),* 960 *(for q prime and $\equiv \pm 3 \pmod 8$),* 720 *(for q prime and $\equiv \pm 1 \pmod{12}$),* 360 *(for q prime, $\neq 7$ and $\equiv \pm 5 \pmod{12}$) and* 2520 *(for $q = 7$).*

The groups of orders 1920 and 960 are groups of symplectic type. The groups of orders 360, 720 and 2520 are isomorphic to A_6, $A_6.2$ and A_7 respectively. A parabolic congruence consists of a self-polar line together with all of the self-polar lines that meet it, so the stabilizer of a parabolic congruence is the stabilizer of a self-polar line. A hyperbolic congruence is the set of self-polar lines that meet a pair of skew polar lines, so the stabilizer of a hyperbolic congruence is the stabilizer of a pair of skew polar lines. An elliptic congruence is a regular spread. The first quadric listed is determined by a pair of skew self-polar lines that is stabilized by the subgroup. The stabilizer of the second quadric in $Sp_4(q)$ has structure $U_2(q^2).2$. These geometric configurations can be described nicely in terms of the Klein quadric that we discuss in the next section. Indeed Hirschfeld defines these configurations in this manner in [27].

Mwene published two papers on the maximal subgroups of $PSL_4(q)$. The first ([45]), on even q, is geometric and group-theoretic in equal measure. The second ([46]), on odd q, is much more algebraic, without a complete determination of maximal subgroups: we don't list the results here.

Theorem 2.9 (Mwene) *The maximal subgroups of $PSL_4(q)$ for q even are as follows:*

(a) *the stabilizer of a point, having order $q^6(q^3-1)(q^2-1)(q-1)$;*

(b) *the stabilizer of a plane, having order $q^6(q^3-1)(q^2-1)(q-1)$;*

(c) *the stabilizer of a line, having order $q^6(q-1)^3(q+1)^2$;*

(d) *the stabilizer of a tetrahedron, having order $24(q-1)^3$ (for $q>4$);*

(e) *the stabilizer of a pair of mutually skew lines, having order $2q^2(q-1)^3(q+1)^2$ (for $q>2$);*

(f) *the stabilizer of a pair of mutually skew imaginary lines, that is, lines in $PG(3,q^2)$, having order $2q^2(q^2-1)(q^2+1)(q+1)$;*

(g) *$PSL_4(q_0)$ where q is a prime power of q_0;*

(h) *$PSp_4(q)$;*

(i) *$PSU_4(q)$ when q is a square;*

(j) *A_7 (for $q=2$).*

2.4 Klein quadric and exceptional isomorphisms

Lines of $PG(3,q)$ are represented as points of $PG(5,q)$. This is a particular case of a construction of the Grassmannian of lines. The representation can be given explicitly in terms of Plücker co-ordinates: given a co-ordinate system for $PG(3,q)$, choose any two points, (x_0,x_1,x_2,x_3) and (y_0,y_1,y_2,y_3), on a given line and calculate the values $p_{ij}=x_iy_j-x_jy_i$; the point $(p_{01},p_{02},p_{03},p_{12},p_{31},p_{23})$ lies on the quadric $\mathcal{K}$ given by $X_0X_5+X_1X_4+X_2X_3=0$. The fundamental property of this representation of lines of $PG(3,q)$ is that two lines intersect if and only if the corresponding points lie on a line on $\mathcal{K}$. A consequence is that a plane on $\mathcal{K}$ corresponds either to the set of lines of $PG(3,q)$ passing through a point or to the set of lines of $PG(3,q)$ lying in a plane, and we thus have two families of planes on $\mathcal{K}$; two planes in one family are either equal or meet in a point; two planes from different families are either disjoint or meet in a line. From this it can be deduced that a line on $\mathcal{K}$ corresponds to the set of lines of $PG(3,q)$ lying in a plane and passing through a common point.

The geometric construction leads to a group isomorphism between $PSL_4(q)$ and $P\Omega_6^+(q)$. It is possible to examine the Plücker co-ordinate construction and find that the lines of a general linear complex of $PG(3,q)$ correspond to points of $\mathcal{K}$ lying in a non-isotropic hyperplane of $PG(5,q)$, and from this there arises an isomorphism between $PSp(4,q)$ and $P\Omega(5,q)$. There is a natural embedding of $PSL_2(q^2)$ in $PSp_4(q)$ permuting a regular spread (q^2+1 lines) of the general linear complex;

it turns out that these lines correspond to points of $\mathcal{K}$ lying in a 3-dimensional subspace of $PG(5,q)$, and this leads to an isomorphism between $PSL_2(q^2)$ and $P\Omega_4^-(q)$. If q is a square, and we consider just the lines of a Hermitian surface $\mathcal{H}(3,q)$, then the corresponding points lie in a subgeometry $PG(5,\sqrt{q})$ of $PG(5,q)$ and the intersection of $PG(5,\sqrt{q})$ with $\mathcal{Q}^+(5,q)$ is an elliptic quadric $\mathcal{Q}^-(5,\sqrt{q})$. From this there arises an isomorphism between $PSU_4(q)$ and $P\Omega_6^-(\sqrt{q})$. Finally for q odd, consider a regulus $\mathcal{R}$ in $PG(3,q)$ together with its opposite regulus $\mathcal{R}'$. Then the lines in $\mathcal{R}$ correspond to $q+1$ points on $\mathcal{K}$ that are pairwise non-orthogonal but that are all orthogonal to the $q+1$ points corresponding to the lines in $\mathcal{R}'$. It follows that $\mathcal{R}$ and $\mathcal{R}'$ determine orthogonal, non-degenerate planes, π and π' of $PG(5,q)$. The subgroup of $PSL_4(q)$ that permutes the lines in $\mathcal{R}$ but fixes each line in $\mathcal{R}'$ acts as $PSL_2(q)$, but at the same time the corresponding subgroup of $P\Omega_6^+(q)$ acts as the identity on π' and as $P\Omega(3,q)$ on π: $PSL_2(q)$ is isomorphic to $P\Omega_3(q)$.

Returning for a moment to $PSp_4(q)$, we have noted that it is the group of a general linear complex whose lines correspond to points of $\mathcal{K}$ lying in a non-isotropic hyperplane of $PG(5,q)$. Let us suppose that P is the point of $PG(5,q)\setminus\mathcal{K}$ such that the hyperplane is $P^\perp$, and write $\mathcal{Q}_4 = P^\perp \cap \mathcal{K}$ (a parabolic quadric). A parabolic congruence is the set of lines in $PG(3,q)$ corresponding to the points of $\mathcal{Q}_4$ polar to a point of $\mathcal{Q}_4$. Consider a point $R \notin \mathcal{K}$ such that the line PR is non-isotropic. If PR is a secant line, then it contains two non-polar points of $\mathcal{K}$ neither of which lie in $\mathcal{Q}_4$, so they correspond to polar non-isotropic lines of $PG(3,q)$; the points of $\mathcal{Q}_4 \cap R^\perp$ are polar to both points of PR and the corresponding lines of $PG(3,q)$ form a hyperbolic congruence. If PR is an external line, then $\mathcal{Q}_4 \cap R^\perp$ is an elliptic quadric with q^2+1 pairwise non-polar points, so corresponds to q^2+1 skew self-polar lines of $PG(3,q)$, i.e., a regular spread. If l is a secant line in $P^\perp$, then $l^\perp \cap \mathcal{Q}_4$ consists of $q+1$ points of a conic, each polar to the two points of $\mathcal{Q}_4$ lying in l, so they correspond to $q+1$ self-polar lines of $PG(3,q)$, each meeting a pair of skew, self-polar lines: the $q+3$ lines described all lie on a quadric in $PG(3,q)$. If l is an external line in $P^\perp$, then $l^\perp \cap \mathcal{Q}_4$ again consists of $q+1$ points of a conic, but no longer polar to a pair of points in $\mathcal{Q}_4$; however, if one worked over $GF(q^2)$ it would be possible to see these points as corresponding to lines on a quadric in $PG(3,q^2)$ and so there is a sense in which the stabilizer in $PSp_4(q)$ of this set of lines stabilizes a quadric.

If we now look at the results of Dickson (for $n=2$) and Mitchell (for $n=4$) we can deduce the following results.

Corollary 2.10 *The maximal subgroups of $P\Omega_3(q)$ ($q>3$ odd) acting on a conic $\mathcal{C}$ are as follows:*

(a) *the stabilizer of a point external to $\mathcal{C}$ for $q \geq 13$;*

(b) *the stabilizer of a point internal to $\mathcal{C}$ for $q \neq 7,9$;*

(c) *the stabilizer of a point of $\mathcal{C}$;*

(d) *$P\Omega_3(q_0)$, where q is an odd prime power of q_0;*

(e) *$PSO_3(q_0)$, where $q=q_0^2$;*

(f) S_4 *when* $q \equiv \pm 1 \pmod 8$, *with either* q *prime, or* $q = p^2$ *and* $3 < p \equiv \pm 3 \pmod 8$;

(g) A_4 *when* $q \equiv \pm 3 \pmod 8$, *with* $q > 3$ *prime;*

(h) A_5 *when* $q \equiv \pm 1 \pmod{10}$, *with either* q *prime, or* $q = p^2$ *and* $p \equiv \pm 3 \pmod{10}$.

Corollary 2.11 *The maximal subgroups of* $P\Omega_4^-(q)$ *are as follows:*

(a) *the stabilizer of a secant line for* $q > 3$;

(b) *the stabilizer of a spread of lines, half of which are secant and the other half external;*

(c) *the stabilizer of a point on the elliptic quadric;*

(d) $P\Omega_4^-(q_0)$, *where* q *is an odd prime power of* q_0;

(e) *the stabilizer of a point off* $\mathcal{Q}$;

(f) S_4 *when* q *is prime and* $3 < q \equiv \pm 3 \pmod 8$;

(g) A_5 *when* q *is prime and* $q \equiv \pm 3 \pmod{10}$.

$PSL_2(q^2)$ acts on the $q^2 + 1$ lines of the regular spread of $PG(3, q)$ as though the lines were points of $PG(1, q^2)$, so a pair of lines in the spread corresponds to a pair of non-orthogonal points on $\mathcal{Q}$, i.e., to a secant line (and its orthogonal complement which is an external line).

A point of $\mathcal{Q}$ corresponds to a line of the spread.

Corollary 2.12 *The maximal subgroups of* $P\Omega_5(q)$ *for odd* q *are as follows:*

(a) $P\Omega_5(q_0)$, *where* q *is an odd prime power of* q_0;

(b) $PSO_5(q_0)$, *where* $q = q_0^2$;

(c) *the stabilizer of a line on* $\mathcal{Q}$;

(d) *the stabilizer of a point on* $\mathcal{Q}$;

(e) *the stabilizer of a point off* $\mathcal{Q}$ *orthogonal to a hyperbolic quadric;*

(f) *the stabilizer of a point off* $\mathcal{Q}$ *orthogonal to an elliptic quadric;*

(g) *the stabilizer of a secant line (for* $q > 3$*);*

(h) *the stabilizer of an external line (for* $q > 3$*);*

(i) *the stabilizer of a partial ovoid of size* $q + 1$ *(for* $p > 3, q > 7$*), isomorphic to* $PSL_2(q)$;

(j) *groups of orders* 1920 *(for* q *prime and* $\equiv \pm 1 \pmod 8$*),* 960 *(for* q *prime and* $\equiv \pm 3 \pmod 8$*),* 720 *(for* q *prime and* $\equiv \pm 1 \pmod{12}$*),* 360 *(for* q *prime,* $\neq 7$ *and* $\equiv \pm 5 \pmod{12}$*) and* 2520 *(for* $q = 7$*).*

The subgroups of orders 1920 and 960 stabilize a pentagon of pairwise orthogonal points off $\mathcal{Q}$. A line on $\mathcal{Q}$ corresponds to a point together with its polar plane in $PG(3,q)$, a point on $\mathcal{Q}$ corresponds to a self-polar line in $PG(3,q)$ which in turn determines a parabolic congruence, a point off $\mathcal{Q}$ corresponds to a hyperbolic or elliptic congruence, a secant line contains two points of $\mathcal{Q}$ and corresponds to a pair of skew self-polar lines of $PG(3,q)$, an external line is orthogonal to a conic that corresponds to a partial spread of $PG(3,q)$ determined by a subgroup $U_2(q^2)$ of $PSp_4(q)$.

Corollary 2.13 *The maximal subgroups of $P\Omega_6^+(q)$ for q even are as follows:*

(a) *the stabilizer of a plane on $\mathcal{K}$;*

(b) *the stabilizer of a point on $\mathcal{K}$;*

(c) *the stabilizer of three pairwise orthogonal secant lines spanning $PG(5,q)$ (for $q > 4$);*

(d) *the stabilizer of a secant line (for $q > 2$);*

(e) *the stabilizer of an external line;*

(f) *$P\Omega_6^+(q_0)$ where q is a prime power of q_0;*

(g) *the stabilizer of a point off $\mathcal{K}$;*

(h) *$P\Omega_6^-(q_0)$ where $q = q_0^2$;*

(i) *A_7 (for $q = 2$).*

A plane on $\mathcal{K}$ corresponds to either a point or a plane of $PG(3,q)$, a point on $\mathcal{K}$ corresponds to a line of $PG(3,q)$, the three pairwise orthogonal secant lines each contain two points of $\mathcal{K}$ corresponding to opposite lines of a tetrahedron in $PG(3,q)$, a secant line contains two points of $\mathcal{K}$ that correspond to a pair of skew lines of $PG(3,q)$, the stabilizer of an external line fixes a pair of points on $\mathcal{Q}_6^+(5,q^2)$ (i.e., imaginary points), the stabilizer of a point off $\mathcal{K}$ corresponds to $PSp_4(q)$, $P\Omega_6^-(q_0)$ is isomorphic to $PSU_4(q)$.

3 Aschbacher's Theorem

In the early 1980s, several papers appeared determining some classes of maximal subgroups of classical groups. We shall refer to these later. In 1984, Michael Aschbacher published his paper, "On the maximal subgroups of the finite classical groups" ([1]), in which he identified eight classes of subgroups $\mathcal{C}_1 - \mathcal{C}_8$ and showed that an arbitrary subgroup either lies in a maximal subgroup belonging to one of $\mathcal{C}_1 - \mathcal{C}_8$ or satisfies several significant constraints. All but one of the classes can be described geometrically, but the proof of Aschbacher's Theorem is group-theoretic. It is worth spending a short time on the shape of the proof and the way in which the eight classes arise.

In the statement of Aschbacher's Theorem, the groups studied are *almost simple*, meaning that they lie between a simple group and its automorphism group, and this means that the classical groups in this context are subgroups of $P\Gamma_n(q)$. However the description of the classes of subgroups is substantially in terms of matrix groups acting on vector spaces. For the purposes of exposition, we shall regard the classical group G as being one of $SL_n(q)$, $Sp_n(q)$, $\Omega_n(q)$ and $SU_n(q)$, and we shall write $\bar{G}$ for the projective equivalent. Thus $\bar{G}$ is non-abelian simple except for: $PSL_2(2) = PSp_2(2)$, $PSL_2(3) = PSp_2(3)$, $PSp_4(2)$, $P\Omega_2(q)$, $P\Omega_3(3)$, $P\Omega_4^+(q)$ and $PSU_3(4)$. These cases are thus omitted from the following discussion. We shall consider a subgroup M of G which is assumed to contain the centre of G. We shall write A for the non-degenerate alternating form given by $A(x, y) = xJ_Sy^T$ preserved by $Sp_n(q)$ (with J_S as in the introduction) and C for the non-degenerate Hermitian form given by $C(x, y) = x\bar{y}^T$ preserved by $SU_n(q)$; the non-degenerate symmetric bilinear form B and the quadratic form Q preserved by $\Omega_n(q)$ are as given in the introduction. Given any subspace U of $V = V(n, q)$, the *orthogonal complement* (with respect to a form $(,)$) is given by $U^\perp = \{v \in V : (u, v) = 0 \text{ for all } u \in U\}$ (in projective terms, this is the *polar space*). To say that a form is non-degenerate simply means that $V^\perp = \{0\}$. A subspace U is said to be *totally isotropic* (self-polar in projective terms) if $U \subseteq U^\perp$ and *non-isotropic* if $U \cap U^\perp = \{0\}$; when we have a quadratic form, U is *totally singular* if $Q(u) = 0$ for all $u \in U$, we shall use this term to be synonymous with 'totally isotropic' for the alternating and Hermitian forms.

The first possibility is that M is reducible, i.e., it stabilizes (globally) a non-trivial proper subspace U. If G has an associated form, then M also stabilizes $U^\perp$ and thus $U \cap U^\perp$. It follows that M lies in the stabilizer of a smaller subspace unless $U \cap U^\perp = U$ or $\{0\}$. This means that we need only consider totally isotropic or non-isotropic subspaces. In the case of $\Omega_n(q)$ with q even, a totally isotropic subspace that is not totally singular has a unique totally singular subspace of codimension 1 that would also be stabilized by M, and that means that we need only consider 1-dimensional 'non-singular', totally isotropic subspaces. There is a further complication when U is non-isotropic, where there is the possibility that G contains elements that switch U and $U^\perp$; in this case U is said to be *isometric* to $U^\perp$ and the stabilizer of U is properly contained in the stabilizer of $\{U, U^\perp\}$. The class $\mathcal{C}_1$ consists of subgroups that stabilize a non-isotropic subspace not isometric to its orthogonal complement, a totally singular subspace or a 1-dimensional non-singular, totally isotropic subspace.

If M is not reducible, then clearly it is irreducible. One might consider the possibility that M is irreducible but not absolutely irreducible, but it turns out that it is more fruitful to consider a normal subgroup N of M.

It is possible that M is irreducible, but that there is a non-trivial subgroup N that is reducible (non-trivial here means that N is not contained in the centre of G). Two quite different possibilities arise. The first is that we can choose an appropriate subspace U stabilized by N and find that the images of U under M form a set whose direct sum is V. In this case M stabilizes this set of subspaces and is an *imprimitive* subgroup of G. If G has an associated form, then the subspaces are all either non-isotropic and pairwise orthogonal, or totally singular with n even and such that U can be chosen to have dimension $n/2$. The class $\mathcal{C}_2$ consists of the stabilizers of such direct sum decompositions. The second possibility is that if we choose U as small as possible (so that N acts irreducibly on U), say of dimension m, then V can be expressed as the direct sum of subspaces $U_1, U_2, \ldots, U_{n/m}$ fixed by N in such a way that the elements of N can be written as block diagonal matrices $diag(P, P, \ldots, P)$

with P an $m \times m$ matrix. Let F be the set of $m \times m$ matrices that commute with all the Ps arising from N. Then, by Schur's Lemma, F is a division ring and then, by Wedderburn's Theorem, F is a field. We can see that F effectively contains $GF(q)$ as the set of scalar matrices: let r be the degree of F over $GF(q)$. If $r > 1$, then $N \leq SL_{n/r}(F)$ and M can be shown to normalize $SL_{n/r}(F)$; we find here that F lies in the class $\mathcal{C}_3$ that we describe more fully in the next paragraph. If $r = 1$, then the matrices in M have the form $A \otimes B$, where $A \in GL_{n/m}(q)$ and $B \in GL_m(q)$, so that M stabilizes a tensor product structure. If there is a form on V, then each factor of the tensor product admits a form. In terms of maximal subgroups, it is only necessary to consider groups arising from $GL_{n_1} \otimes GL_{n_2}$ $(n_1 < n_2)$, $U_{n_1} \otimes U_{n_2}$ $(n_1 < n_2)$, $Sp_{n_1} \otimes O^{\epsilon}_{n_2}$ (q odd, $n_2 \geq 3$), $Sp_{n_1} \otimes Sp_{n_2}$ $(n_1 < n_2)$ and $O^{\epsilon_1}_{n_1} \otimes O^{\epsilon_2}_{n_2}$ ($(n_1, \epsilon_1) \neq (n_2, \epsilon_2)$, q odd, $n_2 \geq 3$) as subgroups of $SL_n(q)$, $SU_n(q)$, $Sp_n(q)$, $\Omega_n(q)$ and $\Omega_n(q)$ respectively; some of these restrictions are present in order to avoid the possibility of a larger group that includes switching of the factors. The class of groups just described is $\mathcal{C}_4$. In geometric terms, the structure in $PG_{n-1}(q)$ on which such groups act is a *Segre variety*, sometimes written S_{n_1,n_2}.

It is possible that M is irreducible, and that there is a non-trivial subgroup N that is irreducible but not absolutely irreducible (this includes the case where M is not absolutely irreducible). If we consider the set C of $n \times n$ matrices that commute with N (in the manner of the previous paragraph), then C is a field containing $GF(q)$. The group N is *absolutely irreducible* if $C = GF(q)$. In the current setting, where N is not absolutely irreducible, C is isomorphic to $GF(q^r)$ for some $r \geq 2$ and $m = n/r$ is an integer; if an appropriate basis is chosen for V, then the elements in N become $m \times m$ block matrices, where the blocks are $r \times r$ matrices representing elements of $GF(q^r)$. Conjugation by elements of M then corresponds to conjugation by an element of N, extended by an automorphism of $GF(q^r)$ that fixes $GF(q)$. In other words M lies in $(G \cap GL_{n/r}(q^r)).Gal_{GF(q)}(GF(q^r))$. If r is composite, say $r = st$ with $s, t > 1$, then M lies inside the larger group $(G \cap GL_{n/s}(q^s)).Gal_{GF(q)}(GF(q^s))$. Hence there is a restriction that r must be prime; we now have the class $\mathcal{C}_3$ that preserve an overfield structure. In geometric terms, if $n > r$ then V can be viewed as an n/r-dimensional vector space over $GF(q^r)$, the set of 1-dimensional subspaces of which become a set of r-dimensional subspaces of $V(n, q)$ preserved by M. Thus M stabilizes a spread of $(n/r) - 1$-subspaces of $PG(n-1, q)$. If $n = r$, then $GL_{n/r}(q^r)$ is just the (cyclic) multiplicative group of $GF(q^n)$, in other words it is the Singer cyclic subgroup of $GL_n(q)$; in this case M normalizes the Singer cyclic subgroup of G.

If M is irreducible, then the remaining possibility is that every non-trivial normal subgroup N is absolutely irreducible. In this case we can choose N such that $\bar{N}$ (the image of N in $\bar{G}$) is a minimal normal subgroup of $\bar{M}$. It is known that such a subgroup $\bar{N}$ is one of: non-abelian simple; a product of isomorphic non-abelian simple groups; an elementary abelian r-group for some prime r distinct from p. Let us take the elementary abelian r-group case: this is the case where there appears little geometric structure in $PG(n-1, q)$, despite the name (N is an extraspecial r-group of symplectic type); this is the class $\mathcal{C}_6$.

If $\bar{N}$ is a product of (at least two) isomorphic non-abelian simple groups, then N becomes a central product of isomorphic groups $N_1 N_2 \ldots N_r$ and each N_i has to be reducible. The setting is quite similar to that for $\mathcal{C}_4$ described above, but regarding

N as the absolutely irreducible subgroup of G and N_1 the reducible normal subgroup. We find that N preserves a tensor product structure with r factors (that are isometric when a form is present) permuted by M: this is the class $\mathcal{C}_7$. In geometric terms, this class corresponds once more to a Segre variety, $S_{m,m,\ldots,m}$ (where $m = (n/r)-1$).

Finally consider the possibility that $\bar{N}$ is non-abelian and simple. Two examples come to mind. The first is that N is defined over a subfield of $GF(q)$: these are the subgroups in the class $\mathcal{C}_5$ and in geometric terms are groups that preserve a sub-geometry. The other is that N cannot be defined over a subfield but that it fixes a form: this is the class $\mathcal{C}_8$. These examples do not exhaust the possibilities for N, but they do significantly restrict the the characteristics of such a group if it has not already been addressed. We thus arrive at Aschbacher's Theorem:

Theorem 3.1 (Aschbacher) *Let F_0 be a simple classical group and suppose that $F_0 \leq F \leq Aut(F_0)$. If $F_0 \cong P\Omega^+(8,q)$ assume that F contains no triality automorphism. Let H be a proper subgroup of F such that $F = HF_0$. Then either H is contained in one of $\mathcal{C}_1 - \mathcal{C}_8$ or the following hold:*

(a) *$H_0 \leq H \leq Aut(H_0)$ for some non-abelian simple group H_0;*

(b) *Let L be the full covering group of H_0 and let V be the natural vector space on which L acts (such that the projective image of L is precisely H_0), then L is absolutely irreducible on V;*

(c) *The representation of L on V is defined over no proper subfield of $GF(q)$;*

(d) *If L fixes a form on V, then F_0 is the group $PSL_n(q), PSp_n(q), P\Omega_n(q)$ or $PSU_n(q)$ corresponding to the form.*

In geometrical terms, Aschbacher's Theorem states that a subgroup of a simple classical group acting on $PG(n-1,q)$ stabilizes one of the following configurations: a subspace, a set of m subspaces of dimension $(n/m)-1$ spanning $PG(n-1,q)$, a spread of subspaces, a Segre variety, a sub-geometry, a polar space or a quadric, or it normalizes a Singer cycle, or it normalizes an r-group of symplectic type, or it is almost simple.

Liebeck and Seitz ([41]) have produced an alternative proof of Aschbacher's Theorem as a corollary to a version of the theorem for algebraic groups, but this does not amount to a geometric proof.

4 Progress towards a geometric classification for $\mathcal{C}_1$ to $\mathcal{C}_8$

In this section we chart the progress towards proving maximality or otherwise of subgroups in the classes $\mathcal{C}_1$ to $\mathcal{C}_8$ in a geometric manner. It is appropriate to note the work of Shangzhi Li who has obtained extensive results by matrix methods, but also to state that these are not geometrical. It is also appropriate to note that Aschbacher's Theorem itself can be used, because an overgroup of a prospective subgroup would itself either lie inside a maximal subgroup of one of the eight classes or be almost simple. This is precisely the approach taken in [40], where knowledge of the representations of finite groups of Lie type, of symmetric and alternating groups, and of sporadic simple groups enables Kleidman and Liebeck to obtain a classification of the maximal subgroups among the classes $\mathcal{C}_1$ to $\mathcal{C}_8$. It should be noted, however, that the statement of their result is for $n \geq 13$, with smaller cases referred to a book by Kleidman that has not, to date, been published. A statement of Kleidman's results appears in his thesis ([39]), but no proof is furnished and one or two errors have been identified. In his paper, Aschbacher identifies more restrictions than we have noted in the previous section and Kleidman and Liebeck cover similar ground; we give details, where appropriate, below. Many of the papers cited below state results over arbitrary commutative fields, but we restrict statements to finite fields. To date there have been no geometrical proofs of maximality for subgroups in $\mathcal{C}_4$ or $\mathcal{C}_7$, the stabilizers of Segre varieties, although it is reasonable to anticipate such proofs. The normalizers of symplectic type groups ($\mathcal{C}_6$) are harder to fathom geometrically and they tend to be relatively small subgroups, so a complete geometric proof of maximality appears some way off (there is a partly geometric proof in one case, but it requires a deep result from Group Theory).

4.1 $\mathcal{C}_1$

The groups in this section are stabilizers of subspaces. We denote by G the classical group $SL_n(q)$, $Sp_n(q)$, $\Omega_n^\epsilon(q)$ or $SU_n(q)$ and by H the stabilizer of an r-dimensional subspace U of $V(n,q)$. We have already noted that in the cases $G \neq SL_n(q)$, the subspace has to be totally isotropic or non-isotropic, and that in the case $G = \Omega_n^\epsilon(q)$, the subspace has to be either totally singular, or 1-dimensional with q even. We denote the Witt dimension by ν, this being the dimension of a maximal totally singular subspace of $V(n,q)$. The maximal subgroups in this class are completely determined. The corresponding results for $\bar{G}$ follow very easily from the theorems below.

Theorem 4.1 ([30]) *If $G = SL_n(q)$, then H is a maximal subgroup of G.*

Theorem 4.2 ([30]) *If $G = Sp_n(q)$, $\Omega_n^\epsilon(q)$ or $SU_n(q)$, and if U is totally singular, then H is a maximal subgroup of G, except when $G = \Omega_n^\epsilon(q)$ and $(n,\nu) = (2,1)$, $n = 2\nu = 2r+2$ or $(n,\nu,q) = (4,2,2)$.*

In the first exceptional case, there are only two totally singular 1-dimensional subspaces and G fixes them both, so $H = G$. In the second case, U lies in two $r+1$-dimensional totally singular subspaces and H fixes both of them, so H lies in the stabilizer of a larger subspace. The third case is one that we exclude from our discussion because $\Omega_4^+(2)$ is not simple, although H is in fact maximal when $r = 2$. Note that in [30] results are stated for $SO_n(q)$ rather than $\Omega_n(q)$, but the methods are readily adapted to $\Omega_n(q)$.

Theorem 4.3 ([31]) *If $G = Sp_n(q)$, $\Omega_n^\epsilon(q)$ or $SU_n(q)$, and if U is non-isotropic but not isometric to $U^\perp$, then H is a maximal subgroup of G, except in the following cases:*

(a) *$G = \Omega_n^\epsilon(q)$ and*

$$(n,r,\nu,\nu_1,\nu_2,q) = (\geq 6,2,\nu,1,\nu_2,2),(6,2,2,0,2,2),(3,1,1,0,1,\leq 11),$$
$$(3,1,1,0,0,\leq 9),(4,2,1,0,1,3),(5,2,2,0,1,3).$$

(b) *$G = SU_n(q)$ and $(n,r,q) = (3,1,4)$.*

It happens again that [31] gives results for $SO_n(q)$ rather than $\Omega_n(q)$. However the methods can be adapted and further exceptions arise only when $n = 3$ (and here $\Omega_3(q)$ is isomorphic to $PSL_2(q)$). In fact the arguments in [31] can be simplified significantly for finite fields by noting that H is self-normalizing and generated by elements leaving fixed a subspace of codimension 2.

Theorem 4.4 ([32]) *If $G = \Omega_n^\epsilon(q)$ with q even and if U is non-singular with $r = 1$, then H is a maximal subgroup of G, except when $(n, q, \nu) = (4, 2, 2)$.*

In the exceptional case, there is a unique anisotropic 2-dimensional subspace W containing U, and H preserves the direct sum decomposition $W \oplus W^\perp$. Thus H lies inside a subgroup of class $\mathcal{C}_2$ (and inclusion is proper).

4.2 $\mathcal{C}_2$

In this section H is the stabilizer of a direct sum decomposition into subspaces of dimension r. If $G = Sp_n(q)$, $\Omega_n^\epsilon(q)$ or $SU_n(q)$, then the subspaces have to be isometric and either totally isotropic or non-isotropic: we write them as $U_1, U_2, \ldots, U_s$. Moreover if U_i is totally isotropic, then $s = 2$; if U_i is non-isotropic, then $U_1, U_2, \ldots, U_s$ are pairwise orthogonal. The maximal subgroups in this class are almost completely determined and the corresponding results for $\bar{G}$ follow very easily.

Theorem 4.5 ([28, 29, 33]) *If $G = SL_n(q)$, then H is a maximal subgroup of G, except when $(n, r, q) = (4, 2, 3)$, $(n, 2, 2)$, $(odd > 3, 1, odd)$ or $(\geq 3, 1, 2$ or $4)$.*

Theorem 4.6 ([35]) *If $G = Sp_n(q)$ (with q odd), $\Omega_n^+(q)$ (with n a multiple of 4) or $SU_n(q)$, and if U_i is totally isotropic, then H is a maximal subgroup of G, except when $G = Sp_n(q)$ and $(n, q) = (4, 3)$, $G = \Omega_n^+(q)$ and $(n, q) = (4, \leq 11)$ or $(8, 2)$, and $G = SU_n(q)$ and $(n, q) = (4, 4$ or $9)$.*

We note that [35] deals with $SO_n^+(q)$ rather than $\Omega_n^+(q)$, but that the methods are readily adapted for $n > 4$. In the isomorphism from $P\Omega_4^+(q)$ to $PSL_2(q) \times PSL_2(q)$, the copies of $PSL_2(q)$ act independently on the two sets of $q+1$ totally singular lines. The subgroup H may be seen to correspond to $PSL_2(q) \times D_{q-1}$ and is therefore maximal precisely when D_{q-1} is maximal in $PSL_2(q)$. We should also explain that when $G = Sp_n(q)$ and q is even, H lies inside an orthogonal group, and when $G = \Omega_n^+(q)$ with r odd, H stabilizes each of U_1 and U_2.

Theorem 4.7 ([31, 33, 34, 36, 24]) *If $G = Sp_n(q)$, $\Omega_n^\epsilon(q)$ or $SU_n(q)$, and $r \geq 2$, and if each U_i is non-isotropic, then H is a maximal subgroup of G, except when*

(a) *$G = Sp_n(q)$ and $(n, r, q) = (4, 2, 3)$ or $(n, 2, 2)$;*

(b) *$G = \Omega_n^\epsilon(q)$ and $(r, \nu_1, q) = (2, 1, \leq 5)$, $(3, 1, 3)$, or $(4, 2, 2)$, $(n, r) = (4, 2)$, or $(r, \nu_1, q) = (2, 0, 3)$.*

(c) *$G = SU_n(q)$ and $(r, q) = (2, 4)$.*

We have excluded here the stabilizer of a simplex (i.e., when $r = 1$) when $G = \Omega_n^\epsilon(q)$ or $SU_n(q)$. In [24], Dye has results for $O_n(q)$ and $U_n(q)$ but the methods are not readily applicable when $\Omega_n^\epsilon(q)$ and $SU_n(q)$. In [31] and [34] (but not in [36]) results are stated for $SO_n(q)$ rather than $\Omega_n(q)$, but again the methods are readily adapted to $\Omega_n(q)$.

4.3 $\mathcal{C}_3$

In this section we describe the subgroups in terms of their actions on $PG(n-1, q)$: each stabilizes a spread of subspaces. Throughout the section, $n = rm$ with r prime and $m \geq 2$, and H is the stabilizer of a spread of $r-1$-spaces. Complete results are available for some groups but not for others. Little has been done from a geometric viewpoint regarding the normalizers of Singer cyclic subgroups.

The first group considered is $PSL_n(q)$. All maximal subgroups in the class are determined and are given by the following result.

Theorem 4.8 ([20]) *H is the normalizer of $PSL_m(q^r)$ and is a maximal subgroup of $PSL_n(q)$.*

Next we consider $PSp_n(q)$. There is one set of subgroups corresponding to the $PSL_n(q)$ case, but there is an additional set of subgroups arising from the embedding of $U_m(q^2)$ in $Sp_n(q)$ when $r = 2$. In the first case, the subspaces in the spread are all totally isotropic, but in the second case the spread is a mixed spread of totally isotropic and non-isotropic lines. All maximal subgroups in the class are determined and are given as follows.

Theorem 4.9 ([17, 18, 19]) *Suppose that n is a multiple of 4 when $r = 2$. Then H is the normalizer of $PSp_m(q^r)$ is a maximal subgroup of $PSp_n(q)$.*

Theorem 4.10 ([4]) *For $m \geq 2$ and $n = 2m$, the H is the image of the normalizer of $U_m(q^2)$ in $Sp_n(q)$ and is a maximal subgroup of $PSp_n(q)$ except for $(m, q) = (2, 3)$.*

Now consider $P\Omega_n(q)$. There are results when $r = 2$, although it is known (from [40] for example) that there are maximal subgroups for all prime $r > 2$ corresponding to the first set below (i.e., $\Omega_m(q^r) \leq \Omega_n(q)$). In the first result, the spread is a mixed spread of totally isotropic and non-isotropic lines. In the second result, the spread of $PG(n-1, q)$ is again a mixed spread but the totally singular lines partition the quadric so that the subgroup may be seen as the stabilizer of a spread of a quadric.

Theorem 4.11 ([6]) *Suppose that $n = 2m$ with $m \geq 3$ and q odd. Then H is the normalizer of $P\Omega_m(q^2)$ and is a maximal subgroup of $P\Omega_n(q)$.*

Theorem 4.12 ([21]) *Suppose that $n = 2m$ with $m \geq 3$. Then H is the image of the normalizer of $U_m(q^2) \cap \Omega_n(q)$ in $\Omega_n(q)$ and is a maximal subgroup of $P\Omega_n(q)$.*

There are currently no results for $PSU_n(q)$, although it is known (from [40] for example) that there are maximal subgroups for all prime $r > 2$.

4.4 $\mathcal{C}_5$

There are not many results for groups in this class. Here we take q_0 such that $q = q_0^r$: an immediate restriction is that r must be prime. It is known that there maximal subgroups for all $r \geq 2$ of the same type, and a result for $PSL_n(q)$ has been announced by Cossidente and Siciliano to the effect that the normalizer of $PSL_n(q_0)$ is maximal in $PSL_n(q)$. The only published results are the following, covering the only exceptions to the 'same type' pattern. They both arise from commuting polarities of $PG(n-1, q)$, studied extensively in [48]. In the first case, the commuting polarities are symplectic and unitary and in the second case they are orthogonal and unitary. In both cases the subgeometry consists of the points fixed by the product of the polarities.

Theorem 4.13 ([7]) *Suppose that n is even and $q = q_0^2$. Then the normalizer of $PSp_n(q_0)$ is a maximal subgroup of $PSU_n(q)$.*

Theorem 4.14 ([5]) *Suppose that $n \geq 3$ and that $q = q_0^2$ is odd. The normalizer of $P\Omega_n^\epsilon(q_0)$ is a maximal subgroup of $PSU_n(q)$, except when $(n, q) = (3, 3)$ or $(3, 5)$ and when $(n, q, \epsilon) = (4, 3, +)$.*

4.5 $\mathcal{C}_8$

The groups in this class are essentially stabilizers of forms, sometimes up to a scalar factor. In practice this means the normalizer of one classical group lying inside another. With one exception this amounts to studying $PSp_n(q)$, $P\Omega_n^\epsilon(q)$ or $PSU_n(q)$ as subgroups of $PSL_n(q)$. The geometric structure can be viewed as the polar space structure: the set of totally isotropic or totally singular subspaces. In the cases of $P\Omega_n^\epsilon(q)$ or $PSU_n(q)$ this equivalent to stabilizing the set of points of a quadric or Hermitian variety; in the case of $PSp_n(q)$ it is equivalent to stabilizing the self-polar lines. In the following the given subgroup is the stabilizer of the relevant polarity. The maximal subgroups in this class are completely determined.

Theorem 4.15 ([23, 37, 38])

(a) *For $n > 2$ and even, the normalizer of $PSp_n(q)$ is a maximal subgroup of $PSL_n(q)$.*

(b) *When q is odd, the normalizer of $P\Omega_n(q)$ is a maximal subgroup of $PSL_n(q)$ except when $n = 2$ and $q \leq 11$.*

(c) *The normalizer of $PSU_n(q)$ is a maximal subgroup of $PSL_n(q)$.*

The only other situation that needs to be addressed is when q is even and one considers $P\Omega_n^\epsilon(q) \leq PSL_n(q)$. Here the bilinear form is an alternating form and $PO_n(q)$ is a subgroup of $PSp_n(q)$, and the subgroup of $PSp_n(q)$ that we need is the stabilizer of a quadric.

Theorem 4.16 ([22]) *When q is even, the normalizer of $P\Omega_n^\epsilon(q)$ is a maximal subgroup of $PSp_n(q)$, except when $n = 2$ and $\epsilon = -$.*

In fact Dye only considers considers quadratic forms with positive Witt index. However, the only remaining case is that of the normalizer of $P\Omega_2^-(q)$, i.e., $D_{2(q+1)}$, as a subgroup of $PSL_2(q)$, and this is maximal unless $q = 2$ (where the two groups are equal).

5 The Class $\mathcal{S}$: subgroups not lying inside groups in $\mathcal{C}_1$ – $\mathcal{C}_8$

By Aschbacher's Theorem, a subgroup of a simple finite classical group that does not lie inside one of the classes $\mathcal{C}_1$ to $\mathcal{C}_8$ is almost simple and its pre-image in $SL_n(q)$ is absolutely irreducible. Such a subgroup lies in a class that Aschbacher denotes by $\mathcal{S}$. However, there arises within $\mathcal{S}$ a class of subgroups that is worth picking out and labelling as $\mathcal{C}_9$. Early attention paid to these groups in the context of $\mathcal{S}$ appears in [49], although not in a geometric fashion. The existence of the groups comes out of Steinberg's Tensor Product Theorem ([50]). Nevertheless, the groups can be described relatively easily in geometric terms: they stabilize the intersection of a Segre variety and a sub-geometry. In [8], the terms *Hermitian Veronesean* and *Twisted Hermitian Veronesean* are used for these sets in the context of $PG(8, q)$.

5.1 The Class $\mathcal{C}_9$: Twisted Tensor Products

Consider the group $G = GL_m(q^t)$ acting on an m-dimensional vector space V_1 over $GF(q^t)$ and let $\psi : GF(q^t) \to GF(q^t)$, $x \mapsto x^q$, be the Frobenius automorphism of $GF(q^t)$. Given a basis, $x_1, x_2, \ldots, x_m$ for V_1, for any matrix A and any vector $v \in V_1$, we denote by A^ψ and v^ψ the matrix and vector obtained by raising each coefficient to the power q. The tensor product $V_t = V_1 \otimes V_1 \otimes \cdots \otimes V_1$ (with t components) admits an action whereby $v_1 \otimes v_2 \otimes \cdots \otimes v_t$ is mapped to $v_1 A \otimes v_2 A^\psi \otimes \cdots \otimes v_t A^{\psi^{t-1}}$. We thus have a representation ρ of G on V_t that happens to be absolutely irreducible.

This representation of G can be written over $GF(q)$, which means to say that there is a basis for V_t with respect to which the elements of G act as matrices of $GL_n(q)$ (where $n = \dim V_t$). In order to see this, let $\phi : V_t \to V_t$ be the map given by $\lambda u_1 \otimes u_2 \otimes \cdots \otimes u_t \to \lambda^q u_t \otimes u_1 \otimes \cdots \otimes u_{t-1}$, with each u_i being one of $x_1, x_2, \ldots, x_m$, extended linearly over $GF(q)$. The set V of all vectors in V_t that are fixed by ϕ is fixed by G and is a $GF(q)$–subspace of V_t. Moreover $GF(q)$–linearly independent vectors in V are linearly independent over $GF(q^t)$ and we conclude that V has dimension $n = m^t$ over $GF(q)$. Let $\Omega = \{1, 2, \ldots, t\}$ and let $c = (1234\ldots t)$, a cyclic permutation of Ω; we can consider the action of c on the set of partitions of Ω into m (possibly empty) subsets. For each orbit, Δ say, of c on these partitions, choose an element $\mathcal{P}$ of Δ (i.e., a partition of Ω into m subsets, $\mathcal{O}_1, \mathcal{O}_2, \ldots, \mathcal{O}_m$ say) and a vector $u = u_1 \otimes u_2 \otimes \cdots \otimes u_t$, with each u_i being one of $v_1, v_2, \ldots, v_m$ and equalling v_j if and only if $i \in \mathcal{O}_j$. Let s be the length of Δ. Then the vectors $\sum_{j=1}^{s} \phi^{j-1}(\lambda u)$, as λ ranges over $GF(q^s)$, span a $GF(q)$–subspace $V(\Delta)$ of V_t of dimension s, fixed by ϕ vectorwise. A basis for $GF(q^s)$ over $GF(q)$ gives rise to a basis for $V(\Delta)$. The direct sum of such subspaces gives a $GF(q)$-subspace of dimension $n = m^t$ and this is precisely V.

This representation of G is not necessarily faithful, for scalar matrices in $GL_m(q^t)$ can act trivially on V_t. However the associated representation of $PGL_m(q^t)$ on $PG(n-1, q)$ is faithful. In geometric terms, $PG(n-1, q)$ is a subgeometry of

$PG(n-1,q^t)$ fixed by $PGL_m(q^t)$. For any $0 \neq v \in V_1$, consider a vector $\mathbf{v} = v \otimes v^{\psi} \otimes \cdots \otimes v^{\psi^{t-1}} \in V_t$. We see that $\mathbf{v}$ is fixed by ϕ so lies in V. Moreover $\lambda\mathbf{v} = (\lambda + \lambda^q + \cdots + \lambda^{q^{t-1}})\mathbf{v}$ with $\lambda + \lambda^q + \cdots + \lambda^{q^{t-1}} \in GF(q)$. Hence points of $PG(m-1,q^t)$ are mapped to points of $PG(n-1,q)$ and the image of $PG(m-1,q^t)$ is a configuration fixed by $PGL_m(q^t)$. Moreover, whenever there is a configuration of points in $PG(m-1,q^t)$ fixed by a subgroup H of $PGL_m(q^t)$, there is a corresponding configuration in $PG(n-1,q)$ fixed by H. We shall say more on possible configurations shortly. If $g \in GL_m(q^t)$ has determinant 1, then the same is true for $\rho(g) \in GL_n(q)$, so $PSL_m(q^t)$ is embedded in $PSL_n(q)$.

Suppose that f_1 is a non-degenerate form on V_1, being one of: alternating, symmetric bilinear or Hermitian. Then f_1 determines a matrix A_1 so that $f_1(u,v) = uA_1v^T$ and we can regard f_1^{ψ} as being the form corresponding to A_1^{ψ}. In most cases the basis for V_1 can be chosen in a natural manner so that A_1 has coefficients in $GF(q)$ and then $A_1^{\psi} = A_1$. Now let f be the form on V_t given by

$$f(u_1 \otimes \cdots \otimes u_t, w_1 \otimes \cdots \otimes w_t) = \prod_{i=1}^{t} f_1^{\psi^{i-1}}(u_i, w_i),$$

where u_i, w_i are arbitrary vectors in V_1. It is not difficult to show that f is non-degenerate. If f_1 is symmetric bilinear, or if f_1 is alternating and t is even, then f is symmetric bilinear. If f_1 is alternating and t is odd, then f is alternating. If q is even, then the terms alternating and symmetric bilinear are equivalent; here there is a unique quadratic form Q on V_t such that f is the bilinear form associated with Q and such that $Q(u_1 \otimes \cdots \otimes u_t) = 0$ for all $u_i \in V_1$. It is not difficult to show that the restrictions of f and Q to V take values in $GF(q)$: the vectors $\mathbf{v}$ span V and $f(\mathbf{u},\mathbf{v}) = \prod_{i=1}^{t} f_1^{\psi^{i-1}}(u^{\psi^{i-1}}, w^{\psi^{i-1}}) = \prod_{i=1}^{t} f_1(u,w)^{\psi^{i-1}} \in GF(q)$. If H is a subgroup of $GL_m(q^t)$ preserving f_1, then H preserves f and Q. Thus we have embeddings $PO_m(q^t)$ in $PO_n(q)$, $PSp_m(q^t)$ in $PO_n(q)$ when t is even and/or q is even, and $PSp_m(q^t)$ in $PSp_n(q)$ when t and q are odd. When q is even we have $PSp_m(q^t) \leq PO_n(q) \leq PSp_n(q)$. Given points of $PG(m-1,q^t)$ represented by vectors u, v we see that $f(\mathbf{u},\mathbf{v}) = 0$ if and only if $f_1(u,v) = 0$. It follows that a partial ovoid in a polar space in $PG(m-1,q^t)$ is embedded as a partial ovoid of a polar space in $PG(n-1,q)$, and that a subgroup of $PGL_m(q^t)$ fixing the former is embedded as a subgroup of $PGL_n(q)$ fixing the latter.

If we now consider $U_m(q^{2t})$, then f is given by

$$f(u_1 \otimes \cdots \otimes u_t, w_1 \otimes \cdots \otimes w_t) = \prod_{i=1}^{t} f_1^{\psi^{2(i-1)}}(u_i, w_i).$$

This is an Hermitian form on V_t. This time V is defined over $GF(q^2)$. The restriction of f to V takes values in $GF(q^2)$. However this restriction is an Hermitian form only if t is odd; if t is even, then the restriction becomes a symmetric bilinear form and $PU_m(q^{2t})$ becomes a subgroup of $O_n(q^2)$. Steinberg's Tensor Product Theorem leads us to believe that for t even $\rho(U(n,q^t))$ is not absolutely irreducible. Indeed for the case $t=2$ it is known that $\rho(U(n,q^2))$ is reducible, for it follows from [11, Theorem 43.14] that $\rho(U(n,q^2))$ fixes all vectors in a 1-dimensional subspace of $V(n^2,q^2)$; moreover the restriction of the Hermitian form f to $V_q = W$ is actually a symmetric bilinear form so $\rho(U(n,q^2))$ is a subgroup of $O(n^2,q)$ (for q odd) or $Sp(n^2,q)$ (for q even).

If t is composite, say $t = rs$, then we can consider initially the map from $GL_m((q^r)^s)$ to $GL_{m^s}(q^r)$ followed by the map from $GL_{m^s}(q^r)$ to $GL_n(q)$. This explains most of the restrictions below to t prime. The same argument applied to the map from $GL_m((q^r)^{2s})$ to $GL_{m^s}(q^{2r})$ followed by the map from $GL_{m^s}(q^{2r})$ to $GL_n(q^2)$ explains the unitary restrictions.

Now we consider a twisted version of the above. Given a matrix A, we denote by A^* the matrix $(A^T)^{-1}$. For t even, there is an action on V_t (taking V_1 as being defined over $GF(q^{2t})$) whereby $v_1 \otimes v_2 \otimes \cdots \otimes v_t$ is mapped to $v_1 A \otimes v_2 A^{*\psi} \otimes v_3 A^{\psi^2} \otimes \cdots \otimes v_t A^{*\psi^{t-1}}$. Once again we have a representation, this time denoted ρ^*, of $GL_m(q^{2t})$ on V_t that happens to be absolutely irreducible and realizable over $GF(q^2)$. If $t > 2$, say $t = 2s$ with $s > 1$, then we have $(A \otimes A^{\psi^2} \otimes \cdots \otimes v_t A^{*\psi^{2s-2}}) \otimes (A \otimes A^{\psi^2} \otimes \cdots \otimes v_t A^{*\psi^{2s-2}})^{*\psi} \in GL_{m^s}(q^4) \leq GL_n(q^2)$. Thus $t = 2$ is the only case that we need to consider. We actually consider the action of G given by $v_1 \otimes v_2$ being mapped to $v_1 A^* \otimes v_2 A^{\psi}$. A non-degenerate Hermitian form is defined by $(u \otimes v, w \otimes z) = (uz^{\psi T}).(w^{\psi} v^T)$ and this is preserved by $\rho^*(g) = (g^T)^{-1} \otimes g^{\psi}$ for all $g \in G$. Hence $\rho^*(G) \leq U_n(q^2)$.

We have observed in [9] that $PSL(2,2^t)$ fixes an ovoid in $PG(2^t-1,2)$ which is also a polygon admitting A_{2^t+1} as an automorphism group, so that $PSL(2,2^t) < A_{2^t+1} < P\Omega^+(2^t,2)$ and that the alternating group is the full stabilizer of the ovoid. The representation here of the alternating group is precisely that arising from the fully deleted permutation module. In the case of the embedding of $P\Omega(3,3^t)$ in $P\Omega(3^t,3)$ (see [8]), the configuration is a rational curve and a partial ovoid in $PG(3^t-1,3)$. The partial ovoid has A_{3^t+1} as its stabilizer, acting on the points of the partial ovoid, and once again the representation is equivalent to that arising from the deleted permutation module.

In summary the inclusions considered as $\mathcal{C}_9$ subgroups are as follows, with t prime and $n = m^t$:

(a) $PSL_m(q^t) \leq PSL_n(q)$ with $m \geq 3$;

(b) $PSp_m(q^t) \leq PSp_n(q)$ with m even, t odd and q odd;

(c) $PSp_m(q^t) \leq P\Omega_n^+(q)$ with m even, t odd and q even;

(d) $PSp_m(q^2) \leq P\Omega_n^{\epsilon}(q)$ with m even and $\epsilon = (-1)^{m/2}$;

(e) $P\Omega_m(q^t) \leq P\Omega_n(q)$ with m odd and q odd;

(f) $P\Omega_m(q^t) \leq P\Omega_n(q)$ with m even, $m \geq 6$, t odd and q odd;

(g) $P\Omega_m(q^2) \leq P\Omega_n^{\epsilon}(q)$ with m even, $m \geq 6$, q odd and $\epsilon = (-1)^{m/2}$;

(h) $PSU_m(q^{2t}) \leq PSU_n(q^2)$ with $m \geq 3$ and t odd;

(i) $PSL_m(q^4) \leq PSU_n(q^2)$ with $m \geq 3$ and $t = 2$.

Most of the subgroups listed above are maximal. This is proved by Schaffer in the theorem given below, but the proof is very far from geometrical, being based on the representation theory of the groups involved. We should add that Schaffer has the restriction tq odd in (e) above, although there appears no reason to exclude $t = 2$.

Theorem 5.1 ([47]) *The subgroups listed above are maximal except in the following cases, with a maximal overgroup as indicated.*

(a) $PSp_2(2^t) \leq A_{2^t+1} \leq P\Omega^+_{2t}(2)$ *with t odd;*

(b) $PSp_2(q^t) \leq PSp_{2t}(q) \leq P\Omega^+_{2t}(q)$ *with t odd and q even;*

(c) $PSp_4(q^t) \leq PSp_{4t}(q) \leq P\Omega^+_{4t}(q)$ *with q even;*

(d) $P\Omega_3(3^t) \leq A_{3^t+1} \leq P\Omega_{3^t}(3)$.

As noted above, points of $PG(m-1, q^t)$ are mapped to points of $PG(n-1, q)$ and so various configurations in $PG(m-1, q^t)$ are mapped to configurations in $PG(n-1, q)$. In particular, partial ovoids of a polar space in $PG(m-1, q^t)$ are mapped to partial ovoids of a polar space in $PG(n-1, q)$ having the same size. In what might be regarded as a degenerate case of this, when $m = 2$ the points of a projective line are mapped to points of a partial ovoid of a quadric in $PG(2^t - 1, q)$. When q is even, the size of this partial ovoid meets the Blokhuis-Moorhouse bound (see [2]). In terms of the embedding of groups, this is the excepted case $PSp_2(q^t) \leq P\Omega^+_{2t}(q)$ with t odd and q even. This embedding is studied in detail in [9], along with the embedding of $PSp_4(q^t)$ in $PSp_{4t}(q) \leq P\Omega^+_{4t}(q)$ with q even. This latter case sees an ovoid of $PG(3, q^t)$ mapped to a partial ovoid of $PG(4^t - 1, q)$, again the size of this partial ovoid meets the Blokhuis-Moorhouse bound. In [10] we shall show that the intermediate subgroups noted by Schaffer for these two cases can be described geometrically.

5.2 Other groups in $\mathcal{S}$

It is difficult to assess the general pattern of almost simple groups that might appear as subgroups of a simple finite classical group and that lie in $\mathcal{S}$ but not $\mathcal{C}_9$. When listing these groups we often give the simple group rather than its slightly larger normalizer, and we omit most details as to which values of q apply.

Among the small dimensional groups for which complete details were given in Section 2 (at least of maximal subgroups) we have:

- $A_5 \leq PSL_2(q)$;
- $A_5 \leq P\Omega_3(q)$;
- $PSL_3(2)$, A_6, A_7 all subgroups of $PSL_3(q)$;
- $PSL_3(2)$, A_6, A_7 all subgroups of $PSU_3(q)$;
- $A_6, A_7, PSL_2(q)$ all subgroups of $PSp_4(q)$ for odd q;
- $A_7 \leq PSL_4(2)$;
- $A_5 \leq P\Omega^-_4(q)$;

- $A_6, A_7, PSL_2(q)$ all subgroups of $P\Omega_5(q)$;
- $A_7 \leq P\Omega_6^+(2)$.

We have already commented on the status of the lists in Kleidman's thesis ([39]). However the lists do provide useful evidence. In the lists below, we give the simple groups rather than their normalizers, and note that these go as far as $n = 11$.

The first list comprises specific simple groups rather than families of simple groups, and we omit the classical groups they belong to. Suffice to say that these are all subgroups of $PSL_n(q)$ for some $n \leq 11$ and in many cases for infinitely many values of q.

- A_m for $m = 6, 7, 9, 10, 11, 12, 13$;
- $PSL_2(r)$ for $r = 8, 11, 13, 17, 19, 23$;
- $PSL_3(3)$; $PSL_3(4)$;
- $PSU_3(9)$; $PSU_4(4)$; $PSU_4(9)$; $PSU_5(4)$;
- $PSp_6(2)$;
- $M_{11}, M_{12}, M_{22}, M_{24}$;
- J_2; J_3;
- $P\Omega_8^+(2)$;
- $Sz(8)$.

The second list comprises families of simple groups that appear to arise as subgroups in $\mathcal{S}$, excluding those known to lie in $\mathcal{C}_9$.

- $PSL_3(q) \leq PSL_6(q)$;
- $PSL_3(q), PSL_4(q), PSL_5(q)$ all subgroups of $PSL_{10}(q)$;
- $PSU_3(q) \leq PSU_6(q)$;
- $PSU_3(q), PSU_4(q), PSU_5(q)$ all subgroups of $PSU_{10}(q)$;
- $Sz(q) \leq PSp_4(q)$ for q even;
- $G_2(q), PSL_2(q)$ both subgroups of $PSp_6(q)$;
- $PSL_2(q) \leq PSp_{10}(q)$;
- $G_2(q) \leq P\Omega_7(q)$;
- $PSL_3(q)$, $PSU_3(q^2)$, $P\Omega_7(q)$, $P\Omega_8^-(\sqrt{q})$, ${}^3D_4(\sqrt[3]{q})$ all subgroups of $P\Omega_8^+(q)$;
- $PSL_3(q), PSU_3(q^2)$ both subgroups of $P\Omega_8^-(q)$;
- $PSL_2(q) \leq P\Omega_9(q)$;

- $PSp_4(q) \leq P\Omega_{10}^{-}(q)$;
- $PSp_4(q) \leq P\Omega_{10}^{+}(q)$;
- $PSL_2(q) \leq P\Omega_{11}(q)$.

References

[1] M. Aschbacher, On the maximal subgroups of the finite classical groups, *Invent. Math.* **76** (1984), 469–514.

[2] A. Blokhuis, G.E. Moorhouse, Some p-ranks related to orthogonal spaces, *J. Algebraic Combin.* **4** (1995), 295–316.

[3] J.H. Conway, R.T. Curtis, S.P. Norton, R.A. Parker, R.A. Wilson, *Atlas of Finite Groups*, Clarendon Press, Oxford, 1985.

[4] A. Cossidente, O.H. King, Maximal subgroups of finite symplectic groups stabilizing spreads of lines, *J. Algebra* **258** (2002), 493–506.

[5] A. Cossidente, O.H. King, Maximal orthogonal subgroups of finite unitary groups, *J. Group Theory* **7** (2004), 447–462.

[6] A. Cossidente, O.H. King, Maximal subgroups of finite orthogonal groups stabilizing spreads of lines, preprint.

[7] A. Cossidente, O.H. King, On some maximal subgroups of unitary groups, *Comm. Algebra* **32** (2004), 989–995.

[8] A. Cossidente, O.H. King, Embeddings of finite classical groups over field extensions and their geometry, *Adv. Geom.* **2** (2002), 13–27.

[9] A. Cossidente, O.H. King, Twisted tensor product group embeddings and complete partial ovoids on quadrics in $PG(2^t-1,q)$, *J. Algebra* **273** (2004), 854–868.

[10] A. Cossidente, O.H. King, On twisted tensor product group embeddings and the spin representation of symplectic groups, preprint.

[11] C.W. Curtis and I. Reiner, *Representation Theory of Finite Groups and Associative Algebras*, Wiley, New York 1962.

[12] L.E. Dickson, *Linear groups, with an Exposition of the Galois Field Theory*, Teubner, Leipzig, 1901.

[13] J. Dieudonné, *Sur les Groupes Classiques*, Actualités Sci. Indust. No. 1040, Hermann, Paris, 1948.

[14] J. Dieudonné, *La Géometrie des Groupes Classiques*, 2nd edition, Springer–Verlag, Berlin, New York, 1963.

[15] R.H. Dye, Partitions and their stabilizers for line complexes and quadrics, *Ann. Mat. Pura Appl.* (4) **114** (1977), 173–194.

[16] R.H. Dye, A maximal subgroup of $PSp_6(2^m)$ related to a spread, *J. Algebra* **84** (1983), 128–135.

[17] R.H. Dye, Maximal subgroups of symplectic groups stabilizing spreads, *J. Algebra* **87** (1984), 493–509.

[18] R.H. Dye, Maximal subgroups of symplectic groups stabilizing spreads II, *J. London Math. Soc.* **40** (1989), 215–226.

[19] R.H. Dye, Maximal subgroups of $PSp_{6n}(q)$ stabilizing spreads of totally isotropic planes, *J. Algebra* **99** (1986), 191–209.

[20] R.H. Dye, Spreads and classes of maximal subgroups of $\mathrm{GL}_n(q)$, $\mathrm{SL}_n(q)$, $\mathrm{PGL}_n(q)$ and $\mathrm{PSL}_n(q)$, *Ann. Mat. Pura Appl.* (4) **158** (1991), 33–50.

[21] R.H. Dye, Maximal subgroups of finite orthogonal groups stabilizing spreads of lines, *J. London Math. Soc.* (2) **33** (1986), 279–293.

[22] R.H. Dye, On the maximality of the orthogonal groups in the symplectic groups in characteristic two, *Math. Z.* **172** (1980), 203–212.

[23] R.H. Dye, Maximal subgroups of $\mathrm{GL}_{2n}(K)$, $\mathrm{SL}_{2n}(K)$, $\mathrm{PGL}_{2n}(K)$ and $\mathrm{PSL}_{2n}(K)$ associated with symplectic polarities, *J. Algebra* **66** (1980), 1–11.

[24] R.H. Dye, Maximal subgroups of the finite orthogonal and unitary groups stabilizing anisotropic subspaces, *Math. Z.* **189** (1985), 111–129.

[25] R.H. Dye, A geometric characterization of the special orthogonal groups and the Dickson invariant, *J. London Math. Soc.* (2) **15** (1977), 472–476.

[26] R.W. Hartley, Determination of the ternary collineation groups whose coefficients lie in the $\mathrm{GF}(2^n)$, *Ann. of Math.* **27** (1925/26), 140–158.

[27] J.W.P. Hirschfeld, *Finite Projective Spaces of Three Dimensions*, Oxford University Press, Oxford, 1991.

[28] J.D. Key, Some maximal subgroups of $PSL(n,q), n \geq 3, q = 2^r$, *Geom. Dedicata* **4** (1975), 377–386.

[29] J.D. Key, Some maximal subgroups of certian projective unimodular groups, *J. London Math. Soc.* (2) **19** (1979), 291–300.

[30] O.H. King, On some maximal subgroups of the classical groups, *J. Algebra* **68** (1981), 109–120.

[31] O.H. King, Maximal subgroups of the classical groups associated with non–isotropic subspaces of a vector space, *J. Algebra* **73** (1981), 350–375.

[32] O.H. King, Maximal subgroups of the orthogonal group over a field of characteristic two, *J. Algebra* **76** (1982), 540–548.

[33] O.H. King, Imprimitive maximal subgroups of the general linear, special linear, symplectic and general symplectic groups, *J. London Math. Soc.* (2) **25** (1982), 416–424.

[34] O.H. King, Imprimitive maximal subgroups of the orthogonal, special orthogonal, unitary and special unitary groups, *Math. Z.* **182** (1983), 193–203.

[35] O.H. King, Imprimitive maximal subgroups of the symplectic, orthogonal and unitary groups, *Geom. Dedicata* **15** (1984), 339–353.

[36] O.H. King, Imprimitive maximal subgroups of finite orthogonal groups, *Geom. Dedicata* **21** (1986), 341–348.

[37] O.H. King, On subgroups of the special linear group containing the special orthogonal group, *J. Algebra* **96** (1985), 178–193.

[38] O.H. King, On subgroups of the special linear group containing the special unitary group, *Geom. Dedicata* **19** (1985), 297–310.

[39] P.B. Kleidman, The maximal subgroups of the low–dimensional classical groups, Ph.D. Thesis, Cambridge 1987.

[40] P.B. Kleidman, M. Liebeck, *The Subgroup Structure of the Finite Classical Groups*, LMS Lecture Note Series 129, Cambridge University Press, Cambridge 1990.

[41] M.W. Liebeck, G.M. Seitz, On the subgroup structure of classical groups, *Invent. Math.* **134** (1998), 427–453.

[42] H.H. Mitchell, Determination of the ordinary and modular ternary linear groups, *Trans. Amer. Math. Soc.* **12** (1911), 207–242.

[43] H.H. Mitchell, The subgroups of the quaternary abelian linear group, *Trans. Amer. Math. Soc.* **15** (1914), 377–396.

[44] E.H. Moore, The subgroups of the generalized finite modular group, *Dicennial Publications of the University of Chicago* **9** (1904), 141–190.

[45] B. Mwene, On the subgroups of the group $PSL_4(2^m)$, *J. Algebra* **41** (1976), 79–107.

[46] B. Mwene, On some subgroups of $PSL(4,q)$, q odd, *Geom. Dedicata* **12** (1982), 189–199.

[47] M. Schaffer, Twisted tensor product subgroups of finite classical groups, *Comm. Alg.* **27** (1999), 5097-5166.

[48] B. Segre, Forme e geometrie hermitiane con particolare riguardo al caso finito, *Ann. Mat. Pura Appl.* (4) **70** (1965), 1–201.

[49] G.M. Seitz, Representations and maximal subgroups of finite groups of Lie type, *Geom. Dedicata* **25** (1988), 391–406.

[50] R. Steinberg, Representations of algebraic groups, *Nagoya Math. J.* **22** (1963), 33–56.

[51] A. Wiman, Bestimmung aller Untergruppen einer doppelt unendlichen Reihe von einfachen Gruppen, *Stockh. Akad. Bihang* **25** (1899), 1–47.

School of Mathematics and Statistics
University of Newcastle
Newcastle Upon Tyne
NE1 7RU, U.K.
O.H.King@ncl.ac.uk

Constructing combinatorial objects via cliques

Patric R. J. Östergård

Abstract

Many fundamental combinatorial objects, including balanced incomplete block designs and error-correcting codes, can be constructed and classified via cliques in certain problem-specific graphs. Various such objects are here identified and surveyed, and the utilization of clique algorithms in the construction of these is considered. Occasionally the type of problem admits a formulation as an instance of the exact cover problem, which, for computational reasons, is even more desirable.

1 Introduction

Cliques and *independent sets* are two of the most fundamental concepts in graph theory. A clique in a graph $G = (V, E)$ is a subset of vertices $V' \subseteq V$ that induces a complete graph. (A *complete graph* is a graph where all vertices are mutually adjacent.) An independent set, on the other hand, is a subset of vertices $V' \subseteq V$ that induces an empty graph. Obviously, a clique in a graph G is an independent set in the complement graph $\overline{G}$, and vice versa, so without loss of generality one may focus on just one of these concepts. Note that cliques are occasionally defined as complete subgraphs rather than sets; we choose the latter alternative, which is much more convenient.

In the current work we study combinatorial objects that can be viewed as *set systems* (but when discussing these objects later, they will generally not be treated using the set system formulation). A set system is a collection of subsets of a given set X, $\mathcal{S} = \{S_1, S_2, \ldots, S_m\}$, $S_i \subseteq X$, which has some additional specific properties. Such set systems correspond to cliques in a graph if the additional properties can be formulated via

(1) conditions that any set S_i must fulfill, such sets are called *admissible*; and

(2) conditions that any pair of sets, $\{S_i, S_j\}$, $i \neq j$, must fulfill, such pairs of sets are called *compatible*.

(Observe that the condition on pairs of sets is symmetric.) Now a graph can be constructed whose vertices correspond to the admissible sets and whose edges correspond to the compatible set pairs. Carter, in his Ph.D. thesis [16] in 1975, recognized that many problems of constructing combinatorial objects are of the above mentioned type, and was probably not the first one to do so.

For some problems, a formulation of this kind is straightforward and comes more or less directly from its definition, but in other cases some insight into the problem is needed.

Often one is interested in the largest set system, that is, the largest clique in the graph that was just constructed. Such a clique is called a *maximum clique*. A clique that is not a proper subset of another clique is said to be a *maximal clique*.

Algorithms for clique problems are treated in Section 2. Some of the problems to be encountered can also be stated in terms of the closely related exact cover problem; if this is the case, a substantially more efficient algorithm can be utilized. Only exact algorithms are considered here. Stochastic algorithms, on the other hand, are useful if one wants to find large cliques quickly, but they cannot tell whether the cliques found are maximum cliques.

In Section 3 we enter a journey through a variety of problems with a clique formulation. For all but the smallest instances it is not advisable to apply a clique algorithm directly, but it should be used as one component of a problem-specific machinery; these issues are covered in Section 4. If clique algorithms are used for classifying combinatorial objects up to isomorphism rather than constructing single examples, an even more involved apparatus is required [38].

It is possible that we have missed some important problem classes in this survey—which inevitably is biased towards the research topics of the author—but the choice of problems hopefully gives a clear picture of the central position of clique problems in combinatorics. Cliques corresponding to subobjects in objects—there are many problems of this kind in, for example, geometry [54, 78]—are not considered here (but this distinction is obviously rather vague); this also excludes extremal graphs in Ramsey theory from the scope of this survey.

2 Algorithms

We shall now briefly look at algorithms for clique and exact cover problems. Application of these algorithms to construction and classification problems is discussed in Section 4 after the treatment of combinatorial objects in Section 3.

2.1 Clique Algorithms

Finding cliques in graphs is in general a computationally hard problem, the following version of the problem being NP-complete [25].

Input: A graph $G = (V, E)$ and an integer k.
Question: Is there a clique in G of size k?

It is also known that the problem of approximating the maximum clique size is hard [28]. The fact that we often, in classifying combinatorial objects, not only need one maximum clique but *all* maximum cliques may look discouraging in view of these complexity results. However, albeit infeasible for general parameters, problem instances that one encounters are solvable with some effort until a limit that can be pushed surprisingly far. The limit is pushed by using (1) tailored clique algorithms and (2) tailored approaches utilizing properties of the objects under consideration.

A variety of results have been published on exact algorithms for the maximum clique problem—that is, algorithms that always find a maximum clique, in contrast to stochastic algorithms that try to find as large a clique as possible. The efficiency of the published algorithms very much depends on the type of instances. Two major types of exact algorithms have been considered: enumerative branch-and-bound algorithms and branch-and-cut algorithms utilizing solutions of relaxed linear programs.

A basic branch-and-bound algorithm, which despite its simplicity performs reasonably well, is published in [15]. A more advanced branch-and-bound algorithm is presented in [62, 65]; an implementation of this algorithm in C, called Cliquer, has been distributed [58]. Several published branch-and-bound algorithms depend on using colorings for bounding [82]; since coloring is an expensive procedure, tuning these algorithms is a nontrivial task.

In a branch-and-cut algorithm, the integrality conditions of an integer linear program (ILP) are relaxed and the optimal solution of the relaxation is used to cut branches of the search tree. The solution is also used to choose the variable to branch on; with 0-1 variables only, the variable whose value is closest to $\frac{1}{2}$ could be chosen.

To be able to apply branch-and-cut algorithms to an ILP for the maximum clique problem, let x_i be a variable that tells whether a vertex i in a graph $G = (V, E)$ is in a particular set (1) or not (0). Via the definition that a clique is a set of vertices no two of which are nonadjacent, the size of a maximum clique is

$$\begin{aligned}&\max \textstyle\sum_{i=1}^{n} x_i,\\ &\text{subject to}\\ &x_i + x_j \leq 1 \text{ for } \{i, j\} \notin E,\\ &x_i \in \{0, 1\}.\end{aligned}$$

An alternative—better, for computational reasons—ILP formulation is

$$\begin{aligned}&\max \textstyle\sum_{i=1}^{n} x_i,\\ &\text{subject to}\\ &\textstyle\sum_{i \in S} x_i \leq 1,\ \ S \in \mathcal{S},\\ &x_i \in \{0, 1\},\end{aligned}$$

where $\mathcal{S}$ is the collection of all maximal independent sets. With such an ILP formulation, the standard method of solving its LP relaxation can be used as a part of the algorithm. Whereas the global maxima of the proposed ILP formulations coincide, this is (usually) not the case for their relaxations. The latter formulation gives a smaller (or equal) maximum cost, which in turn is an upper bound on the size of a maximum clique. A solution also gives information for the branching step.

Even though the latter formulation is better, the number of maximal independent sets is in general prohibitively large. However, since the main idea of the ILP formulation is to obtain an upper bound on the size of a maximum clique via the LP relaxation, one may as well use a subset of the maximal independent sets to get a (possibly) slightly worse upper bound. In any case, the number of inequalities will be fairly large. Since most LP algorithms are able to handle many variables better than they can handle many inequalities, one should think about considering the *dual* problem instead; see [14], which also contains several other interesting ideas for solving maximum clique problems in this manner.

2.2 Exact Cover Algorithms

A problem closely related to clique problems is the exact cover problem, which is also NP-complete [25].

Input: A collection of subsets $\mathcal{S} = \{S_1, S_2, \ldots, S_m\}$ of a universal set U.
Question: Is there a subcollection $\mathcal{T} \subseteq \mathcal{S}$ that forms a partition of U?

Instances where all S_i have the same size can be reduced directly to the problem of finding cliques of a given size. Construct a graph with one vertex for each S_i, and add edges between vertices that correspond to disjoint sets. Then the exact cover problem has a solution exactly when there is a (maximum) clique of size $|U|/|S_i|$ in the constructed graph.

For computational reasons, it is important to identify exact cover problems and not try to solve these with a clique algorithm. It is a very strong requirement that all elements of U must be in some chosen set S_i, we say that they have to be *covered.* Whereas cliques can often be viewed as packings, exact covers are simultaneously packings and coverings.

There is a single state-of-the-art backtracking algorithm for the exact cover problem. The core of this algorithm is a heuristic that minimizes the branching factor on each level of the search tree [26, 40, 54]. Knuth [40] observes that a proper use of links in the basic data structure of the search leads to additional speed-up.

3 Combinatorial Objects as Cliques

We shall now present families of combinatorial objects that can be viewed as (maximum) cliques or exact covers. A common feature of many of these objects is the fact that they can be viewed as various types of packings. A good example of a packing is an error-correcting code. The connection between error-correcting codes and cliques is emphasized in [83].

3.1 Unrestricted Error-Correcting Codes

An unrestricted *code* is a set $C \subseteq \mathbf{Z}_q^n$, where $\mathbf{Z}_q = \{0, 1, \ldots, q-1\}$; the elements in $\mathbf{Z}_q^n$ are called *words* and those in C are called *codewords*. The parameter n is the *length* of the code and gives the number of *coordinates*. The *(Hamming) weight* of a word is the number of nonzero coordinates. The *(Hamming) distance* between two codewords $c, c' \in C$, denoted by $d_H(c, c')$, is the number of coordinates in which they differ. The *minimum distance* of a code C is

$$d(C) := \min\{d_H(c, c') : c, c' \in C, c \neq c'\}.$$

A code $C \subseteq \mathbf{Z}_q^n$ with cardinality M and minimum distance at least d is called an $(n, M, d)_q$ code, and the maximum cardinality of a code $C \subseteq \mathbf{Z}_q^n$ with minimum distance at least d is denoted by $A_q(n, d)$.

A code with minimum distance at least d is a clique in the following graph. Let there be one vertex in V for each word in $\mathbf{Z}_q^n$ and an edge between two vertices exactly when the Hamming distance between the corresponding codewords is at least d. This graph is called the *Hamming graph* $H_q(n, d)$.

A code C with minimum distance d is able to correct up to $d' = \lfloor (d-1)/2 \rfloor$ errors. Namely, since Hamming distance is a metric, we get by the triangle inequality that if at most d' errors have occurred in the coordinates of $c \in C$, then the distance from the erroneous word to any codeword in $C \setminus \{c\}$ is at least $d - d' > d'$. For odd minimum distance d, the spheres with radius d' around codewords must be nonoverlapping; error-correcting codes with odd minimum distance may therefore be viewed as packings of such spheres.

The graph $\overline{H_q(n,2)}$ is the q-ary n-dimensional hypercube, or just the q-ary n*-cube*. Since $\mathrm{Aut}(G) \cong \mathrm{Aut}(\overline{G})$, it is not difficult to see that the automorphism group of the q-ary n-cube—which has order $(q!)^n n!$, this was apparently first proved by Winkler [93]—is a subgroup of $\mathrm{Aut}(H_q(n,d))$. The order of $\mathrm{Aut}(H_q(n,d))$ has been studied, for example, in [13, 57].

For an overall approach for obtaining unrestricted codes efficiently utilizing a clique algorithm, see Section 4 (in particular, Subsection 4.2). We here list a few results in the literature that have been achieved in this manner. In [67], the long-standing open problem of determining $A_2(10,3)$ and $A_2(11,3)$ is settled, and the optimal codes are classified. Other bounds and classification results for unrestricted error-correcting codes that are obtained utilizing—explicitly or implicitly—a clique algorithm include [34, 47, 48] (ternary codes) and [60, 61] (binary/ternary mixed codes).

3.2 Various Error-Correcting Codes

Unrestricted error-correcting codes form a very general class of codes. Sometimes we are interested in a subclass of codes with some particular property or in error-correcting codes with a different metric.

A graph of a subclass of codes is formed by defining the set of admissible codewords accordingly. *Constant weight codes* form one important class of codes; a word (of length n) is then admissible exactly when its weight takes a prescribed value, w. The maximum cardinality of a code $C \subseteq \mathbf{Z}_q^n$ with minimum distance at least d and constant weight w is denoted by $A_q(n,d,w)$.

Most studies of constant weight codes concern binary codes. The graph for binary constant weight codes, formed analogously to the unrestricted case, is called a *Johnson graph* $J(n,d,w)$. In the construction of binary constant weight codes in [12, Sect. XII], a clique algorithm is not utilized but can be easily incorporated. See also [11]. Clique algorithms have been used for ternary constant weight codes [70], ternary constant composition codes (constant weight codes where the number of occurrences is specified for each value) [89], ternary equidistant codes [8], and permutation codes (constant composition codes in $\mathbf{Z}_q^q$ with each value of $\mathbf{Z}_q$ occurring once in each codeword) [18].

By considering error-correcting codes for other metrics than the Hamming metric, one arrives at a wide variety of clique problems. There is no possibility of covering all conceivable situations, so we restrict the consideration to some codes and metrics related to problems in communication theory.

The type of errors that can occur when transmitting digital symbols are modeled by the particular *communication channel*. The channel depicted in Figure 1 is called the Z-channel; alternatively, the symbol 0 could be the symbol subjected to possible distortion. (For real-world channels, one is usually able to attach probabilities to the edges of the model, describing more precisely the behavior of the channel.)

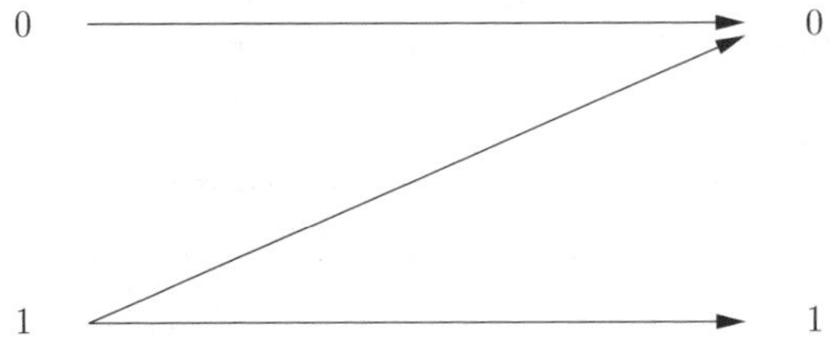

Figure 1: The Z-channel

Codes used for the Z-channel are called *asymmetric codes*. Certain channels behave either by distorting 1s only—as in Figure 1—or by distorting 0s only, in such a way that only one (but we do not know which one) of these situations occurs in any transmitted codeword; the codes used for such channels are called *unidirectional codes*. See [22] for more information about codes of these types.

In order to obtain a clique formulation for constructing the largest possible asymmetric and unidirectional codes, the following definitions are useful. For two words, x and y, let $N(x, y)$ denote the number of coordinates where x is 0 and y is 1, so $d_H(x, y) = N(x, y) + N(y, x)$. The asymmetric distance is defined by

$$d_a(x, y) := \max\{N(x, y), N(y, x)\}$$

and the unidirectional distance is defined by

$$d_u(x, y) := \begin{cases} d_H(x, y) & \text{if } d_a(x, y) = d_H(x, y), \\ 2d_a(x, y) - 1 & \text{otherwise.} \end{cases}$$

It is known, and not difficult to prove, that a code with minimum asymmetric distance d is able to correct up to $d - 1$ asymmetric errors. A code with minimum unidirectional distance d is able to correct up to $\lfloor (d-1)/2 \rfloor$ unidirectional errors.

Codes for other possible channels, which are often not as easily depicted as the Z-channel, include deletion-correcting codes (symbols in a codeword may be lost) [84], transposition-correcting codes (adjacent symbols may be transposed) [14], and variants of these. See [14] for some results on codes discussed in this section obtained by using clique algorithms.

3.3 Set Packings in Hamming Spaces

There are several types of problems that are not clique (or exact cover) problems as such, but that can be divided into a number of subproblems, all of which are clique (or exact cover) problems. One such problem is discussed in this section, another in Subsection 3.4.

For unrestricted error-correcting codes with minimum distance at least $d = 3$, the spheres of radius 1 around the codewords are nonoverlapping. The problem of finding such codes can then be formulated as a set packing problem. Let

$$A = \{00\cdots 0, 100\cdots 0, 0100\cdots 0, \ldots, 00\cdots 01\}$$

consist of the all-zero vector and the vectors of weight 1. Then we want to find the largest code C such that all sums $c+a$ for $c \in C$ and $a \in A$ differ.

In the general case, for two sets $C, A \subseteq \mathbf{Z}_2^r$ with the property that all sums $c+a$ for $c \in C$ and $a \in A$ differ, the inequality $|C||A| \leq 2^r$ holds. (In vector arithmetic in $\mathbf{Z}_2^r$, the coordinate values are determined modulo 2.) We may let $C = \{00\cdots 0\}$ and $A = \mathbf{Z}_2^r$ to reach this upper bound, so we need introduce some further restrictions to arrive at a packing problem that does not admit degenerate solutions. One possible restriction is to fix the size of one of the sets. The following theorem, which is closely related to [68, Theorem 2], shows that there is a clear motivation behind such a restriction.

Theorem 3.1 *Let $C, A \subseteq \mathbf{Z}_2^r$ such that all sums $c+a$ for $c \in C$ and $a \in A$ differ. Then $A_2(|A|-1, 3) \geq |C|2^{|A|-r-1}$ and, by symmetry, $A_2(|C|-1, 3) \geq |A|2^{|C|-r-1}$.*

A more general result for codes with arbitrary minimum distance is stated in [68, Theorem 1]. We shall now elaborate on the construction that leads to Theorem 3.1; see [68] for more details. It is not difficult to see that after adding any vector simultaneously to all vectors in C or A, we still have a packing. Without loss of generality, we may therefore assume that $00\cdots 0 \in A$. Form a matrix $\mathbf{H}$ with the vectors in $A \setminus \{00\cdots 0\}$ as columns. For the sake of clarity, we consider only the case when $\mathbf{H}$ has full rank. Then $\{x : \mathbf{H}x^T \in C\}$ has the desired size and it just remains to prove the minimum distance. The code obtained consists of $|C|$ cosets of the linear code with parity check matrix $\mathbf{H}$ and size $2^{|A|-r-1}$.

As an example, let us consider the case $r = 7$. Since $10 \cdot 10 < 2^7 = 128$ it might be tempting to conjecture that there be a packing with $|C| = |A| = 10$. The question whether such a packing exists was in fact posed in [31], and answered in the negative in [59]. By Theorem 3.1 and the related discussion, this would imply $A_2(9, 3) \geq 40$ and the existence of a corresponding code consisting of 10 cosets of a linear code of size 4. However, the unique code [49] attaining $A_2(9, 3) = 40$ cannot be expressed as 10 cosets of a linear code of size 4.

Having fixed the size of one of the sets, say A, the set packing problem does not admit a useful formulation via graphs and cliques. However, we may utilize the fact that A corresponds to a parity check matrix of a linear code and choose A from the set of all nonisomorphic linear codes with the given parameters. See [64] for a classification of linear codes.

Codes corresponding to sets that attain the upper bound $|C||A| \leq 2^r$ are of special interest, namely, by Theorem 3.1, they lead to codes with the same parameters as Hamming codes (only the linear code with these parameters is called a Hamming code). Now $|C| = 2^i$ and $|A| = 2^{r-i}$, and we know that such packings exist for all $0 \leq i \leq r$. The existence problem for such packings where $00\cdots 0 \in C \cap A$ and both C and A have full rank has been studied since these are fundamental building blocks of perfect codes. This problem was finally resolved in [71]—we do not go into details here but refer the reader to that paper, where an exact cover formulation plays a central role: after fixing A it is required that the sets $c+A$ with $c \in C$ partition $\mathbf{Z}_2^r$.

Utilizing the more general result for error-correcting codes with arbitrary minimum distance, [68, Theorem 1], codes proving $A_2(21,9) \geq 64$ and $A_2(22,9) \geq 80$ were found in [66] via a clique approach.

3.4 The Two-User Binary Adder Channel

A ubiquitous multi-user communication channel is depicted in Figure 2, for the case of two users. Two (or more in the general case) users transmit binary symbols in a synchronized way, and the output of the channel is the sum (in $\mathbf{Z}$) of the inputs.

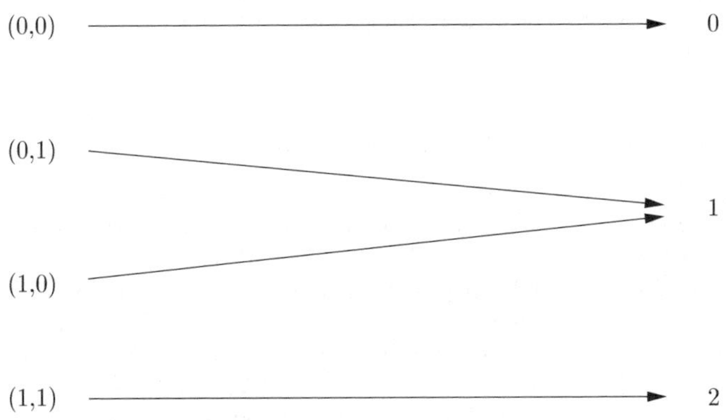

Figure 2: A two-user channel

Since a received 1 cannot be unambiguously decoded, information should be transmitted in the form of codewords of a prescribed length n. Let C and D denote the codes of the two users. Then decoding can be carried out unambiguously exactly when

$$c_1 + d_1 \neq c_2 + d_2$$

for all distinct $c_1, c_2 \in C$ and distinct $d_1, d_2 \in D$. Alternatively, one may write

$$c_1 - c_2 \neq d_2 - d_1. \tag{3.1}$$

(Recall that the operations are carried out in $\mathbf{Z}$.) The speed—the *rate* in bits per transmission—at which information can be transmitted from the respective users is $R_1 = (\log_2 |C|)/n$ and $R_2 = (\log_2 |D|)/n$, and the total rate $R = R_1 + R_2$ should be maximized.

If one of the codes, C or D, is known, then (3.1) leads to a maximum-clique formulation for determining the maximum size of the other code [32]. In the constructed graph, there is one vertex for each binary word of length n, and compatible vertices are given by (3.1). This approach is used in [24] to determine the best possible two-user codes of length $n \leq 5$ by first classifying all possible codes of length

n (that is, subsets of $\mathbf{Z}_2^n$) and then solving the corresponding instances of the maximum clique problem. The same approach is also used in [55], where record-breaking codes of length $n = 6$ are found in a partial search.

3.5 The Zero-Error Capacity of Graphs

So far we have looked at error-correcting codes for several types of channels. We shall here consider yet another channel, which has received a lot of attention. Consider the channel in Figure 3, where input i leads to output $i-1$ or $i+1$ (modulo 5).

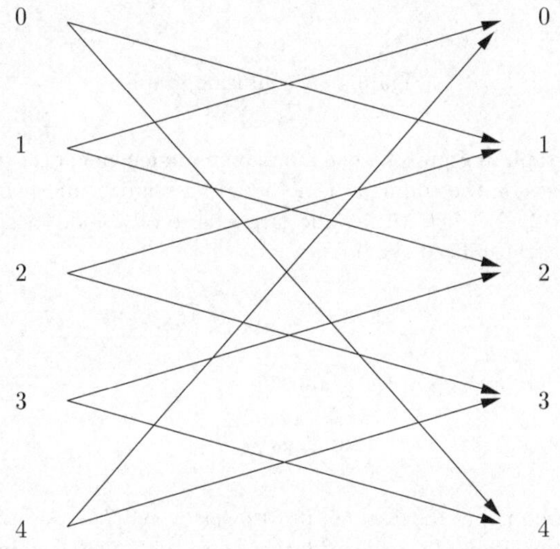

Figure 3: A communication channel

For a received symbol, there is ambiguity regarding the transmitted symbol. We may now construct a graph with one vertex for each input, and an edge between inputs that may lead to the same output. The *confusion graph* for the channel in Figure 3 is shown in Figure 4.

If we do not use all possible symbols for transmission, but only the symbols that cannot be confused, then all received symbols can be uniquely decoded. A set of symbols cannot be confused if and only if they form an independent set in the corresponding confusion graph. To transmit information at as high a rate as possible, it is desirable to use symbols corresponding to a maximum independent set (which has size 2 in our example).

Instead of transmitting information one symbol at a time, one may group symbols together and use codewords with length greater than 1. If the confusion graph is $G = (V, E)$, then the nth confusion graph is the strong product $G^n = G \times G \times \cdots \times G$ (n times) having vertex set $V = \{(v_1, v_2, \ldots, v_n)\} : v_i \in V\}$ and an edge between two vertices $u = (u_1, u_2, \ldots, u_n)$ and $v = (v_1, v_2, \ldots, v_n)$ if and only if, for all i, $u_i = v_i$ or $u_i v_i \in E$.

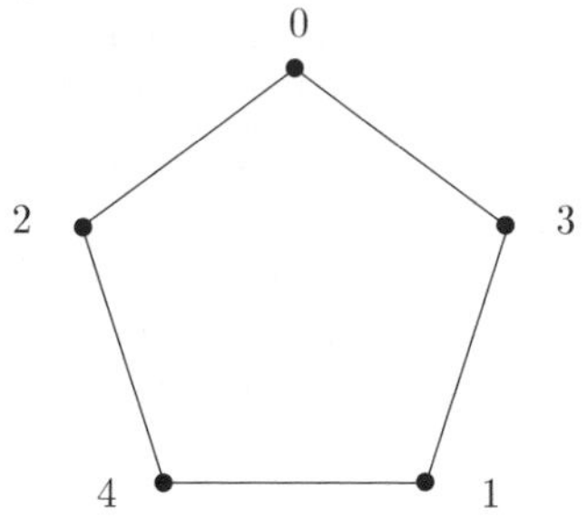

Figure 4: A confusion graph

If G is the graph in Figure 4, then a maximum independent set in $G \times G$ has size 5, which improves on the coding scheme based on sending information one symbol at a time ($5 > 2 \cdot 2$). Let M_n be the largest size of a code in G^n. The rate of information per transmitted symbol is

$$\frac{\log_2 M_n}{n} = \log_2 \sqrt[n]{M_n}\,,$$

and the *zero-error capacity* of the graph G is

$$\sup_{n \geq 1} \log_2 \sqrt[n]{M_n}\,.$$

In a celebrated paper, Lovász [50] proves that when $G \cong C_5$, as in our example, the zero-error capacity is $\sqrt{5}$, achieved by the aforementioned maximum independent set in $G \times G$.

One may study any confusion graph, but the cyclic graphs C_q have been the main focus of interest. We already saw how one arrives at instances of the maximum independent set problem. We shall now give an alternative formulation, which brings out its nature as a packing problem, cf. [3].

We label the vertices of the graph C_q in consecutive order $0, 1, \ldots, q-1$, and the vertices of a strong product by the corresponding n-tuples (words). Two vertices of such a strong product are nonadjacent if and only if there exists a coordinate of the corresponding words whose Lee distance

$$d_L(a, b) := \min\{|a-b|, q-|a-b|\}$$

is at least 2. In other words, this means that we are packing n-dimensional cubes of size $2 \times 2 \times \cdots \times 2$ in the n-dimensional torus of size $q \times q \times \cdots \times q$ (where the codewords denote the positions of the centers of the cubes, or of any other specified part of the cubes). Obviously, the case q even is trivial, since then the torus can be tiled with the cubes (and we have an exact cover).

The search for good codes of this type via large independent sets has been studied for $q > 5$ in [91, 92].

3.6 Keller's Conjecture

The solution to the problem in the previous section is obvious when q is even. One may, however, impose further requirements on the tiling by forbidding complete common $(n-1)$-dimensional faces (that is, facets) of any two cubes. This additional restriction is related to Keller's cube-tiling conjecture.

Keller [39] conjectured that any tiling of the n-dimensional Euclidean space by translates of a finite n-dimensional cube contains a pair of cubes that share a complete common facet. A tiling of an n-dimensional torus with n-dimensional cubes that do not share complete facets can easily be extended to a tiling of the whole space with the same property. By finding explicit such tilings, Lagarias and Shor [41] obtained a counterexample to Keller's conjecture for dimensions $n \geq 10$; Mackay [51] later improved this to $n \geq 8$. For $n \leq 6$ it is known that Keller's conjecture holds [74, 75], so dimension 7 is the sole unsettled case.

The problem of finding tilings forbidding certain pairs of tiles (here, cubes) is not quite an exact cover problem. If an algorithm for the exact cover problem is used, it should be modified slightly to take into account the forbidden pairs of tiles. Such an approach should in any case be preferred to a clique approach. On the other hand, in the small dimensions where Keller's conjecture holds, a clique algorithm can be used to determine the best possible packings of a torus with cubes that do not share complete facets [20].

3.7 Balanced Incomplete Blocks Designs

A *balanced incomplete block design*, briefly BIBD, is a pair $(V, \mathcal{B})$ where V is a v-set (of *points*) and $\mathcal{B}$ is a collection of k-subsets of V (called *blocks*) such that any 2-subset of points is contained in exactly λ blocks. A BIBD with these parameters is called a 2-(v, k, λ) design. The number of blocks in a BIBD is denoted by b and the number of blocks in which each point occurs is denoted by r. A counting argument gives

$$vr = bk \quad \text{and} \quad r(k-1) = \lambda(v-1). \tag{3.2}$$

From the definition of BIBDs, viewing these as collections of blocks, one does not in general arrive at a clique problem. Namely, when $\lambda > 1$, the occurrence of a certain block in the BIBD does not have any implication on whether another block may occur or not. For $\lambda = 1$, however, any two blocks must intersect in either 0 or 1 point. This condition (leading to a definition of compatible blocks) is indeed sufficient for constructing BIBDs.

Theorem 3.2 *A set of b k-element blocks out of a v-set of points with no two blocks intersecting in more than one point, where the parameters are interrelated by* (3.2), *is a* 2-$(v, k, 1)$ *design.*

Proof Every k-element block covers $\binom{k}{2}$ pairs, which are not covered by other blocks. Therefore $b\binom{k}{2} = bk(k-1)/2$ pairs of points are contained in $1 = \lambda$ block and, by (3.2),

$$bk(k-1)/2 = vr(k-1)/2 = \lambda v(v-1)/2 = \binom{v}{2},$$

so each 2-subset of points is contained in exactly $1 = \lambda$ block. □

Since, when $\lambda = 1$, each pair of points is to occur in exactly one block, we may view the construction problem as an instance of the exact cover problem instead. Then the universal set U contains all 2-subsets of points, $|U| = \binom{v}{2}$, and there are $\binom{v}{k}$ sets S_i, each of which cover $\binom{k}{2}$ elements of the universal set. This approach, which is used in [36] to classify 2-$(19, 3, 1)$ designs (Steiner triple systems of order 19), may be generalized to t-designs with $\lambda = 1$; t-designs form a generalization of BIBDs where any t-subset (rather than 2-subset) of points is contained in exactly λ blocks.

Instead of considering BIBDs as sets of blocks—which correspond to columns in incidence matrices—we may consider them as sets of vectors corresponding to rows in incidence matrices. Since every point occurs in r blocks (out of b blocks), the vectors with length b and weight r are admissible in this setting. Since every pair of points must occur in exactly λ blocks, two admissible vectors are compatible exactly when they have λ 1s in common or, in other words, (real) inner product λ.

If we form a graph based on these definitions of admissible and compatible vectors, and search for cliques of size v, it is not obvious that those cliques will correspond to BIBDs. Namely, we have so far neglected the condition that every block of a BIBD must contain k points. However, the following theorem shows that we indeed get BIBDs.

Theorem 3.3 *Let v, k, λ, b, and r be the parameters of a putative BIBD. If there are v vectors of length b and weight r whose pairwise inner product is λ, then these vectors have exactly k 1s in each of the coordinates.*

Proof We count in two ways all pairs of 1s that occur in the same coordinate. If we look at all pairs of vectors, the sum is $\lambda v(v-1)/2$ since all pairwise inner products are λ.

Next we look at each coordinate separately. If we would have k 1s in each of the coordinates, then the sum would be $bk(k-1)/2$ and coincide with the former sum by (3.2). Even if we do not yet know the number of 1s in the respective coordinates, we know that the total number of 1s in all vectors is $vr = bk$. We shall now see that the bk 1s must be evenly distributed among the b coordinates. Namely, if they are not evenly distributed, then we can decrease the number of pairs of 1s as follows.

Take two coordinates with k_1 and k_2 1s, respectively, such that $k_1 \geq k_2 + 2$ (such coordinates must exist if the 1s are not evenly distributed). Then if one 1 is moved from the former coordinate (corresponding to k_1) to the latter (ignoring any other properties), then the sum of pairs of 1s in these two coordinates becomes

$$\binom{k_1 - 1}{2} + \binom{k_2 + 1}{2} = \binom{k_1}{2} + \binom{k_2}{2} - (k_1 - k_2 - 1)$$

and thereby decreases. Repeating this procedure finally leads to the even distribution of 1s, which then necessarily has the unique distribution of 1s with the desired sum. □

The assertion of Theorem 3.3 is directly related to the *second Johnson bound* for constant weight codes, see [12, Sect. III]. We may look at the vectors as codewords of a binary code of length b, constant weight r, and minimum distance $2(r-\lambda)$. In fact, it turns out that whenever a BIBD exists, the corresponding constant weight code is optimal, $A_2(b, 2(r-\lambda), r) = v$.

Clique algorithms played, implicitly, a central role in settling the long-standing open question of whether there exists a projective plane of order 10—that is, a 2-$(111, 11, 1)$ design—a monumental achievement completed by Lam and his colleagues in the late 1980s [42, 45, 90]. Spence used clique algorithms in several studies to search for and classify various BIBDs [85, 87, 88]. Other studies that build on clique algorithms include [29, 35, 43, 63, 79].

Several other types of combinatorial objects can be constructed via designs or codes.

Hadamard matrices of order $4n$—matrices of size $4n \times 4n$ with elements from $\{-1, 1\}$ so that the (real) inner product of any two rows is 0—can be constructed with a clique approach in a straightforward manner; one may also use the correspondence between Hadamard matrices and various objects including 2-$(4n-1, 2n-1, n-1)$ symmetric designs and resolvable 3-$(4n, 2n, n-1)$ designs, cf. [86, 87].

Latin squares can be efficiently constructed by viewing these as certain triple systems and using an exact cover algorithm [38] (in the same manner as Steiner triple systems [36], mentioned earlier). A Latin square of side n has the property that each row and each column contains all elements from $\mathbf{Z}_n$ exactly once. Such a Latin square can be transformed into a collection of triples with one element from each of the sets U, V, W with $|U| = |V| = |W| = n$ telling, for each cell of the square, its row, its column, and its symbol. Each pair of elements from different sets U, V, W should be contained in exactly one triple. This can also be viewed as a group divisible design of type n^3. For example:

$$\begin{array}{|cc|} \hline 0 & 1 \\ 1 & 0 \\ \hline \end{array} \quad \leftrightarrow \quad \{\{u_0, v_0, w_0\}, \{u_0, v_1, w_1\}, \{u_1, v_0, w_1\}, \{u_1, v_1, w_0\}\}$$

The interested reader is referred to [38] for more details.

Two Latin squares, L_1 and L_2, are orthogonal if for each pair of symbols, (i, j), there is a unique cell in which L_1 has symbol i and L_2 has symbol j. A collection of k *mutually orthogonal Latin squares* (MOLS) corresponds to a $(k+2)$-tuple system that is a generalization of the aforementioned representation of Latin squares; just add more sets of elements. Incidentally, many other kinds of combinatorial objects may in fact be viewed as various kinds of set systems in a similar manner. Computationally, however, construction is very hard for k-tuple systems with $k \geq 4$.

If we are given a Latin square L and want to find one or all orthogonal squares, we may proceed as follows. A *transversal* of a Latin square is a subset of cells with exactly one representative from each row, column, and symbol. Any symbol of a Latin square orthogonal to L must occur in cells that form a transversal of L. Hence, we are seeking to partition the cells of L into n transversals. Having found the transversals, we may easily form an instance of the exact cover problem. A clique approach is used to solve this problem in [53]; see also [73]. The transversals may also be found using exact cover: the rows, the columns, and the symbols have to be covered using n cells.

3.8 Resolutions of Designs

A *resolution* of a design is a partition of the blocks into *parallel classes*, which in turn partition the points. Finding one or all (or proving nonexistence of) resolutions of a given design can be achieved with algorithms solving two instances of the exact cover problem, as follows.

If the universal set U is the set of points and the blocks form the subsets S_i, then the solutions to such an instance of the exact cover problem are the possible parallel classes. Having computed the possible parallel classes in this way, consider an exact cover instance where the universal set U' has the blocks of the design as elements, and form for each of the possible parallel classes a set S'_i consisting of the blocks in that parallel class. In [19] and [69], the first instance and both instances, respectively, are solved with a clique approach; this is possible since all subsets have the same set size.

Since a block may occur several times in a design—if this does not happen, the design is said to be *simple*—it is important to consider labelled blocks and parallel classes in solving the instances of exact cover. Blocks occurring many times introduce symmetry and may slow down the algorithm considerably; [69] discusses one possible course of action in this situation.

Alternatively, resolutions of designs may be constructed from scratch. We restrict the following discussion to resolutions of BIBDs. Then it is possible to make use of the result that a resolution of a 2-(qk, k, λ) design corresponds to an $(r, qk, r - \lambda)_q$ code, where $r = \lambda(qk-1)/(k-1)$ [81]. It turns out that a code with such parameters is necessarily equidistant (that is, with all pairwise distances between codewords coinciding) and equireplicate (that is, with the coordinate values evenly distributed in each coordinate). To construct such codes, we can use a clique approach as discussed in Subsection 3.1, with the further requirement that the codes be equidistant; see [34]. By a theorem analogous to Theorem 3.3 one may ignore the fact that the codes must be equireplicate; on the other hand, this fact provides a means for speeding up the search.

For a given even number of points $2v$, resolutions of the unique 2-$(2v, 2, 1)$ design are of particular interest since these correspond to 1-factorizations of the complete graph of order $2v$. A 1-*factor* of a graph $G = (V, E)$ is a 1-regular spanning subgraph (in other words, a set of v edges of G whose set of endpoints is V). A 1-*factorization* of a graph is a partition of the edge set into 1-factors.

Indeed, in the study of 1-factorizations of complete graphs in the guise of resolutions of 2-$(2v, 2, 1)$ designs, we may consider the corresponding $(2v - 1, 2v, 2v - 2)_v$ codes instead, as discussed earlier. More generally, a classification of 1-factorizations of r-regular graphs of order $2v$—recall that a complete graph of order $2v$ is $(2v - 1)$-regular—may be done via equireplicate $(r, 2v, r - 1)_v$ codes [37].

The fastest known approach for classifying 1-factorizations of complete graphs is achieved by formulating the problem in terms of certain triple systems and using an exact cover algorithm [37]. We have already seen such an approach for Steiner triple systems and Latin squares; only the type of triple system is different here. Now we have two sets U and V with $|U| = 2v - 1$ and $|V| = 2v$, such that the elements of U correspond to the parallel classes and the elements of V to the vertices of the graph. All triples intersect U in one element and V in two elements, indicating in what parallel class a certain edge (pair of elements) resides. We have a 1-factorization when each pair of elements from V and each pair with one element from U and another from V occurs in exactly one triple. This is a group divisible design of type $(2n-1)^1 1^{2n}$. For example, a 1-factorization of the complete graph of order 4 is given by

$$\{\{u_1, v_1, v_2\}, \{u_1, v_3, v_4\}, \{u_2, v_1, v_3\},$$
$$\{u_2, v_2, v_4\}, \{u_3, v_1, v_4\}, \{u_3, v_2, v_3\}\}.$$

For further details, [37] should be consulted.

In a similar manner, as described in [38], the construction of resolutions of any given design can be formulated as a single instance of the exact cover problem. Let $p_1, p_2, \ldots, p_v$ be the points of the BIBD, and let $B_1, B_2, \ldots, B_b$ be the blocks (represented as sets of points). Furthermore, let $P_1, P_2, \ldots, P_r$ be a set of distinct labels for the putative parallel classes. The task is now to find an exact cover for all sets of the form $\{P_l, p_i\}$, where $l \in \{1, 2, \ldots, r\}$ and $i \in \{1, 2, \ldots, v\}$, and all sets $\{B_j\}$, where $j \in \{1, 2, \ldots, b\}$, using $(k+2)$-sets of the form

$$\{P_l\} \cup \{B_j\} \cup \{p_i : p_i \in B_j\},$$

where $l \in \{1, 2, \ldots, r\}$ and $j \in \{1, 2, \ldots, b\}$. Each $(k+2)$-set covers the subsets that occur in it.

Connecting to the discussion at the end of Subsection 3.7, finding the Latin squares orthogonal to a given square that is viewed as a triple system is actually a problem of finding resolutions of that triple system. For example, parallel classes of the triple system correspond to transversals of the Latin square. The reader is encouraged to work out the details of the formulation of an instance of the exact cover problem that can be used to find the resolutions.

A clique algorithm is used as a part of the approach in [27], where near resolutions of designs are classified via a mapping to a special type of codes.

3.9 Partitions of Codes and Designs

Clique and exact cover algorithms may be used to find packings or partitions in a certain space by codes or designs. Such packings and partitions are useful, for example, in the construction of larger objects. This is also an interesting problem in its own right, and can in some cases be viewed as a coloring problem [30].

Formulation of problems of these types as clique and exact cover problems is straightforward. In the former case, the admissible objects (codes, designs) correspond to vertices and we have edges between two objects if and only they are nonoverlapping. In the latter case, on the other hand, the universal set consists of all points in the ambient space and the admissible objects form a set system where the subsets are codes or designs, that is, sets of codewords or blocks. A partition of all k-element blocks into t-(v, k, λ) designs is of particular interest in design theory and is called a *large set* of disjoint designs.

We shall here mention a few references where the aforementioned approach has been used or can be used. Mutually disjoint designs and large sets of designs have been considered in many papers, including [1, 46, 54]. Disjoint constant weight codes have been studied in [14]; see also [83].

The binary $(2^k - 1)$-cube can be partitioned into codes with the same parameters as Hamming codes. Phelps [76] showed that there are 11 inequivalent partitions of the binary 7-cube into such codes. We shall now see how this can be done using an exact cover algorithm on two levels.

First, we find all labelled codes of the given type in $\mathbf{Z}_2^7$. For these parameters the Hamming code C is the unique code up to equivalence, so by the orbit-stabilizer theorem there are

$$\frac{(q!)^n n!}{|\operatorname{Aut}(C)|} = \frac{2^7 7!}{2688} = 240$$

codes, as we know that $|\operatorname{Aut}(C)| = 2688$; see [52, p. 400] and [72, Remark 3]. The obvious exact cover approach, cf. Subsection 3.3, finds these 240 codes instantly. Even a suboptimal clique approach finds the codes in a fraction of a second.

In the second step we have an instance of exact cover, where the universal set is the set of vertices of the binary 7-cube, which should be partitioned using the codes found in the first step. For the instance under consideration, a clique approach is also feasible. There are 27360 solutions but after isomorph rejection only 11 inequivalent partitions remain.

In the list of open problems, in Subsection 4.3, a partitioning problem for the binary 8-cube is mentioned.

3.10 Objects with Prescribed Automorphisms

Occasionally, one is not interested in all objects with certain parameters, but only in those with some prescribed automorphisms. Moreover, prescribing automorphisms is a common tool in trying to construct large objects whose existence is unknown.

Automorphisms can be prescribed for any of the objects considered in Section 3, within the group of symmetries of the ambient space. For example, for BIBDs the automorphisms are elements of the symmetric group S_v, and for unrestricted codes they are elements of the wreath product $S_q \wr S_n$.

The prescribed automorphism group partitions all elements into orbits. For each orbit it must be checked whether it is *admissible*, that is, if all pairs of its elements are compatible. Two orbits are *compatible* if all pairs of elements from the two sets are compatible (but it suffices to compare one element of one orbit with all elements of the other).

One can now construct a graph with one vertex for each admissible orbit and an edge between each pair of compatible orbits. However, in general all orbits do not have the same size, so we have arrived at an instance of the *maximum-weight clique* problem, where weights (here, the orbit sizes) are associated with the vertices and we want to find one or all cliques of maximum weight. The program Cliquer [58] is able to handle weighted graphs, using the algorithm from [62].

For exact cover instances, on the other hand, no modifications to the basic algorithm are necessary. The action of the prescribed automorphism group merges some of the elements in U and some of the sets in $\mathcal{S}$ into orbits. It must be checked though that the sets in an orbit are pairwise disjoint, otherwise the orbit is not admissible and should be removed.

Designs with prescribed automorphisms have been studied in this manner—explicitly using an exact cover algorithm—in, for example, [5, 6, 33, 54]; this feature is also included in the DISCRETA tool for constructing t-designs [7]. The possibility of constructing error-correcting codes with prescribed automorphisms using maximum-weight clique algorithms is discussed in [12, Sect. X] and [83]. Chu and Colbourn [17] used a clique algorithm in determining the maximum size of certain cyclic constant weight codes. Unrestricted nonbinary codes are considered in [9, 10].

Note that the construction related to Theorem 3.1 and the more general result in [68, Theorem 1] is in fact about codes with certain automorphisms. Namely, as we then get codes that consist of cosets of a linear code, we may instead say that we have a code with certain translational symmetries (given by the linear code).

4 Applying Clique and Exact Cover Algorithms

In some rare cases, to the best of our knowledge, one can do no better than apply a clique or a maximum cover algorithm directly to construct or classify objects. In most cases, however, a more clever approach leads to an essential increase in speed. We shall now present some general guidelines for developing such approaches.

4.1 Some General Guidelines

Only the very smallest instances of the problems discussed here can be solved quickly by applying, depending on the particular problem, a clique algorithm or an exact cover algorithm directly. For somewhat larger instances—in particular, when one aims at proceeding beyond what has been done before—a more sophisticated approach is unavoidable. Generally, such approaches try to

(1) reduce the search space by taking symmetries into account,

(2) use known properties of the final objects to prune the search, and

(3) construct the final object via subobjects.

If only (1) is incorporated, ignoring (2) and (3), this could in principle be done by developing maximum clique algorithms that make use of the automorphisms of the graph related to the problem instance—and analogously for exact cover algorithms. Perhaps somewhat surprisingly, we have not encountered algorithms of this kind in the literature. If this approach is used to classify objects rather than to find one maximum clique or exact cover, then one should be very careful since an automorphism of the graph need not in general correspond to a symmetry of the original search space; this matter is related to the concept of (strong) reconstructability [2].

If a combinatorial object is constructed from scratch, at each step of the build-up the full automorphism group—or stabilizer—of the partial object reveals the prevalent symmetries. In the beginning of a search, when the partial object is the empty set, the full automorphism group coincides with that of the ambient space; this group often acts transitively on the elements of the search space, and there is then only one choice for the first element to be added. For example, the first word of an unrestricted error-correcting code may be chosen as the all-zero word $00\cdots0$, and the first row of an incidence matrix of a BIBD may be taken as r 1s followed by 0s.

There are two main frameworks for building up a complete set of nonisomorphic combinatorial objects: the method of canonical representatives [23, 77] and the canonical construction path method [56]. A proper level for abandoning isomorph rejection—to focus on finding cliques or exact covers—is when the number of partial objects gets prohibitively large or when most partial objects tend to have a very small automorphism group. See also [38].

So far we have paid no attention to the issue of correctness. If single combinatorial objects are constructed, they can often be checked even by hand. For a census of objects, however, this issue is intricate and out of the scope of this survey. In any case, it should be stressed that a wide range of techniques are applicable, ranging from utilization of independently developed computer programs to techniques based on double counting [36, 44].

We shall now have a brief look at how some of the objects listed in this survey can be constructed via subobjects.

4.2 Constructing Objects via Subobjects

Combinatorial objects can often be efficiently constructed or classified by using intermediate subobjects in two or more stages. Isomorph rejection should be carried out after each stage to speed up the search, and, in the case of classification, always for a complete object. For this task, the methods mentioned in Subsection 4.1 can be used, or, if the number of mutually nonisomorphic objects is small, pairwise comparison may be carried out.

The choice of parameters and types of subobjects is crucial; a successful choice leads to relatively few nonisomorphic subobjects that determine a great part of the final object. (Unfortunately, not all problems admit such subobjects.) We shall here look at a few examples related to the objects considered in Section 3.

Consider the codes C_i, $i \in \mathbf{Z}_q$, obtained by partitioning a punctured $(n, M, d)_q$ code according to the value in the deleted coordinate. Since

$$\sum_{i=0}^{q-1} |C_i| = M,$$

at least one of the subcodes must have cardinality at least $\lceil M/q \rceil$. Since all codes C_i have minimum distance at least d, one may therefore start a classification of $(n, M, d)_q$ codes from the set of $(n-1, M', d)_q$ codes with $M' \geq \lceil M/q \rceil$. This argument may be repeated recursively.

Analogously, one may classify constant weight codes via shorter codes. Then, one may use the fact that a code with length n, cardinality M, and constant weight w has Mw 1s in the codewords. Hence, there must exist a coordinate with at least

$$\lceil M(n-w)/n \rceil$$

0s, and a coordinate with at least

$$\lceil Mw/n \rceil$$

1s. These bounds are in fact related to the *first Johnson bound* for constant weight codes, see [12, Sect. III].

For the objects that we have viewed as triple systems—Steiner triple systems, Latin squares, and 1-factorizations of complete graphs—a good choice of an intermediate subobject is the set of all triples that intersect a given triple [36, 37]. From such a subobject, the rest of the triple system can be determined using an exact cover algorithm.

Balanced incomplete block designs, viewed as $v \times b$ incidence matrices, are conveniently classified by first classifying $v' \times b$ submatrices—in fact, it often suffices to consider only a certain subset of such matrices—and searching for the missing $v - v'$ rows with a clique algorithm. Similarly, to classify resolutions of BIBDs in the framework of codes, one first classifies the subcodes of a certain size and type and then one completes the codes by searching for desired cliques.

4.3 Interesting Open Problems

For most types of problems considered in Section 3 there is a smallest unsolved case—with respect to construction or classification—which is a challenge to researchers in this field. We conclude this survey paper by bringing forward a few particularly interesting open problems.

(1) There is room for many new bounds and classification results for binary constant weight codes. See [12, 38].

(2) Linear and semidefinite programming (with problem-dependent inequalities, different from those in Subsection 2.1) have been used to find upper bounds on the sizes of error-correcting codes of various types [4, 21, 80]. Can those approaches be combined with a computer search to efficiently find the maximum size of error-correcting codes, that is, the size of maximum cliques? In particular, determine whether $A_2(16,7) = 36$ or $A_2(16,7) = 37$.

(3) Since $A_2(8,3) = 20$, a partitioning of the binary 8-cube into one-error-correcting codes needs at least $\lceil 256/20 \rceil = 13$ codes. There is a partitioning with 14 codes, but is there one with 13 codes? See [30, Sect. 9.7] and [94].

(4) A comparative study of clique algorithms should be carried out, since the principles behind the most efficient algorithms are not yet fully understood. Moreover, there is still place for new ideas and further development of the algorithms. Perhaps a combination of the idea in [65] and pruning based on vertex coloring [82] could prove efficient.

(5) Develop clique and exact cover algorithms for problem instances with nontrivial automorphisms.

Acknowledgements

The author thanks Petteri Kaski and Alfred Wassermann for valuable comments. The research was supported by the Academy of Finland under Grants No. 100500 and No. 202315.

References

[1] M. Araya, More mutually disjoint Steiner systems $S(5,8,24)$, *J. Combin. Theory Ser. A* **102** (2003), 201–203.

[2] L. Babai, Automorphism groups, isomorphism, reconstruction, in *Handbook of Combinatorics, Vol. II* (eds. R.L. Graham, M. Grötschel & L. Lovász), North-Holland, Amsterdam (1995), pp. 1447–1540.

[3] L.D. Baumert, R.J. McEliece, E. Rodemich, H.C. Rumsey, Jr., R. Stanley & H. Taylor, A combinatorial packing problem, in *Computers in Algebra and Number Theory (Proc. SIAM-AMS Symposium in Applied Mathematics, New York, 1970)* (eds. G. Birkhoff & M. Hall, Jr.), American Mathematical Society, Providence (1971), pp. 97–108.

[4] M.R. Best, A.E. Brouwer, F.J. MacWilliams, A.M. Odlyzko & N.J.A. Sloane, Bounds for binary codes of length less than 25, *IEEE Trans. Inform. Theory* **24** (1978), 81–93.

[5] A. Betten, R. Laue, S. Molodtsov & A. Wassermann, Steiner systems with automorphism groups PSL(2, 71), PSL(2, 83), and $\mathrm{P\Sigma L}(2, 3^5)$, *J. Geom.* **67** (2000), 35–41.

[6] A. Betten, R. Laue & A. Wassermann, A Steiner 5-design on 36 points, *Des. Codes Cryptogr.* **17** (1999), 181–186.

[7] A. Betten, R. Laue & A. Wassermann, DISCRETA: A tool for constructing t-designs, in *Computer Algebra Handbook: Foundations, Applications, Systems* (eds. J. Grabmeier, E. Kaltofen & V. Weispfenning), Springer, Berlin (2003), pp. 372–375.

[8] G. Bogdanova, Ternary equidistant codes and the maximum clique problem, in *Groups and Graphs (Proc. European Women in Mathematics International Workshop on Groups and Graphs, Varna, 2002)* Institute of Mathematics and Informatics, Bulgaria (2002), pp. 15–18.

[9] G.T. Bogdanova, A.E. Brouwer, S.N. Kapralov & P.R.J. Östergård, Error-correcting codes over an alphabet of four elements, *Des. Codes Cryptogr.* **23** (2001), 333–342.

[10] G.T. Bogdanova & P.R.J. Östergård, Bounds on codes over an alphabet of five elements, *Discrete Math.* **240** (2001), 13–19.

[11] I. Bouyukliev, Maximal cliques in graphs and some new upper bounds for constant-weight codes, in *Groups and Graphs (Proc. European Women in Mathematics International Workshop on Groups and Graphs, Varna, 2002)* Institute of Mathematics and Informatics, Bulgaria (2002), pp. 19–22.

[12] A.E. Brouwer, J.B. Shearer, N.J.A. Sloane & W.D. Smith, A new table of constant weight codes, *IEEE Trans. Inform. Theory* **36** (1990), 1334–1380.

[13] S.A. Burr, On the automorphism group of a power of a hypercube, in *Graph Theory, Combinatorics, and Applications, Vol. 1* (eds. Y. Alavi, G. Chartrand, O.R. Oellermann & A.J. Schwenk), Wiley, New York (1991), pp. 187–192.

[14] S. Butenko, P. Pardalos, I. Sergienko, V. Shylo & P. Stetsyuk, Estimating the size of correcting codes using extremal graph problems, in *Optimization: Structure and Applications* (eds. E. Hunt & C.E.M. Pearce), Kluwer, Boston, in press.

[15] R. Carraghan & P.M. Pardalos, An exact algorithm for the maximum clique problem, *Oper. Res. Lett.* **9** (1990), 375–382.

[16] J.L. Carter, On the existence of a projective plane of order ten, Ph.D. thesis, Mathematics Department, University of California, Berkeley, 1975.

[17] W. Chu & C.J. Colbourn, Optimal $(n, 4, 2)$-OOC of small orders, *Discrete Math.* **279** (2004), 163–172.

[18] W. Chu, C.J. Colbourn & P. Dukes, Constructions for permutation codes in powerline communications, *Des. Codes Cryptogr.* **32** (2004), 51–64.

[19] M.J. Colbourn, Algorithmic aspects of combinatorial designs: A survey, *Ann. Discrete Math.* **26** (1985), 67–136.

[20] K. Corrádi & S. Szabó, A combinatorial approach for Keller's conjecture, *Period. Math. Hungar.* **21** (1990), 95–100.

[21] P. Delsarte, Bounds for unrestricted codes, by linear programming, *Philips Research Reports* **27** (1972), 272–289.

[22] G. Fang and H.C.A. van Tilborg, Bounds and constructions of asymmetric or unidirectional error-correcting codes, *Appl. Algebra Engrg. Comm. Comput.* **3** (1992), 269–300.

[23] I.A. Faradžev, Constructive enumeration of combinatorial objects, in *Problèmes combinatoires et théorie des graphes (Univ. Orsay, Orsay, 1976) Colloq. Internat. CNRS*, 260, Paris, (1978), pp. 131–135.

[24] D.F. Fuccio, Codes for the two-user binary adder channel, M.Sc. thesis, Department of Electrical and Communications Engineering, Helsinki University of Technology, Espoo, 2003.

[25] M.R. Garey & D.S. Johnson, *Computers and Intractability: A Guide to the Theory of NP-Completeness*, Freeman, New York (1979).

[26] S.W. Golomb & L.D. Baumert, Backtrack programming, *J. ACM* **12** (1965), 516–524.

[27] H. Haanpää & P. Kaski, The near resolvable 2-$(13, 4, 3)$ designs and thirteen-player whist tournaments, *Des. Codes Cryptogr.*, in press.

[28] J. Håstad, Clique is hard to approximate within $n^{1-\epsilon}$, *Acta Math.* **182** (1999), 105–142.

[29] S.K. Houghten, L.H. Thiel, J. Janssen & C.W.H. Lam, There is no $(46, 6, 1)$ block design, *J. Combin. Des.* **9** (2001), 60–71.

[30] T.R. Jensen & B. Toft, *Graph Coloring Problems*, Wiley, New York (1995).

[31] G.A. Kabatyanskii & V.I. Panchenko, Unit sphere packings and coverings of the Hamming space (in Russian), *Problemy Peredachi Informatsii* **24** (4) (1988), 3–16. (English translation in *Problems Inform. Transmission* **24** (1988), 261–271.)

[32] T. Kasami, S. Lin, V.K. Wei & S. Yamamura, Graph theoretic approaches to the code construction for the two-user multiple-access binary adder channel, *IEEE Trans. Inform. Theory* **29** (1983), 114–130.

[33] P. Kaski, Isomorph-free exhaustive generation of designs with prescribed groups of automorphisms, preprint.

[34] P. Kaski & P.R.J. Östergård, There exists no $(15, 5, 4)$ RBIBD, *J. Combin. Des.* **9** (2001), 357–362.

[35] P. Kaski & P.R.J. Östergård, Miscellaneous classification results for 2-designs, *Discrete Math.* **280** (2004), 65–75.

[36] P. Kaski & P.R.J. Östergård, The Steiner triple systems of order 19, *Math. Comp.* **73** (2004), 2075–2092.

[37] P. Kaski & P.R.J. Östergård, One-factorizations of regular graphs of order 12, *Electron. J. Combin.* **12** (1) (2005), R2.

[38] P. Kaski & P.R.J. Östergård, *Classification Algorithms for Codes and Designs*, preprint.

[39] O.H. Keller, Über die lückenlose Einfüllung des Raumes mit Würfeln, *J. Reine Angew. Math.* **163** (1930), 231–248.

[40] D.E. Knuth, Dancing links, in *Millennial Perspectives in Computer Science* (eds. J. Davies, B. Roscoe & J. Woodcock), Palgrave Macmillan, Basingstoke (2000), pp. 187–214.

[41] J.C. Lagarias & P.W. Shor, Keller's cube-tiling conjecture is false in high dimensions, *Bull. Amer. Math. Soc. (N.S.)* **27** (1992), 279–283.

[42] C.W.H. Lam, The search for a finite projective plane of order 10, *Amer. Math. Monthly* **98** (1991), 305–318.

[43] C.W.H. Lam, G. Kolesova & L. Thiel, A computer search for finite projective planes of order 9, *Discrete Math.* **92** (1991), 187–195.

[44] C.W.H. Lam & L. Thiel, Backtrack search with isomorph rejection and consistency check, *J. Symbolic Comput.* **7** (1989), 473–485.

[45] C.W.H. Lam, L.H. Thiel & S. Swiercz, A computer search for a projective plane of order 10, in *Algebraic, Extremal and Metric Combinatorics* (eds. M.-M. Deza, P. Frankl & I.G. Rosenberg), Cambridge University Press, Cambridge (1988), pp. 155–165.

[46] R. Laue, S. Magliveras & A. Wassermann, New large sets of t-designs, *J. Combin. Des.* **9** (2001), pp. 40–59.

[47] M.J. Letourneau & S.K. Houghten, Optimal ternary $(10, 7)$ error-correcting codes, *Congr. Numer.* **155** (2002), 71–80.

[48] M.J. Letourneau & S.K. Houghten, Optimal ternary $(11, 7)$ and $(14, 10)$ error-correcting codes, *J. Combin. Math. Combin. Comput.* **51** (2004), 159–164.

[49] S. Litsyn & A. Vardy, The uniqueness of the Best code, *IEEE Trans. Inform. Theory* **40** (1994), 1693–1698.

[50] L. Lovász, On the Shannon capacity of a graph, *IEEE Trans. Inform. Theory* **25** (1979), 1–7.

[51] J. Mackey, A cube tiling of dimension eight with no facesharing, *Discrete Comput. Geom.* **28** (2002), 275–279.

[52] F.J. MacWilliams & N.J.A. Sloane, *The Theory of Error-Correcting Codes*, North-Holland, Amsterdam (1977).

[53] B.M. Maenhaut & I.M. Wanless, Atomic Latin squares of order eleven, *J. Combin. Des.* **12** (2004), 12–34.

[54] R. Mathon, Searching for spreads and packings, in *Geometry, Combinatorial Designs and Related Structures* (eds. J.W.P. Hirschfeld, S. S. Magliveras & M.J. de Resmini), Cambridge University Press, Cambridge (1997), pp. 161–176.

[55] M. Mattas & P.R.J. Östergård, A new bound for the zero-error capacity region of the two-user binary adder channel, preprint.

[56] B.D. McKay, Isomorph-free exhaustive generation, *J. Algorithms* **26** (1998), 306–324.

[57] Z. Miller & M. Perkel, A stability theorem for the automorphism groups of powers of the n-cube, *Australas. J. Combin.* **10** (1994), 17–28.

[58] S. Niskanen & P.R.J. Östergård, Cliquer user's guide, version 1.0, Technical Report T48, Communications Laboratory, Helsinki University of Technology, Espoo, 2003.

[59] P.R.J. Östergård, On the structure of optimal error-correcting codes, *Discrete Math.* **179** (1998), 285–287.

[60] P.R.J. Östergård, On binary/ternary error-correcting codes with minimum distance 4, in *Applied Algebra, Algebraic Algorithms and Error-Correcting Codes* (eds. H. Imai, S. Lin & A. Poli), Springer, Berlin (1999), pp. 472–481.

[61] P.R.J. Östergård, Classification of binary/ternary one-error-correcting codes, *Discrete Math.* **223** (2000), 253–262.

[62] P.R.J. Östergård, A new algorithm for the maximum-weight clique problem, *Nordic J. Comput.* **8** (2001), 424–436.

[63] P.R.J. Östergård, A 2-$(22, 8, 4)$ design cannot have a 2-$(10, 4, 4)$ subdesign, *Des. Codes Cryptogr.* **27** (2002), 257–260.

[64] P.R.J. Östergård, Classifying subspaces of Hamming spaces, *Des. Codes Cryptogr.* **27** (2002), 297–305.

[65] P.R.J. Östergård, A fast algorithm for the maximum clique problem, *Discrete Appl. Math.* **120** (2002), 195–205.

[66] P.R.J. Östergård, Two new four-error-correcting codes, *Des. Codes Cryptogr.*, in press.

[67] P.R.J. Östergård, T. Baicheva & E. Kolev, Optimal binary one-error-correcting codes of length 10 have 72 codewords, *IEEE Trans. Inform. Theory* **45** (1999), 1229–1231.

[68] P.R.J. Östergård & M.K. Kaikkonen, New single-error-correcting codes, *IEEE Trans. Inform. Theory* **42** (1996), 1261–1262.

[69] P.R.J. Östergård & P. Kaski, Enumeration of 2-$(9, 3, \lambda)$ designs and their resolutions, *Des. Codes Cryptogr.* **27** (2002), 131–137.

[70] P.R.J. Östergård & M. Svanström, Ternary constant weight codes, *Electron. J. Combin.* **9** (1) (2002), R41.

[71] P.R.J. Östergård & A. Vardy, Resolving the existence of full-rank tilings of binary Hamming spaces, *SIAM J. Discrete Math.* **18** (2004), 382–387.

[72] P.R.J. Östergård & W.D. Weakley, Constructing covering codes with given automorphisms, *Des. Codes Cryptogr.* **16** (1999), 65–73.

[73] E.T. Parker, Computer investigation of orthogonal Latin squares of order ten, in *Experimental Arithmetic, High-Speed Computing and Mathematics (Proc. 15th AMS Symposium in Applied Mathematics, Chicago, Atlantic City, 1962)* (eds. N.C. Metropolis, A.H. Taub, J. Todd & C.B. Tompkins), American Mathematical Society, Providence (1963), pp. 73–81.

[74] O. Perron, Über lückenlose Ausfüllung des n-dimensionalen Raumes durch kongruente Würfel, I, *Math. Z.* **46** (1940), 1–26.

[75] O. Perron, Über lückenlose Ausfüllung des n-dimensionalen Raumes durch kongruente Würfel, II, *Math. Z.* **46** (1940), 161–180.

[76] K.T. Phelps, An enumeration of 1-perfect binary codes, *Australas. J. Combin.* **21** (2000), 287–298.

[77] R.C. Read, Every one a winner; or, How to avoid isomorphism search when cataloguing combinatorial configurations, *Ann. Discrete Math.* **2** (1978), 107–120.

[78] G.F. Royle, An orderly algorithm and some applications in finite geometry, *Discrete Math.* **185** (1998), 105–115.

[79] C.J. Salwach & J.A. Mezzaroba, The four biplanes with $k = 9$, *J. Combin. Theory Ser. A* **24** (1978), 141–145.

[80] A. Schrijver, New code upper bounds from the Terwilliger algebra, preprint.

[81] N.V. Semakov & V.A. Zinov'ev, Equidistant q-ary codes with maximal distance and resolvable balanced incomplete block designs (in Russian), *Problemy Peredachi Informatsii* **4** (2) (1968), 3–10. (English translation in *Problems Inform. Transmission* **4** (2) (1968), 1–7.)

[82] E.C. Sewell, A branch and bound algorithm for the stability number of a sparse graph, *INFORMS J. Comput.* **10** (1998), 438–447.

[83] N.J.A. Sloane, Unsolved problems in graph theory arising from the study of codes, *Graph Theory Notes N. Y.* **18** (1989), 11–20.

[84] N.J.A. Sloane, On single-deletion-correcting codes, in *Codes and Designs (Columbus, OH, 2000)* (eds. K.T. Arasu & A. Seress), de Gruyter, Berlin (2002), pp. 273–291.

[85] E. Spence, The complete classification of symmetric $(31, 10, 3)$ designs, *Des. Codes Cryptogr.* **2** (1992), 127–136.

[86] E. Spence, Classification of Hadamard matrices of order 24 and 28, *Discrete Math.* **140** (1995), 185–243.

[87] E. Spence, Construction and classification of combinatorial designs, in *Surveys in Combinatorics, 1995* (ed. P. Rowlinson), Cambridge University Press, Cambridge (1995), pp. 191–213.

[88] E. Spence, The complete classification of Steiner Systems $S(2, 4, 25)$, *J. Combin. Des.* **4** (1996), 295–300.

[89] M. Svanström, P.R.J. Östergård & G.T. Bogdanova, Bounds and constructions for ternary constant-composition codes, *IEEE Trans. Inform. Theory* **48** (2002), 101–111.

[90] L. Thiel, C. Lam & S. Swiercz, Using a CRAY-1 to perform backtrack search, in *Supercomputing '87: Supercomputer Design, Performance Evaluation and Performance Education (Proc. 2nd International Conference on Supercomputing, San Fransisco, 1987), Vol. III* (eds. L.P. Kartashev & S.I. Kartashev), International Supercomputing Institute, St. Petersburg, FL (1987), pp. 92–99.

[91] A. Vesel, The independence number of the strong product of cycles, *Comput. Math. Appl.* **36** (7) (1998), 9–21.

[92] A. Vesel & J. Žerovnik, Improved lower bound on the Shannon capacity of C_7, *Inform. Process. Lett.* **81** (2002), 277–282.

[93] P.M. Winkler, Isometric embedding in products of complete graphs, *Discrete Appl. Math.* **7** (1984), 221–225.

[94] G.M. Ziegler, Coloring Hamming graphs, optimal binary codes, and the 0/1-Borsuk problem in low dimensions, in *Computational Discrete Mathematics* (ed. H. Alt), Springer, Berlin (2001), pp. 159–171.

Dept. of Electrical and Communications Engineering
Helsinki University of Technology
P.O. Box 3000, 02015 TKK, Finland
patric.ostergard@hut.fi

Flocks of circle planes

Tim Penttila

Abstract

Flocks of finite circle planes—inversive, Minkowski and Laguerre planes—are surveyed, including their connections with projective planes, generalised quadrangles and ovals.

1 Circle planes

In the last thirty years, there has been considerable activity in the study of flocks of circle planes, originally by Thas, Walker and Fisher, but later, after Kantor, Payne and Thas had established connections between flocks of Laguerre planes and generalised quadrangles in the 1980s, by many authors. Their importance lies mainly in their connections with projective planes and generalised quadrangles.

The *circle planes* are the inversive, Minkowski and Laguerre planes, defined below. Their study received impetus when Benz published his book [9] devoted to them in 1973. They are related to ovoids, sharply 3-transitive sets and ovals, respectively.

1.1 Inversive planes

An *inversive plane*, I, is an incidence structure with a finite number of points and circles with the following properties.

(1) Every 3 distinct points are incident with a unique circle.

(2) Every circle has $n+1>2$ points incident with it.

(3) There are n^2+1 points.

The integer n is called the *order* of I.

Example 1.1 The *classical inversive plane* $I(q)$ has as its points the points of an elliptic quadric E of $\mathrm{PG}(3,q)$ and as its circles the non-tangent plane sections of E. It has order q, and automorphism group $P\Gamma O^-(4,q)$. See [27] for more on elliptic quadrics. □

An *ovoid* of $\mathrm{PG}(3,q)$ is a set Ω of q^2+1 points, no 3 collinear (unless $q=2$, when it is a set of 5 points, no 4 coplanar). This is the maximum size of a set of points of $\mathrm{PG}(3,q)$, no 3 collinear, for $q>2$. An elliptic quadric of $\mathrm{PG}(3,q)$ is an ovoid.

Example 1.2 Each ovoid Ω of $\mathrm{PG}(3,q)$ gives an inversive plane $I(\Omega)$ of order q, with points the points of Ω and circles the secant plane sections of Ω. $I(\Omega)$ is classical if and only if Ω is an elliptic quadric.

An inversive plane isomorphic to $I(\Omega)$ for some ovoid Ω is called *egglike*. All known inversive planes are egglike.

Example 1.3 (The Tits ovoids) (Tits, 1962 [63]) The set of points $\Omega = \{(1, s, t, s^{2^{e+1}} + st + t^{2^{e+1}+2}) : s, t \in \mathrm{GF}(q)\} \cup \{(0,0,0,1)\}$ is an ovoid of $\mathrm{PG}(3,q)$, for $q = 2^{2e+1}$, $e > 0$, with stabiliser in $PGL(4,q)$ being the Suzuki group $Sz(q)$. It is not an elliptic quadric.

Theorem 1.4 (Barlotti, 1955 [7]; Panella, 1955 [39]) *Every ovoid of* $\mathrm{PG}(3,q)$*, q odd, is an elliptic quadric.*

Theorem 1.5 (Bose, 1947 [11]; Seiden, 1950 [55]; O'Keefe–Penttila, 1990 [34], 1992 [35]) *Every ovoid of* $\mathrm{PG}(3,q)$*, $q = 2, 4, 16$, is an elliptic quadric.*

Theorem 1.6 (Fellegara, 1961 [19]; O'Keefe–Penttila–Royle, 1994 [37]) *Every ovoid of* $\mathrm{PG}(3,q)$*, $q = 8, 32$, is an elliptic quadric or a Tits ovoid.*

We boldly venture the following.

Conjecture 1.7 *The only ovoids of* $\mathrm{PG}(3,q)$ *are elliptic quadrics and Tits ovoids.*

Theorem 1.8 (Dembowski, 1964 [17]) *Every inversive plane of even order is egg-like.*

The *residue* of an inversive plane I of order n at a point P is the incidence structure with points the points other than P and lines the circles on P. It is an affine plane of order n.

Theorem 1.9 (Thas, 1994 [60]) *Every inversive plane of odd order with at least one Desarguesian residue is classical.*

The proof of this theorem relies on Theorem 3.2 below.

1.2 Minkowski planes

A *Minkowski plane*, M, is an incidence structure with a finite number of points, lines and circles with the following properties.

(1) Every point is on exactly 2 lines.

(2) Every line has $n + 1 > 2$ points incident with it.

(3) Every 3 points, no 2 collinear, are incident with a unique circle.

(4) Every circle has $n + 1$ points incident with it.

(5) Every line and every circle meet in a unique point.

(6) There are $(n + 1)^2$ points.

The integer n is called the *order* of M.

Example 1.10 The *classical Minkowski plane* $M(q)$ has as its points the points of a hyperbolic quadric H of $\mathrm{PG}(3,q)$, as lines the generators of H and as its circles the non-tangent plane sections of H. It has order q, and automorphism group $P\Gamma O^+(4,q)$. See [27] for more on hyperbolic quadrics.

A set S of permutations of a set X is *sharply 3-transitive* if, whenever (x_1, x_2, x_3) and (y_1, y_2, y_3) are triples of distinct elements of X, there is a unique element σ of S with $x_i^\sigma = y_i$, for all i. The *degree* of S is the cardinality of X. Note that $PGL(2, q)$ is a sharply 3-transitive set of degree $q + 1$ in its action on PG$(1, q)$.

Example 1.11 Each sharply 3-transitive set S of permutations of a set X of degree $n + 1$ gives a Minkowski plane $M(S)$ of order n, with points the elements of X^2, lines the vertical $\{(c, y) : y \in X\}$ and horizontal $\{(x, c) : x \in X\}$ lines, and circles the graphs $\{(x, x^\sigma) : x \in X\}$ of elements σ of S, and conversely, each Minkowski plane of order n gives rise to a sharply 3-transitive set of permutations of degree $n + 1$. The plane $M(S)$ is classical if and only if S is $PGL(2, q)$. See [16].

Example 1.12 (Pedrini, 1966 [45]) Let $q = p^h$, p prime, let n be a non-square in GF(q), and let α be an automorphism of GF(q). The union S_α of $PSL(2, q)$ and the coset of $PSL(2, q)$ in $P\Gamma L(2, q)$ containing $x \mapsto nx^\alpha$ is a sharply 3-transitive set of permutations of PG$(1, q)$, giving rise to a Minkowski plane $M(S_\alpha)$, which is non-classical if $\alpha \neq 1$.

These are all the known Minkowski planes [16].

Theorem 1.13 (Heise, 1974 [25]; Percsy, 1974 [51]) *Every Minkowski plane of even order is classical.*

It follows that every sharply 3-transitive set of permutations of odd degree is a group, isomorphic to $PGL(2, q)$.

The *residue* of a Minkowski plane M of order n at a point P is the incidence structure with points the points not collinear with P and lines the circles on P and the lines not on P. It is an affine plane of order n.

Theorem 1.14 (Chen–Kaerlein, 1973 [12]) *Every Minkowski plane of odd order with at least one Desarguesian residue is classical.*

1.3 Laguerre planes

A *Laguerre plane*, L, is an incidence structure with a finite number of points, lines and circles with the following properties.

(1) Every point is on exactly 1 line.

(2) Every line has $n > 1$ points incident with it.

(3) Every 3 points, no 2 collinear, are incident with a unique circle.

(4) Every circle has $n + 1$ points incident with it.

(5) Every line and every circle meet in a unique point.

(6) There are $n^2 + n$ points.

The integer n is called the *order* of L.

Example 1.15 Let K be a quadratic cone of $\mathrm{PG}(3,q)$ with vertex V. The *classical Laguerre plane* $L(q)$ has as its points the points of K other than V, as lines the generators of K and as its circles the plane sections of K by planes not on V. It has order q, and the automorphism group is the semidirect product of a group of order $q^3(q-1)$ and $P\Gamma O(3,q)$. See [27] for more on quadratic cones.

An *oval* of $\mathrm{PG}(2,q)$ is a set O of $q+1$ points, no 3 collinear. This is the maximum size of a set of points of $\mathrm{PG}(2,q)$, no 3 collinear, for q odd and one less than the maximum for q even, when a set of $q+2$ points, no 3 collinear is called a *hyperoval*. A conic is an oval.

Example 1.16 Each oval O of $\mathrm{PG}(2,q)$ gives a Laguerre plane $L(O)$ of order q as follows. Embed $\mathrm{PG}(2,q)$ as a hyperplane π in $\mathrm{PG}(3,q)$, choose a point V not on π and let K be the cone with vertex V subtended by O. Then $L(O)$ has as points the points of K other than V, as lines the generators of K and as circles the plane sections of K, where the planes do not contain V. The plane $L(O)$ is classical if and only if O is a conic.

A Laguerre plane isomorphic to $L(O)$ for some oval O is called *embeddable*. All known Laguerre planes are embeddable (see [16]).

Theorem 1.17 (Segre, 1955 [53]) *Every oval of* $\mathrm{PG}(2,q)$, *q odd, is a conic.*

There are 10 known infinite families of hyperovals (giving many more infinite families of ovals) in $\mathrm{PG}(2,q)$, q even. For a recent survey, see [47].

Theorem 1.18 (Segre, 1957 [54]; Hall, 1975 [24]; Penttila–Royle, 1994 [48]) *Ovals of* $\mathrm{PG}(2,q)$, $q \le 32$, *are classified.*

The problem of classifying hyperovals of $\mathrm{PG}(2,q)$ for $q > 32$, q even, is an important open problem.

Theorem 1.19 (Quattrocchi, 1995 [52]) *There are no Laguerre planes of order* $n \equiv 2 \pmod 4$ *with* $n > 2$.

The *residue* of a Laguerre plane M of order n at a point P is the incidence structure with points the points not collinear with P and lines the circles on P and the lines not on P. It is an affine plane of order n.

Theorem 1.20 (Chen–Kaerlein, 1973 [12]) *Every Laguerre plane of odd order with at least one Desarguesian residue is classical.*

2 Flocks of inversive planes

A *flock* of an inversive plane of order n is a set of $n-1$ pairwise disjoint circles.

Example 2.1 (Linear flocks of egglike inversive planes) Let Ω be an ovoid of $\mathrm{PG}(3,q)$, and l be a line external to Ω. Then the set of sections of Ω by non-tangent planes on l is a flock of $I(\Omega)$. These flocks are called *linear*.

Theorem 2.2 (Thas, 1973 [56]; Orr, 1976 [38]) *Every flock of an egglike inversive plane is linear.*

3 Flocks of Minkowski planes

A *flock* of a Minkowski plane of order n is a set of $n+1$ pairwise disjoint circles.

Example 3.1 (Linear flocks of classical Minkowski planes) Let H be a hyperbolic quadric of $\mathrm{PG}(3,q)$, and l be a line external to H. Then the set of sections of H by planes on l is a flock of $M(q)$. These flocks are called *linear*.

Examples of non-linear flocks of $M(q)$ are constructed in Thas, 1975 [57] for all odd q, and there are 3 exceptional flocks of classical Minkowski planes for $q = 11, 23, 59$ constructed by various authors, with a nice treatment via root systems in Bader–Durante–Law–Lunardon–Penttila 2003.

Theorem 3.2 (Classification of flocks of classical Minkowski planes) (Bader–Lunardon, 1989 [2]; Thas, 1990 [59]) *Every flock of an classical Minkowski plane is linear, Thas or exceptional.*

A short and elegant proof of this was given by Durante–Siciliano, 2003 [18].

Theorem 3.3 (Thas, 1975 [57]; Walker, 1976 [64]) *Each flock of $M(q)$ gives a translation plane of order q^2.*

Flocks of certain non-classical Minkowski planes were constructed in Bonisoli–Fiori 1988 [10].

4 Flocks of Laguerre planes

A *flock* of an Laguerre plane of order n is a set of n pairwise disjoint circles.

4.1 Examples of flocks

Example 4.1 (Linear flocks of embeddable Laguerre planes) Let O be an oval of $\mathrm{PG}(2,q)$, K be a cone of $\mathrm{PG}(3,q)$ subtended by O with vertex V and l be a line external to K. Then the set of sections of K by planes on l, not on V, is a flock of $L(O)$. These flocks are called *linear*.

Non-linear flocks of $L(q)$ are constructed by the following:

1. Fisher–Thas, 1979 [20] and Walker, 1976 [64] for $q > 2$ and $q \equiv 2 \pmod 3$;
2. Fisher (in Fisher–Thas, 1979 [20]) for $q > 3$ and q odd;
3. Payne, 1985 [41] for $q = 2^h$, $h > 3$ and h odd;
4. Kantor, 1986 [29] for q not prime;
5. Kantor, 1986 for $q \equiv 2$ or $3 \pmod 5$;
6. Gevaert–Johnson, 1987 [22] building on work of Ganley, 1981 [21] for $q > 3$ and $q \equiv 0 \pmod 3$;
7. Gevaert–Johnson, 1988 [22] building on work of Kantor, 1982 [28] for $q > 5$ and $q \equiv 0 \pmod 5$;

8. Cherowitzo–Penttila–Pinneri–Royle, 1996 [15] *Subiaco flocks* for $q > 8$ and $q \equiv 0 \pmod 2$;

9. Penttila, 1998 [46] *Mondello flocks* for $q > 9$ and $q \equiv 1$ or $9 \pmod{10}$;

10. Law–Penttila, 2001 [30] for $q > 9$ and $q \equiv 0 \pmod 3$;

11. Cherowitzo–O'Keefe–Penttila, 2003 [14] *Adelaide flocks* for $q = 4^h$, $h > 2$.

Many examples exist for small field orders. For a recent survey see [31].

Further flocks can be derived from these by the method referred to in Section 4.4 below.

4.2 Projective planes: the Thas–Walker construction

Theorem 4.2 (Thas, unpublished; Walker, 1976 [64]) *Each flock of $L(q)$ gives a translation plane of order q^2.*

Gevaert–Johnson–Thas, 1987 [23] characterise the translation planes that arise.

4.3 Generalised Quadrangles

Theorem 4.3 (Kantor, 1986 [29]; Payne, 1985 [41]; Thas, 1987 [58]) *Each flock of $L(q)$ gives a GQ of order (q^2, q).*

These GQs were geometrically characterised by Thas, 1999 [62] and a geometric construction of the GQs was given. An extension and simplification of this work appears in Barwick–Brown–Penttila [8].

4.4 Derivation of flocks and BLT-sets

A *BLT-set* of a parabolic quadric $Q(4, q)$ in $\mathrm{PG}(4, q)$ is a set B of $q+1$ points, such that no point of the quadric is polar to more than 2 points of B.

Theorem 4.4 (Bader–Lunardon–Thas, 1990 [4]) *Each flock of $L(q)$ gives a BLT-set of $Q(4, q)$, and so q further flocks of $L(q)$. The number of equivalence classes of flocks so obtained is the number of orbits of $P\Gamma O^+(5, q)_B$ on B.*

4.5 Semifield flocks, translation ovoids of $Q(4, q)$, eggs and TGQ

A *semifield* is a division algebra (not necessarily associative). The *middle nucleus* of a semifield is $\{b : (ab)c = a(bc)$, for all a and $c\}$. For a finite semifield, the middle nucleus is a field and the semifield is a vector space over its middle nucleus. The *rank* of a finite semifield is its dimension as a vector space over its middle nucleus.

An *ovoid* of $Q(4, q)$ is a set Ω of $q^2 + 1$ points, no 2 collinear (on a line of the parabolic quadric). An ovoid of $Q(4, q)$ is a *translation ovoid* if there is a point P of Ω and a group G of elations with axis the tangent hyperplane to $Q(4, q)$ at P which acts regularly on $\Omega \backslash \{P\}$.

Theorem 4.5 (Thas, 1997 [61]; Lunardon, 1997 [33]) *Finite semifields of rank 2 over their middle nucleus and translation ovoids of $Q(4, q)$ are equivalent objects and correspond to a class of flocks of $L(q)$ called semifield flocks.*

The GQ arising from a semifield flock is the dual of a translation GQ, and so gives rise to an egg (see [44]). The dual egg gives rise to another (possibly isomorphic) GQ (see [3] for conditions on isomorphism). The known semifield flocks of $L(q)$ are the linear flocks, the Kantor (1986) flocks [29], the Ganley flocks, for $q > 3$ and $q \equiv 0 \pmod 3$, and the Bader–Lunardon–Pinneri flock [3] constructed from the Penttila–Williams ovoid [50], for $q = 3^5$. See also Bader–Ghinelli–Penttila [1] for a computer-free construction of this.

Conjecture 4.6 *The semifield flocks of $L(q)$ are all known.*

4.6 Projective planes: Hyperbolic fibrations

A *hyperbolic fibration* is a partition of the points of $\mathrm{PG}(3, q)$ into two lines and $q - 1$ hyperbolic quadrics.

Theorem 4.7 (Baker–Ebert et al. [6, 5]) *Each flock of $L(q)$ gives a hyperbolic fibration of $\mathrm{PG}(3, q)$, and so 2^{q-1} spreads of $\mathrm{PG}(3, q)$ and 2^{q-1} (possibly isomorphic) projective planes of order q^2.*

Norman Johnson, in his talk at the La Roche conference in 2004, gave a beautiful characterisation of the planes which arise.

4.7 Herds of ovals

Theorem 4.8 (Payne, 1985 [41]; Cherowitzo–Pinneri–Penttila–Royle, 1996 [15]) *Each flock of $L(q)$, q even, gives rise to $q+1$ (possibly equivalent) ovals of $\mathrm{PG}(2, q)$, called a herd of ovals.*

There is a converse in Cherowitzo–Pinneri–Penttila–Royle, 1996. The known flocks of $L(q)$, q even, are the linear, Fisher–Thas–Walker, Payne (and the recoordinatised Payne flocks of Payne, 1992 [42]), Subiaco and Adelaide flocks. The Payne, Subiaco and Adelaide flocks gave rise to previously unknown ovals.

Conjecture 4.9 *The flocks of $L(q)$, q even, are known.*

4.8 Ovals from α-flocks

An oval O of $\mathrm{PG}(2, q)$ is a *translation oval* if there is a point P of O and a group G of elations with axis the tangent line to O at P acting regularly on $O \backslash \{P\}$. Translation ovals of $\mathrm{PG}(2, q)$ were classified by Payne, 1971 [40] and Hirschfeld, 1975 [26].

Theorem 4.10 (Cherowitzo, 1998 [13]) *Each flock of $L(O)$ for O a translation oval of $\mathrm{PG}(2, q)$ gives rise to an oval of $\mathrm{PG}(2, q)$.*

Many examples of flocks of $L(O)$ for O a translation oval appear in Cherowitzo 1998. All known hyperovals of $\mathrm{PG}(2, q)$, q even, arise in this way from flocks of $L(O)$ for O a translation oval, with the single exception of the hyperoval of O'Keefe–Penttila 1992 [36] in $\mathrm{PG}(2, 32)$.

4.9 Classification for fields of small order

Theorem 4.11 (Penttila–Royle, 1998 [49]; Law–Penttila, 2004 [32]) *Flocks of $L(q)$ are classified for $q < 31$ and $q = 32$.*

Flocks of non-classical Laguerre planes have received little attention. Carey Jenkins and Anton Betten have classified the flocks of $L(O)$ for O a Segre–Bartocci oval of $PG(2, 32)$ (personal communication, 2003).

References

[1] L. Bader & D. Ghinelli & T. Penttila, On monomial flocks, *European J. Combin.* **22** (2001), 447–454.

[2] L. Bader & G. Lunardon, On the flocks of $Q^+(3, q)$, *Geom. Dedicata* **29** (1989), 177–183.

[3] L. Bader & G. Lunardon & I. Pinneri, A new semifield flock, *J. Combin. Theory Ser. A* **86** (1999), 49–62.

[4] L. Bader & G. Lunardon & J.A. Thas, Derivation of flocks of quadratic cones, *Forum Math.* **2** (1990), 163–174.

[5] R.D. Baker & G.L. Ebert & T. Penttila, Hyperbolic fibrations and q-clans, *Des. Codes Cryptogr.* **34** (2005), 295–305.

[6] R.D. Baker & G.L. Ebert & K.L. Wantz, Regular hyperbolic fibrations, *Adv. Geom.* **1** (2001), 119–144.

[7] A. Barlotti, Un'estensione del teorema di Segre-Kustaanheimo, *Boll. Un. Mat. Ital. (3)* **10** (1955), 498–506.

[8] S. Barwick & M.R. Brown & T. Penttila, Tetradic sets of elliptic quadrics and generalized quadrangles of order (s, s^2) with property (G), submitted, 2004,

[9] W. Benz, *Vorlesungen ber Geometrie der Algebren. Geometrien von Mbius, Laguerre-Lie, Minkowski in einheitlicher und grundlagengeometrischer Behandlung*, Springer-Verlag, Berlin (1973).

[10] A. Bonisoli & C. Fiori, Sharply 1-transitive subsets of certain permutation groups, *Geom. Dedicata* **26** (1988), 309–314.

[11] R.C. Bose, Mathematical theory of the symmetrical factorial design, *Sankhyā* **8** (1947), 107–166.

[12] Y. Chen & G. Kaerlein, Eine Bemerkung über endliche Laguerre- und Minkowski-Ebenen, *Geom. Dedicata* **2** (1973), 193–194.

[13] W. Cherowitzo, α-flocks and hyperovals, *Geom. Dedicata* **72** (1998), 221–246.

[14] W. Cherowitzo & C.M. O'Keefe & T. Penttila, A unified construction of finite geometries associated with q-clans in characteristic 2, *Adv. Geom.* **3** (2003), 1–21.

[15] W. Cherowitzo & T. Penttila & I. Pinneri & G.F. Royle, Flocks and ovals, *Geom. Dedicata* **60** (1996), 17–37.

[16] A. Delandtsheer, Dimensional linear spaces, *Handbook of incidence geometry*, (ed. F. Buekenhout), North-Holland, Amsterdam (1995), pp. 193–294.

[17] P. Dembowski, Möbiusebenen gerader Ordnung, *Math. Ann.* **157** (1964), 179–205.

[18] N. Durante and A. Siciliano, (B)-geometries and flocks of hyperbolic quadrics, *J. Combin. Theory Ser. A* **102** (2003), 425–431.

[19] G. Fellegara, Gli ovaloidi in uno spazio tridimensionale di Galois di ordine 8, *Atti Accad. Naz. Lincei Rend. Cl. Sci. Fis. Mat. Nat. (8)* **32** (1962), 170–176.

[20] J.C. Fisher & J.A. Thas, Flocks in $\mathrm{PG}(3,q)$, *Math. Z.* **169** (1979), 1–11.

[21] M.J. Ganley, Central weak nucleus semifields, *European J. Combin.* **2** (1981), 339–347.

[22] H. Gevaert & N.L. Johnson, Flocks of quadratic cones, generalized quadrangles and translation planes, *Geom. Dedicata* **27** (1988), 301–317.

[23] H. Gevaert & N.L. Johnson & J.A. Thas, Spreads covered by reguli, *Simon Stevin* **62** (1988), 51–62.

[24] M. Hall, Ovals in the Desarguesian plane of order 16, *Ann. Mat. Pura Appl.* **102** (1975), 159–176.

[25] W. Heise, Minkowski-Ebenen gerader Ordnung, *J. Geometry* **5** (1974), 83.

[26] J.W.P. Hirschfeld, Ovals in Desarguesian planes of even order, *Ann. Mat. Pura Appl.* **102** (1975), 79–89.

[27] J.W.P. Hirschfeld, *Finite projective spaces of three dimensions*, Oxford University Press, New York (1985).

[28] W.M. Kantor, On point-transitive affine planes, *Israel J. Math.* **42** (1982), 227–234.

[29] W.M. Kantor, Some generalized quadrangles with parameters q^2, q, *Math. Z.* **192** (1986), 45–50.

[30] M. Law & T. Penttila, Some flocks in characteristic 3, *J. Combin. Theory Ser. A* **94** (2001), 387–392.

[31] M. Law & T. Penttila, Construction of BLT-sets over small fields, *European J. Combin.* **25** (2004), 1–22.

[32] M. Law & T. Penttila, Classification of flocks of the quadratic cone over fields of order at most 29. Special issue dedicated to Adriano Barlotti, *Adv. Geom.* **Supplement** (2003), 232–244.

[33] G. Lunardon, Flocks, ovoids of Q$(4, q)$ and designs, *Geom. Dedicata* **66** (1997), 163–173.

[34] C.M. O'Keefe & T. Penttila, Ovoids of PG$(3, 16)$ are elliptic quadrics, *J. Geom.* **38** (1990), 95–106.

[35] C.M. O'Keefe & T. Penttila, Ovoids of PG$(3, 16)$ are elliptic quadrics. II., *J. Geom.* **44** (1992), 140–159.

[36] C.M. O'Keefe & T. Penttila, A new hyperoval in PG$(2, 32)$, *J. Geom.* **44** (1992), 117–139.

[37] C.M. O'Keefe & T. Penttila & G.F. Royle, Classification of ovoids in PG$(3, 32)$, *J. Geom.* **50** (1994), 143–150.

[38] W.F. Orr, A characterization of subregular spreads in finite 3-space, *Geom. Dedicata* **5** (1976), 43–50.

[39] G. Panella, Caratterizzazione delle quadriche di uno spazio (tridimensionale) lineare sopra un corpo finito, *Boll. Un. Mat. Ital. (3)* **10** (1955), 507–513.

[40] S.E. Payne, A complete determination of translation ovoids in finite Desarguian planes, *Atti Accad. Naz. Lincei Rend. Cl. Sci. Fis. Mat. Natur. (8)* **51** (1971), 328–331.

[41] S.E. Payne, A new infinite family of generalized quadrangles, *Congr. Numer.* **49** (1985), 115–128.

[42] S.E. Payne, Collineations of the generalized quadrangles associated with q-clans, *Ann. Discrete Math.* **52** (1992), 449–461.

[43] S.E. Payne, The fundamental theorem of q-clan geometry, *Des. Codes Cryptogr.* **8** (1996), 181–202.

[44] S.E. Payne & J.A. Thas, *Finite Generalized Quadrangles*, Pitman, Boston (1984).

[45] C. Pedrini, Gruppi transitivi di sostituzioni e t-reti, *Atti Accad. Naz. Lincei Rend. Cl. Sci. Fis. Mat. Natur. (8)* **40** (1966), 226–232.

[46] T. Penttila, Regular cyclic BLT-sets, *Rend. Circ. Mat. Palermo Suppl.* **53** (1998), 167–172.

[47] T. Penttila, Configurations of ovals, *J. Geom.* **76** (2003), 233–255.

[48] T. Penttila & G.F. Royle, Classification of hyperovals in PG$(2, 32)$, *J. Geom.* **50** (1994), 151–158..

[49] T. Penttila & G.F. Royle, BLT-sets over small fields, *Australas. J. Combin.* **17** (1998), 295–307.

[50] T. Penttila & B. Williams, Ovoids of parabolic spaces, *Geom. Dedicata* **82** (2000), 1–19.

[51] N. Percsy, A characterization of classical Minkowski planes over a perfect field of characteristic two, *J. Geom.* **5** (1974), 191–204.

[52] P. Quattrocchi, Orthogonality in finite Laguerre planes, *Atti Sem. Mat. Fis. Univ. Modena* **43** (1995), 17–24.

[53] B. Segre, Ovals in a finite projective plane, *Canad. J. Math.* **7** (1955), 414–416.

[54] B. Segre, Sulle geometrie proiettive finite, in *Convegno Internazionale: Reticoli e Geometrie Proiettive* Cremonese, Rome (1957), pp. 46–61.

[55] E. Seiden, A theorem in finite projective geometry and an application to statistics, *Proc. Amer. Math. Soc.* **1** (1950), 282–286.

[56] J.A. Thas, Flocks of finite egglike inversive planes, in *Finite Geometric Structures and their Applications* Cremonese, Rome (1973), pp. 189–191.

[57] J.A. Thas, Flocks of non-singular ruled quadrics in $\mathrm{PG}(3,q)$, *Atti Accad. Naz. Lincei Rend. Cl. Sci. Fis. Mat. Natur.* **59** (1975), 83–85.

[58] J.A. Thas, Generalized quadrangles and flocks of cones, *European J. Combin.* **8** (1987), 441–452.

[59] J.A. Thas, Flocks, maximal exterior sets, and inversive planes, *Contemp. Math.* **111** (1990), 187–218.

[60] J.A. Thas, The affine plane $\mathrm{AG}(2,q)$, q odd, has a unique one point extension, *Invent. Math.* **118** (1994), 133–139.

[61] J.A. Thas, Symplectic spreads in $\mathrm{PG}(3,q)$, inversive planes and projective planes, *Discrete Math.* **174** (1997), 329–336.

[62] J.A. Thas, Generalized quadrangles of order (s,s^2). III, *J. Combin. Theory Ser. A* **87** (1999), 247–272.

[63] J. Tits, Ovoides et groupes de Suzuki, *Arch. Math.* **13** (1962), 187–198.

[64] M. Walker, A class of translation planes, *Geom. Dedicata* **5** (1976), 135–146.

School of Mathematics and Statistics
University of Western Australia
WA 6009
Australia
penttila@maths.uwa.edu.au

Judicious partitions and related problems

Alex Scott

Abstract

Many classical partitioning problems in combinatorics ask for a single quantity to be maximized or minimized over a set of partitions of a combinatorial object. For instance, Max Cut asks for the largest bipartite subgraph of a graph G, while Min Bisection asks for the minimum size of a cut into two equal pieces.

In *judicious partitioning problems*, we seek to maximize or minimize a number of quantities simultaneously. For instance, given a graph G with m edges, we can ask for the smallest $f(m)$ such that G must have a bipartition in which each vertex class contains at most $f(m)$ edges.

In this survey, we discuss recent extremal results on a variety of questions concerning judicious partitions, and related problems such as Max Cut.

1 Introduction

A wide variety of combinatorial optimization problems ask for an "optimal" partition of the vertex set of a graph or hypergraph. A good example is the Max Cut problem: given a graph G, what is the maximum of $e(V_1, V_2)$ over partitions $V(G) = V_1 \cup V_2$, where $e(V_1, V_2)$ is the number of edges between V_1 and V_2? Similarly, Min Bisection asks for the minimum of $e(V_1, V_2)$ over partitions $V(G) = V_1 \cup V_2$ with $|V_1| \leq |V_2| \leq |V_1|+1$ (there are k-partite versions Max k-Cut and Min k-Section of both problems).

Both of these problems involve maximizing or minimizing a single quantity over graphs from a certain class. In this survey, we shall discuss a group of problems where several quantities must be maximized or minimized simultaneously. There are a number of variations, but we will group these problems together under the heading of *judicious partitioning problems.*

For example, given a graph G, what is the minimum of $\max\{e(V_1), e(V_2)\}$ over partitions $V(G) = V_1 \cup V_2$? Note the difference here from Max Cut: in Max Cut, we are looking for a partition in which $e(V_1, V_2)$ is large, or equivalently $e(V_1) + e(V_2)$ is small, but we do not care how edges are shared between V_1 and V_2. Here we are seeking a partition in which $e(V_1)$ and $e(V_2)$ are small simultaneously.

In general, judicious partitioning problems seem to be more difficult than similar partitioning problems in which a single quantity is optimized. For instance, in the case of Max Cut it is easy to show that every graph with m edges has a bipartite subgraph with at least $m/2$ edges: a random bipartition, in which each vertex is independently assigned to either class with equal probability, gives a bipartite subgraph with expected size $m/2$, and so there must be some bipartite subgraph of at least this size (and random graphs show that the constant $1/2$ is best possible). But what if we want a *judicious* bipartition, in which both vertex classes contain few edges? In a random bipartition $V(G) = V_1 \cup V_2$ of a graph with m edges, we expect $m/4$ edges in each vertex class; but this does not imply the existence of a good judicious partition, as the quantities $e(V_1)$ and $e(V_2)$ are not independent (for instance, a random bipartition of $K_{1,n-1}$ has, with high probability, about $n/2$ edges on one side and no edges on the other side). Thus we cannot expect to get good bounds for this judicious partitioning problem from a naive random argument.

Indeed, it is far from easy to determine the largest c such that every graph with m edges has a bipartition with at most cm edges in each vertex class. A similar problem holds in the algorithmic context: for Max Cut, a simple greedy algorithm finds a cut of size at least $m/2$ in linear $(O(m+n))$ time; finding a reasonably good judicious partition appears to be rather more complicated.

In general, partitioning problems have two aspects: an extremal problem and an algorithmic problem. The extremal problem asks for bounds on the size of a largest cut. For instance, given $m \geq 1$, what is the largest m' such that every graph with m edges has a cut of size at least m'? The algorithmic problem asks for efficient algorithms (or heuristics) to find large cuts, or else a proof that it is hard (in some appropriate sense) to find a large cut, or to determine whether one exists. We shall focus here on extremal problems, although these are often closely related to algorithmic questions: extremal results often give structural information that can be used to design efficient algorithms, or explain why algorithms perform well or badly.

Judicious partitioning problems arguably share certain features with many real-world problems, where multiple constraints and objectives must be taken into account (for example, many measures of success in circuit layout problems can be thought of as judicious partitioning problems; see [33]). However, real-world problems are often sufficiently complex and specific that their mathematical structure is obscured. The problems considered in this paper are hopefully complex enough to exhibit interesting behaviour, but simple enough to remain mathematically interesting. Even with very simple judicious partitioning problems, many fascinating open questions remain.

The remainder of the paper is split into three sections. In the first two sections, we consider partitioning problems on graphs: we begin with problems with a single constraint, concentrating in particular on Max Cut and its variants, while in section 3, we move on to discuss a variety of judicious partitioning problems. In the final section, we look at partitioning problems on hypergraphs: these are in general much more difficult than the corresponding graph partitioning problems, but some progress has been made, and there are some tantalising conjectures.

A large part of this survey is based on a series of joint papers with Béla Bollobás [9, 10, 11, 12, 13, 14, 15, 16, 17, 18], and much of the remainder is influenced by joint discussions. Any mistakes, of course, are mine.

1.1 Notation

For a graph G and $X \subset V(G)$, we define $G[X]$ to be the subgraph induced by X and set $e(X) = e(G[X])$. For disjoint subsets $X, Y \subset V(G)$, we write

$$e(X,Y) = |\{xy \in E(G) : x \in X, y \in Y\}|$$

for the number of edges between X and Y. If G is a digraph, $e(X,Y)$ denotes the number of edges that are directed from X to Y.

If G is a graph with vertex set V, then a *cut* (V_1, V_2) *of* G is a bipartition $V = V_1 \cup V_2$; a *bisection of* G is a cut with $|V_1| \leq |V_2| \leq |V_1| + 1$. The *size of the cut* (V_1, V_2) is $e(V_1, V_2)$. Similarly, if G is a digraph, a *directed cut of* G is an ordered pair (V_1, V_2) such that $V_1 \cup V_2$ partitions $V(G)$; the *size of the directed cut* is $e(V_1, V_2)$ (i.e. we only count edges from V_1 to V_2).

2 Graph partitions with a single objective

2.1 Max Cut

Probably the best known graph partitioning problem is Max Cut, which asks for the largest bipartite subgraph of a graph G. The extremal version of the problem asks for bounds on the size of a maximum cut in terms of various parameters of G; the algorithmic version, which has an immense literature, asks for algorithms to find a maximum cut or to determine its size, or for proofs that it is in some sense hard to do so. Important surveys of the Max Cut problem can be found in Poljak and Tuza [38] and Laurent [31].

We concentrate here on the extremal version of the problem. For a graph G, we define

$$f(G) = \max_{W \subset V} e(W, V \setminus W)$$

to be the maximal size of a bipartite subgraph of G. For $m \geq 1$, we write

$$f(m) = \min_{e(G)=m} f(G).$$

We will also consider Max Cut for edge-weighted graphs. For a graph G with edge-weighting w, we define

$$f(G) = \max_{W \subset V} w(W, V \setminus W),$$

where $w(W, V \setminus W) = \sum_{u \in W, v \in V \setminus W} w(uv)$ is the weight of the cut $(W, V \setminus W)$. We further define

$$f_w(m) = \min_{w(G)=m} f(G),$$

where the minimum is taken over all graphs whose edges are weighted with positive integers and have total weight m. Equivalently, $f_w(m)$ is the minimum value of $f(G)$ over multigraphs with m edges. Clearly

$$f_w(m) \leq f(m). \tag{2.1}$$

It is easily seen (for instance, by considering random bipartitions) that

$$f_w(m) \geq m/2.$$

In 1973, answering a question of Erdős, Edwards [22, 21] proved that

$$f(m) \geq \frac{m}{2} + \sqrt{\frac{m}{8} + \frac{1}{64}} - \frac{1}{8}. \tag{2.2}$$

The extremal graphs are the complete graphs of odd order.

We will give two proofs of (2.2). For a first proof, we note that (2.2) follows immediately from (2.1) and the following theorem.

Theorem 2.1 *For every m,*

$$f_w(m) \geq \frac{m}{2} + \sqrt{\frac{m}{8} + \frac{1}{64}} - \frac{1}{8}. \tag{2.3}$$

We give a short proof of Theorem 2.1 due to Alon [3] and Hofmeister and Lefmann [28].

Proof Let G be a graph with $w(G) = m$ and $f_w(G)$ minimal. If G contains two vertices x and y that are not adjacent (or for which $w(xy) = 0$), we can compress G by identifying x and y to create a new vertex z. We define

$$w(za) = w(xa) + w(ya)$$

for all $a \in V(G)\backslash\{x, y\}$, and leave other edge-weights unchanged (we define $w(uv) = 0$ if uv is not an edge). Let H be the resulting graph. Any bipartition of H can be extended to a bipartition of G with the same weight by replacing z with x and y, so $f(H) \leq f(G)$. By repeated compressions, we may therefore assume that G is an edge-weighting of the complete graph.

Now let $n = |G|$, and consider a bipartition chosen uniformly at random from all bipartitions of $V(G)$ into sets of size $\lfloor n/2 \rfloor$ and $\lceil n/2 \rceil$. Let's check that this has expected weight at least (2.3). The expected weight of our bipartition is

$$\frac{\lfloor n^2/4 \rfloor}{\binom{n}{2}} m \geq \frac{(n^2-1)/4}{n(n-1)/2} m = \frac{n+1}{2n} m.$$

Defining n' by $m = \binom{n'}{2}$, we see that this is at least

$$\begin{aligned}\frac{m}{2} + \frac{m}{2n} &\geq \frac{m}{2} + \frac{m}{2n'} \\ &= \frac{m}{2} + \frac{n'-1}{4} \\ &= \frac{m}{2} + \sqrt{\frac{m}{8} + \frac{1}{64}} - \frac{1}{8},\end{aligned}$$

where the last equality follows from the fact that $m = \binom{n'}{2}$. □

A similar idea (in this case, contracting colour classes to a single vertex) gives the following simple bound observed by several authors [3, 6, 32, 34].

Theorem 2.2 *For a nonempty graph G with m edges,*

$$f(G) \geq \left(\frac{1}{2} + \frac{1}{2\chi(G)}\right) m.$$

Another useful bound on $f(G)$ was proved by Edwards [22].

Lemma 2.3 *If G is a connected graph then*

$$f(G) \geq \frac{e(G)}{2} + \frac{|G|-1}{4}. \tag{2.4}$$

Short proofs or algorithms can be found in [26, 28, 37, 36, 14]. Note that the bound is sharp for complete graphs of odd order.

Lemma 2.3 also gives a short second proof of (2.2). Identifying a vertex from each component, we may assume that G is connected; if G has n vertices and m edges then it is easily checked (as in the proof of Lemma 2.1) that

$$\frac{n-1}{4} \geq \sqrt{\frac{m}{8} + \frac{1}{64}} - \frac{1}{8},$$

giving the required bound (2.2).

The bound of Edwards (2.2) is sharp for complete graphs of odd order. It came as a surprise when Alon [3], answering a question of Erdős [23], showed that (2.2) can be arbitrarily far from $f(m)$. More specifically, he proved that if $m = n_0^2/2$ (half way between $\binom{n_0}{2}$ and $\binom{n_0+1}{2}$) then

$$f(m) \geq \frac{m}{2} + \sqrt{\frac{m}{8}} + cm^{1/4}; \tag{2.5}$$

on the other hand, unions of complete graphs show that, for every m,

$$f(m) \leq \frac{m}{2} + \sqrt{\frac{m}{8}} + O(m^{1/4}).$$

A further improvement was obtained more recently. Alon and Halperin [4] and Bollobás and Scott [14] independently found a recursion for $f_w(m)$, showing that for sufficiently large n, and any $0 \leq k < n$, if $m = \binom{n}{2} + k$ then

$$f_w(m) = \min\left\{ \left\lfloor \frac{(n+1)^2}{4} \right\rfloor, \left\lfloor \frac{n^2}{4} \right\rfloor + f_w(k) \right\}. \tag{2.6}$$

The bound can be attained by taking either a copy of K_{n+1} with some edges removed, or a copy of K_n together with an extremal multigraph with k edges (see [14] for further discussion of extremal graphs). Since it holds only for sufficiently large n, (2.6) does not determine $f_w(m)$ for every m (although Alon and Halperin [4] conjecture that (2.6) holds for every m).

It was noted in [14], as a consequence of (2.6), that there is a constant C such that

$$|f(m) - f_w(m)| \leq C \quad \text{for all } m \geq 1 \tag{2.7}$$

and so (2.6) determines $f(m)$ to within a constant for every m. In fact, the following conjecture was made both by Alon and Halperin [4] and in [14].

Conjecture 2.4 *$f(m) = f_w(m)$ for every m.*

Note that it follows from (2.6) and (2.7) that the functions $f(m)$ and $f_w(m)$ are not very 'smooth': for instance, although $f(m) \sim f_w(m) \sim m/2$, there are intervals of length $\Omega(m^{1/4})$ on which both functions are constant.

Although we do not know the value of $f_w(m)$ for every m, there is an efficient algorithm that will always find a cut of at least this size.

Theorem 2.5 ([14]) *There is an algorithm that, given a multigraph with m edges, will find a cut of weight at least $f_w(m)$ in linear time.*

If Conjecture 2.4 is true, this would give an algorithm for finding a cut of size at least $f(m)$ in every graph G with m edges. Further discussion, and related results concerning fixed-parameter tractability and weak approximation algorithms, can be find in [14]. For a more general discussion of algorithms, see Laurent [31] and Poljak and Tuza [38].

Let us note briefly that for many classes of graph there are better bounds than $f(m)$. For instance, extending results of Poljak and Tuza [39] and Shearer [44] (see also Erdős [24], where the problem is raised, and Erdős, Faudree, Pach and Spencer [25]), Alon [3] proved that if G is triangle-free then

$$f(G) \geq \frac{m}{2} + cm^{4/5},$$

while for every $m > 0$ there is a triangle-free graph with m edges such that

$$f(G) \leq \frac{m}{2} + c'm^{4/5}.$$

Thus the $m^{1/2}$ term in (2.2) is replaced in this case by an $m^{4/5}$ term. Alon, Bollobás, Krivelevich and Sudakov [1] proved that, for $r \geq 4$, if G has girth at least r then

$$f(G) \geq \frac{m}{2} + cm^{r/(r+1)},$$

and showed that, for every m, there is a graph with m edges and girth at least 5 such that

$$f(G) \leq \frac{m}{2} + c'm^{5/6}.$$

They conjecture that a similar result should hold for every $r \geq 4$. Further results of this type can be found in Alon, Krivelevich and Sudakov [5], and a detailed discussion can be found in Poljak and Tuza [38].

2.2 Max k-cut

Let us write $f_k(G)$ for the maximum number of edges in a k-partite subgraph of G, and define

$$f_k(m) = \min\{f_k(G) : e(G) = m\}.$$

We define $\tilde{f}_k$ to be the same quantity for multigraphs (so $\tilde{f}_2(m) \equiv f_w(m)$). Clearly

$$f_k(m) \geq \tilde{f}_k(m).$$

By considering a random partition into k classes, it is easy to see that

$$f_k(m) \geq \frac{k-1}{k}m,$$

while an analogue to the Edwards bound was proved in [14]:

$$\tilde{f}_k(m) \geq \left(1 - \frac{1}{k}\right) m + \frac{k-1}{2k}\sqrt{2m + 1/4} + \frac{k^2 - 2k + 2}{8}, \tag{2.8}$$

with equality when G is a complete graph.

It was shown in [14] that if $\delta(G) = \omega(n)$, where $\omega(n) \to \infty$ as $n \to \infty$, then

$$f_k(G) \geq \frac{k-1}{k}m + \frac{k-1}{2k}n + o(n),$$

which is sharp except for the $o(n)$ term. A sharper result was conjectured.

Conjecture 2.6 ([14]) *If G is $(k-1)$-connected then*

$$f_k(G) \geq \frac{k-1}{k}m + \frac{k-1}{2k}n + O(1).$$

It is also possible (see [14]) to extend (2.6) to the k-partite context: the functions $\tilde{f}_k(m)$ and $f_k(m)$ turn out to have similar properties to $f_w(m)$ and $f(m)$.

2.3 Max Bisection

The extremal theory for Max Cut is now quite well developed. For Max Bisection, however, much less is known. Bisection problems are more restricted than cut problems, and the available tools are consequently more limited. For instance, the 'repeated contraction' argument used for Theorem 2.1 does not work, as we need to keep track of the number of vertices on each side of the partition.

Let us write $b(G)$ for the maximum of $e(V_1, V_2)$ over all partitions $V(G) = V_1 \cup V_2$ with $|V_1| \leq |V_2| \leq |V_1| + 1$, and define

$$b(m) = \min\{b(G) : e(G) = m\}.$$

By considering random bipartitions, it is clear that

$$b(m) \geq \lceil m/2 \rceil.$$

On the other hand, this bound is clearly achieved by the star $K_{1,n}$. However, it is less clear what happens if the graph is not so sparse.

Problem 2.7 *Fix $\delta > 1$. What is*

$$\min\{b(G) : e(G) = m, \delta(G) \geq \delta\}\,?$$

Natural families of graphs to consider here are the complete bipartite graph $K_{\delta,n}$ and the graphs $K_\delta + E_n$ obtained from $K_{\delta,n}$ by filling in one vertex class.

A similar problem arises if we restrict the number of vertices.

Problem 2.8 *For $\delta \geq 1$. For $m, n \geq 1$, what is*

$$\min\{b(G) : e(G) = m, |G| = n\}\,?$$

An interesting range here is when $n = O(\sqrt{m})$. Considering a random bisection shows that in this range we get $b(G) \geq m/2 + \Omega(\sqrt{m})$. But how close do we get to the Edwards bound (2.2)?

3 Judicious partitions of graphs

3.1 Each class contains few edges

We now turn to considering various types of judicious partition. For a graph G, we define

$$g(G) = \min_{V(G)=V_1 \cup V_2} \max\{e(V_1), e(V_2)\}.$$

For $m \geq 1$, we define

$$g(m) = \max_{e(G)=m} g(G).$$

Thus $g(m)$ is the smallest integer such that every graph G with m edges has a bipartition in which each class contains at most $g(m)$ edges.

It is easily seen that

$$g(m) \leq m/3. \tag{3.1}$$

Indeed, consider a maximum cut $V(G) = V_1 \cup V_2$. Clearly every $v \in V_1$ has

$$|\Gamma(v) \cap V_2| \geq |\Gamma(v) \cap V_1|, \tag{3.2}$$

or else we could increase the size of the cut by moving v from V_1 to V_2. Summing (3.2) over vertices in V_1, we get

$$\begin{aligned} 2e(V_1) &= \sum_{v \in V_1} |\Gamma(v) \cap V_1| \\ &\leq \sum_{v \in V_1} |\Gamma(v) \cap V_2| \\ &= e(V_1, V_2), \end{aligned}$$

and so $e(V_1) \leq m/3$. Similarly, for $v \in V_2$, we have

$$|\Gamma(v) \cap V_1| \geq |\Gamma(v) \cap V_2|, \tag{3.3}$$

and the same argument shows that $e(V_2) \leq m/3$.

Bipartitions such that every $v \in V_1$ satisfies (3.2) and every $v \in V_2$ satisfies (3.3) are called *unfriendly*. As we have noted, it is very easy to see that every finite graph has an unfriendly bipartition, but for infinite graphs the picture is quite different. Aharoni, Milner and Prikry [2] showed that unfriendly bipartitions exist for certain classes of infinite graphs. Shelah and Milner [45] proved the following surprising result.

Theorem 3.1 ([45]) *There is a graph with $(2^\omega)^{(+\omega)}$ vertices and no unfriendly bipartition.*

On the other hand, there is a positive result for partitions into more than two classes. We say that a vertex-partition of a graph G (into any number of classes) is *unfriendly* if every vertex has at least as many neighbours in every other class as in its own class.

Theorem 3.2 ([45]) *For $k \geq 3$, every graph has an unfriendly partition into k sets.*

Returning to finite graphs, the bound $g(m) \leq m/3$ is attained with equality only by K_3 (and this is the only extremal graph). But what happens for larger m? For Max Cut, it is easy to show that $f(m) \sim m/2$, but the asymptotics of $g(m)$ are less easy to determine.

A number of papers gave successively better bounds for $g(m)$. It was shown in [18] that

$$g(m) \leq \frac{m}{4} + cm^{4/5}.$$

Porter [42] determined the correct order of magnitude for the second term, showing that

$$g(m) \leq \frac{m}{4} + \sqrt{2}m,$$

and Porter and Bin Yang [43] showed that

$$g(m) \leq \frac{m}{4} + \sqrt{\frac{8m}{9}}.$$

Finally, a sharp result was proved in [16], which gave the following analogue of (2.2). Surprisingly, it turned out that this could be combined with a partition giving a large cut.

Theorem 3.3 ([16]) *Every graph G with m edges has a bipartition $V(G) = V_1 \cup V_2$ such that*

$$e(V_1, V_2) \geq \frac{m}{2} + \sqrt{\frac{m}{8} + \frac{1}{64}} - \frac{1}{8} \tag{3.4}$$

and, for $j = 1, 2$,

$$e(V_i) \leq \frac{m}{4} + \sqrt{\frac{m}{32} + \frac{1}{256}} - \frac{1}{16}. \tag{3.5}$$

The extremal graphs are the complete graphs of odd order.

Note that (3.5) is exactly half of the Edwards bound (3.4).

Proof We give a sketch of the proof. Let us start with a maximum cut $V(G) = V_1 \cup V_2$, and suppose that $e(V_1) \geq e(V_2)$. If V_1 satisfies (3.5), we are done. Otherwise, we successively move vertices from V_1 to V_2 until we obtain $V_1' \subset V_1$ that satisfies (3.5).

Suppose we have reached a partition (W_1, W_2), where $W_1 \subset V_1$ but W_1 does not satisfy (3.5). We pick $w \in W_1$ with $|\Gamma(w) \cap W_1|$ as small as possible, and move it across. Let (W_1', W_2') be the resulting partition. We claim that W_2' satisfies (3.5) and (W_1', W_2') satisfies (3.4). The theorem will then be proved, as we continue moving vertices until W_1' first satisfies (3.5).

First of all, note that for $v \in V_1$ we have $|\Gamma(v) \cap V_2| \geq |\Gamma(v) \cap V_1|$, since (V_1, V_2) is a maximum cut. Since we are only moving vertices from V_1 to V_2, we have for all $v \in W_1$,

$$|\Gamma(v) \cap W_1| \leq |\Gamma(v) \cap W_2|.$$

Now suppose that

$$e(W_1) = \frac{m}{4} + \alpha,$$

where

$$\alpha > \sqrt{\frac{m}{32} + \frac{1}{256}} - \frac{1}{16}.$$

Recall that w is the vertex we move from W_1 to W_2 and let $\delta = |\Gamma(w) \cap W_1|$. Then

$$\begin{aligned} e(W_1', W_2') &= \sum_{v \in W_1'} |\Gamma(v) \cap W_2'| \\ &= \sum_{v \in W_1'} |\Gamma(v) \cap W_2| + \delta \\ &\geq \sum_{v \in W_1 \backslash w} |\Gamma(v) \cap W_1| + \delta \\ &= (2e(W_1) - \delta) + \delta \\ &= \frac{m}{2} + 2\alpha, \end{aligned}$$

which satisfies (3.4), as $\alpha > \sqrt{m/32 + 1/256} - 1/16$.

Since $\delta = \delta(G[W_1]) = |\Gamma(w) \cap W_1|$, we have

$$e(W_1) \geq \binom{\delta + 1}{2},$$

and so

$$\delta \leq \sqrt{2e(W_1) + \frac{1}{4}} - \frac{1}{2} = \sqrt{\frac{m}{2} + 2\alpha + \frac{1}{4}} - \frac{1}{2}.$$

So

$$\begin{aligned} e(W_2') &= m - e(W_1') - e(W_1', W_2') \\ &\leq m - \left(\frac{m}{4} + \alpha - \delta\right) - \left(\frac{m}{2} + 2\alpha\right) \\ &= \frac{m}{4} - 3\alpha + \delta \\ &\leq \frac{m}{4} - 3\alpha + \sqrt{\frac{m}{2} + 2\alpha + \frac{1}{4}} - \frac{1}{2}, \end{aligned}$$

which is bounded above by (3.5) for $\alpha > \sqrt{m/32 + 1/256} - 1/16$. □

A more general version of the argument produces a 'biased' extension of Theorem 3.3.

Theorem 3.4 ([16]) *Let G be a graph with m edges and let $0 \leq p \leq 1$. There is a partition $V(G) = V_1 \cup V_2$ such that*

$$e(V_1) \leq p^2 m + c(p, m)$$

and

$$e(V_2) \leq (1-p)^2 m + c(p, m),$$

where

$$c(p, m) = p(1-p)\left(\sqrt{m/2 + 1/16} - 1/4\right).$$

We now turn to the k-partite version of the question. Let

$$g_k(G) = \min_{V(G)=V_1\cup\cdots\cup V_k} \max\{e(V_1),\ldots,e(V_k)\}.$$

For $m \geq 1$, we define

$$g_k(m) = \max_{e(G)=m} g_k(G).$$

It was shown in [18] that

$$g_k(m) \leq m\Big/\binom{k+1}{2},$$

which is best possible for the complete graph K_{k+1}. However, as in the bipartite case, it is possible to do much better for larger graphs. It was proved in [18] that

$$g(m) \leq \frac{m}{k^2} + O(m^{4/5}),$$

while Porter [41] showed that if k is a power of 2 then

$$g_k(m) \leq \frac{m}{k^2} + \sqrt{\frac{m}{k}}.$$

Porter [40] showed that, for every k,

$$g_k(m) \leq \frac{m}{k^2} + 4k\sqrt{m},$$

and Porter and Bin Yang [43] showed that, for k a power of 2,

$$g_k(m) \leq \frac{m}{k^2} + \frac{1.31\sqrt{m}}{k}.$$

As in the bipartite case, a sharp result was proved in [16] (using Theorem 3.4 as an essential tool).

Theorem 3.5 ([16]) *Every graph with m edges has a vertex-partition into k sets, each of which contains at most*

$$\frac{m}{k^2} + \frac{k-1}{2k^2}\left(\sqrt{2m+\frac{1}{4}} - \frac{1}{2}\right) \tag{3.6}$$

edges.

The extremal graphs are complete graphs of order $kn+1$. However, unlike Theorem 3.3, it is not known whether a k-partition satisfying (3.6) can be achieved with a large cut.

Problem 3.6 ([13]) *Does every graph with m edges have a partition into k sets that satisfies both (3.6) and (2.8)?*

A starting point for this would be to prove that there is a k-cut that satisfies (3.6) and has size at least $(1 - 1/k)m$. (Although note that when k is a power of 2, we can apply Theorem 3.3 recursively.)

The bounds given by (3.5) and (3.4) are both optimal for complete graphs of odd order. It would be interesting to know the behaviour of

$$\frac{m}{4} + \sqrt{\frac{m}{32}} - g(m).$$

For instance, it seems very likely that $m/4 + \sqrt{m/32} - g(m)$ is unbounded, as conjectured in [13]. Perhaps the following judicious version of Alon's result (2.5) is true.

Conjecture 3.7 *There is some $c > 0$ such that*

$$g(m) < \frac{m}{4} + \sqrt{\frac{m}{32}} - cm^{1/4}$$

for infinitely many values of m.

As with Max Cut, a good starting point might be to consider graphs with m about half way between $\binom{n}{2}$ and $\binom{n+1}{2}$, or else of form $\binom{2t}{2}$.

More generally, it would be desirable to pin down $g(m)$ more precisely.

Problem 3.8 *Is there a recursion for $g(m)$, for m sufficiently large, analogous to (2.6)?*

The same problem arises in the k-partite case: if m is not of form $\binom{2n+1}{2}$, how far out can (3.6) be? For instance, Hofmeister and Lefmann [28] proved that if G has $\binom{kn}{2}$ edges then it has a partition into k vertex classes $V_1, \ldots, V_k$ with $\sum_{i=1}^{k} e(V_i) \leq k\binom{n}{2} = n(kn - k)/2$, beating the trivial bound of $(1/k)\binom{nk}{2} = n(kn - 1)/2$ by $(k + 1)n/2$.

Problem 3.9 ([16]) *Does every graph G with $\binom{kn}{2}$ edges have a vertex partition into k sets, each of which contains at most $\binom{n}{2}$ edges?*

Finally, we note that the proof of Theorem 3.3 gives a polynomial-time algorithm that finds a partition satisfying (3.4) and (3.5) (we start with a partition satisfying (3.2) and (3.3), rather than with a maximum cut). It would be nice to have a judicious version of Theorem 2.5.

Problem 3.10 *Is there a polynomial-time algorithm that takes as input a graph with m edges and finds a partition $V(G) = V_1 \cup V_2$ with $\max\{e(V_1), e(V_2)\} \leq g(m)$?*

3.2 Bounded-degree graphs

For graphs with bounded maximal degree, we can get much stronger results. Indeed, the constant in the linear term in (3.5) can be improved.

The following was shown in [12].

Theorem 3.11 ([12]) *If k is odd, then every graph G with m edges and $\Delta(G) \le k$ has a bipartition $V(G) = V_1 \cup V_2$ with*

$$\max\{e(V_1), e(V_2)\} \le \frac{k-1}{4k}m + \frac{k-1}{4} \tag{3.7}$$

and

$$e(V_1, V_2) \ge \frac{k+1}{2k}m. \tag{3.8}$$

The extremal graphs for (3.7) are of form $(2t+1)K_k \cup sK_{k+1}$. Note that if k is even, then an optimal bound is obtained by applying the theorem with $k+1$ as the degree bound: the extremal graphs are of form $(2t+1)K_{k+1}$.

For graphs that are regular, we can do even better. If the maximal degree is odd we get the following.

Theorem 3.12 ([12]) *Let $k \ge 1$ be odd, and suppose G is a k-regular graph with m edges. Then there is a partition $V(G) = V_1 \cup V_2$ with $|V_1| = |V_2|$ and*

$$\max\{e(V_1), e(V_2)\} \le \frac{k-1}{4k}m.$$

The extremal graphs are of form sK_{k+1}.

If the maximal degree is even, we cannot do much better.

Theorem 3.13 ([12]) *Let $k \ge 2$ be even, and suppose G is a k-regular graph with m edges.*

(a) If $|G|$ is even, there is a partition $V(G) = V_1 \cup V_2$ with $|V_1| = |V_2|$ and

$$\max\{e(V_1), e(V_2)\} \le \frac{k}{4(k+1)}m.$$

The extremal graphs are of form $2tK_{k+1}$.

(b) If $|G|$ is odd, there is a partition $V(G) = V_1 \cup V_2$ with $|V_2| = |V_1| + 1$ and

$$\max\{e(V_1), e(V_2)\} \le \frac{1}{4}\frac{k}{k+1}m + \frac{k}{4}.$$

The extremal graphs are of form $(2t+1)K_{k+1}$.

The extremal graphs for Theorem 3.11, 3.12 and 3.13 are all unions of complete graphs. We should expect to be able to do rather better for graphs without large cliques.

Problem 3.14 ([12]) *For $k \ge 3$, what are the optimal constants c_k and d_k such that every k-regular K_{k+1}-free graph with m edges has $f(G) \ge c_k m$ and $g(G) \ge d_k m$?*

Note that if G is k-regular then any partition $V(G) = V_1 \cup V_2$ with $|V_1| = |V_2|$ has $e(V_1) = e(V_2)$, so Max Bisection gives a bound on $g(G)$. For instance, for $k = 3$, Locke [34] showed that every cubic K_4-free graph has a bisection of size at least

$11e(G)/15$; it follows that $g(G) \leq 2e(G)/15$, which is better than the bound $e(G)/6$ from Theorem 3.12.

For graphs with larger girth, it should be possible to get stronger results (see [13, 19, 29, 34, 46]). One approach here is to combine a good colouring with Theorem 2.2. For graphs without large cliques, there are powerful strengthenings of Brooks' Theorem that provide colourings with fewer than $\Delta(G)$ colours (see Molloy and Reed [35] for a detailed discussion). For triangle-free graphs, Johansson [30] showed that $\chi(G) = O(\Delta/\log\Delta)$; it follows that $f(G) \geq m/2 + O(m\log\Delta/\Delta)$. However, the resulting colouring might be very imbalanced, and so this does not help us with $g(G)$. (Note that the Hajnal-Szemerédi Theorem [27] on balanced colourings does not quite give Theorem 3.12.) Further discussion can be found in [13].

Finally, we note that, for graphs with bounded degrees, there are partitions that satisfy very strong conditions.

Theorem 3.15 ([12]) *Let $k \geq 2$. For every $\Delta \geq 1$ there is a constant $K = K(\Delta)$ such that for every graph G with $\Delta(G) \leq \Delta$ and every sequence $p_1, \ldots, p_k$ of non-negative reals with $\sum_{i=1}^k p_i = 1$, there is a partition $V(G) = \bigcup_{i=1}^k V_i$ such that, for every i,*

$$\big||V_i| - p_i|G|\big| \leq K$$

and

$$|e(V_i) - p_i^2 e(G)| \leq K,$$

and, for every $i \neq j$,

$$|e(V_i, V_j) - p_i p_j e(G)| \leq K.$$

A similar result is proved in [12] for hypergraphs.

3.3 Judicious partitions and maximum cuts

Given a graph G, a trivial bound on $g(G)$ is given by the fact that

$$e(V_1, V_2) \geq m - 2\max\{e(V_1), e(V_2)\}. \tag{3.9}$$

It follows immediately that

$$f(m) + 2g(m) \geq m.$$

and, by Theorem 3.3, we know the value of $f(m) + 2g(m)$ whenever m has form $\binom{n}{2}$. However, it is unclear what happens for other values of m. Related to Conjecture 3.7, we have the following problem.

Problem 3.16 *What is the behaviour of the function $f(m) + 2g(m) - m$?*

This is expressed rather imprecisely, but the aim is to get a very precise description of the behaviour of $f(m) + 2g(m)$.

If we have a graph for which $g(G)$ is particularly small, then (3.9) implies that $f(G)$ is large. So graphs with small judicious partitions have large cuts. It is natural to wonder whether the converse is also true, and Alon, Bollobás, Krivelevich and Sudakov proved the following.

Theorem 3.17 ([1]) *Let G be a graph with m edges and suppose that $f(G) = \frac{m}{2} + \delta$. If $\delta \le m/30$ then*

$$g(G) \le \frac{m}{4} - \frac{\delta}{2} + \frac{10\delta^2}{m} + 3\sqrt{m}$$

and if $\delta \ge m/30$ then

$$g(G) \le \frac{m}{4} - \frac{m}{100}.$$

Note that if $f(G) = m/2 + \delta$ then $g(G) \ge m/4 - \delta/2$, so the theorem is sharp up to the error term $10\delta^2/m + 3\sqrt{m}$, which is $o(\delta)$ provided $\delta = o(m)$. It would be very interesting to have more exact bounds, although this might be quite difficult.

Note that Theorem 3.17 does not help when δ is $O(\sqrt{m})$, as the $3\sqrt{m}$ error term overwhelms the gain of $\delta/2$; it would be good to have sharper bounds in this range.

An extension of Theorem 3.17 for partitions into more than two parts was proved in [10].

3.4 Other norms

Elsewhere we have considered Max Cut, which can be thought of as minimizing the l_1 norm of $(e(V_1), e(V_2))$ over partitions $V(G) = V_1 \cup V_2$; we have also considered the judicious partitioning problem of minimizing the l_∞ norm of $(e(V_1), e(V_2))$. However, we could equally well look for the minimum of other norms. For instance, we have the following natural problem.

Problem 3.18 *What is the maximum of*

$$\min_{V(G)=V_1\cup V_2} e(V_1)^2 + e(V_2)^2$$

over graphs G with m edges?

Similar problems arise when we partition into more than two sets, or consider other norms.

In general, given an invariant μ of graphs, we can ask for the minimum of

$$\mu(G[V_1]) + \mu(G[V_2])$$

or

$$\max\{\mu(G[V_1]), \mu(G[V_2])\}$$

over partitions $V(G) = V_1 \cup V_2$. Writing $f_\mu(G)$ and $g_\mu(G)$ for these two functions, we define corresponding functions

$$f_\mu(m) = \min\{f_\mu(G) : e(G) = m\}$$

and

$$g_\mu(m) = \min\{g_\mu(G) : e(G) = m\}.$$

For instance, if $\mu(G) = e(G)$ then determining $f_\mu(G)$ and $g_\mu(G)$ corresponds to Max Cut and the judicious partitioning problem considered in section 3.

There are many problems, for instance the following.

Problem 3.19 *Let*

$$\mu(H) = \sum_{v \in V(H)} d_H(v)^2.$$

What is $f_\mu(m)$*? What is* $g_\mu(m)$*?*

Of course, many other combinatorial problems can be expressed in this manner.

3.5 Multiple graphs

So far, we have been concerned with problems that involve partitions optimizing several quantities simultaneously for a given graph. We can also consider partitions that optimize a single quantity simultaneously for several different graphs defined on the same vertex set. For instance, the following question is raised in [12].

Problem 3.20 ([12]) *Find the largest integer* $f(m;2)$ *such that for every pair of graphs* G_1, G_2, *with* $e(G_1) = e(G_2) = m$ *and* $V(G_1) = V(G_2)$, *there is a partition* $V(G_i) = V_1 \cup V_2$ *such that*

$$\min\{e_{G_1}(V_1, V_2), e_{G_2}(V_1, V_2)\} \geq f(m;2).$$

A starting point for this question is to determine whether

$$f(m;2) = (1 + o(1))m/2.$$

Of course, there are many interesting variations on the question. For instance, what if there are more than two graphs? What about partitions such that $\max\{e_{G_1}(V_1), e_{G_1}(V_2)\}$ and $\max\{e_{G_2}(V_1), e_{G_2}(V_2)\}$ are both small?

There is also an interesting relationship between the maximum cut in a graph G and the maximum cut of its complement. For a graph G with m edges, we define

$$f^+(G) = f(G) - m/2.$$

The following was proved in [9] (see also [11]).

Theorem 3.21 ([9]) *If* G *is a graph with* n *vertices and* $p\binom{n}{2}$ *edges, where* $4/n \leq p \leq 1 - 4/n$ *then*

$$f^+(G)f^+(\overline{G}) \geq c(p)n^3.$$

We can think of Theorem 3.21 as a result about 2-colourings of the edges of K_n. What can we say if we use more than two colours?

3.6 Judicious bisections

As with the extremal version of Max Bisection, comparatively little is known about judicious partitions in which we demand that the partition is balanced, i.e. the vertex classes are as equal as possible in size. Considering $K_{1,n-1}$ shows that we cannot in general demand a balanced bipartition with fewer than $\lfloor m/2 \rfloor$ edges in each vertex class. However, it may be possible to do better for graphs that are denser.

Problem 3.22 ([13]) *What is the smallest $c(k)$ such that every graph G with m edges and minimal degree k has a bisection with at most $c(k)m$ edges in each vertex class?*

It is easily seen that $c(1) = 1/2$: given G, consider a bipartition $V(G) = V_1 \cup V_2$ chosen uniformly at random from all bipartitions with $|V_1| = |V_2|$ (if $|G|$ is odd then add an isolated vertex). Then $\mathbb{E}e(V_1) < m/4$, and so $\mathbb{P}(e(V_1) \geq m/2) < 1/2$, and similarly for $e(V_2)$. So with positive probability the bipartition has $\max\{e(V_1), e(V_2)\} < m/2$. We deduce that $c(1) \leq 1/2$. On the other hand, as noted above, $K_{1,n-1}$ shows that $c(1) \geq 1/2$.

Considering graphs of form $K_{3,n-3}$ shows that $c(2) \geq c(3) \geq 1/3$.

Conjecture 3.23 ([13]) $c(2) = 1/3$.

A starting point for the problem might be the following conjecture from [13], which asserts that every graph has a bisection that is close to an unfriendly partition.

Conjecture 3.24 ([13]) *Every graph G has a bisection $V(G) = V_1 \cup V_2$ such that*

$$|\Gamma(v) \cap V_1| \leq |\Gamma(v) \cap V_2| + 1 \qquad \forall v \in V_1$$
$$|\Gamma(v) \cap V_2| \leq |\Gamma(v) \cap V_1| + 1 \qquad \forall v \in V_2.$$

If true, this could not be improved, as can be seen by considering graphs of form $K_{2k+1,2l+1}$, where $k \neq l$.

A different condition that might ensure a good judicious bipartition is $\Delta(G) = o(n)$, as suggested in [13]. We make the following conjecture.

Conjecture 3.25 *If G is a graph with n vertices and $\Delta(G) = o(n)$ then there is a bisection $V(G) = V_1 \cup V_2$ such that*

$$\max\{e(V_1), e(V_2)\} \leq (1 + o(1))e(G)/4.$$

3.7 Digraphs

There is also a version of the Max Cut problem for digraphs. For a directed graph D, we define

$$df(D) = \max_{V(D)=V_1 \cup V_2} e(V_1, V_2)$$

and

$$df(m) = \min\{df(D) : e(D) = m\}.$$

It is easily seen that $df(D) \geq f(G)/2$, where G is the underlying multigraph of D (if xy and yx are edges then xy will be a double edge in G). Thus

$$df(m) \geq f_w(m)/2.$$

The behaviour of $df(m)$ is closely related to the behaviour of $f(m)$, as shown in [14].

It is natural to ask what happens if we restrict our attention to *acyclic* directed graphs. Surprisingly, we still get a cut of almost the same size.

Theorem 3.26 ([7]) *For every $m \geq 1$, there is an acyclic digraph D with m edges and*

$$df(D) \geq (1 + o(1))m/4.$$

A large directed cut asks only for $e(V_1, V_2)$ to be large. An interesting judicious partitioning problem arises when we ask for *both* $e(V_1, V_2)$ and $e(V_2, V_1)$ to be large. The star $K_{1,n-1}$ with all edges directed away from the centre shows that we need some degree condition.

Problem 3.27 *Fix $k > 0$. What is the largest constant $c(k)$ such that every digraph with m edges and minimum outdegree at least k has a bipartition $V(D) = V_1 \cup V_2$ with*

$$\min\{e(V_1, V_2), e(V_2, V_1)\} \geq c(k)m?$$

The same problem arises for partitions into more than two parts.

4 Hypergraphs

4.1 Each class contains few edges

As with most combinatorial problems, judicious partitioning problems are typically much harder for hypergraphs than for graphs, and correspondingly much less is known. We give just a brief account, concentrating on two types of problem.

We extend our notation as follows. For a hypergraph H, and $X \subset V(H)$, we write $e(X)$ for the number of edges contained in X. For disjoint subsets X, Y of $V(H)$, we write $e(X, Y)$ for the number of edges contained in $X \cup Y$ and incident with both X and Y. So $e(X, Y) = e(X \cup Y) - e(X) - e(Y)$. Then, for a hypergraph H, we define

$$f(H) = \max_{V(H)=V_1 \cup V_2} e(V_1, V_2),$$

where the maximum is taken over partitions $V(H) = V_1 \cup V_2$, and

$$f^{(k)}(m) = \min\{f(H) : H \text{ a } k\text{-uniform hypergraph}, e(H) = m\}.$$

Similarly,

$$g(H) = \min_{V(H)=V_1 \cup V_2} \max\{e(V_1), e(V_2)\},$$

and

$$g^{(k)}(m) = \max\{g(H) : H \text{ a } k\text{-uniform hypergraph}, e(H) = m\}.$$

The extremal theory is much less developed than for graphs. For instance, it is easy to show that

$$f^{(k)}(m) = (1 - 2^{1-k})m + o(m),$$

but the following is open even for $k = 3$.

Problem 4.1 *Prove a bound for k-uniform hypergraphs analogous to* (2.2).

A natural conjecture is that, for m of form $\binom{n}{k}$, the complete k-uniform hypergraph should be extremal.

For $g^{(k)}(m)$, even less is known, The following result was proved in [17].

Theorem 4.2 ([17]) *Every 3-uniform hypergraph with m edges has a partition into k sets, each of which contains at most*

$$(1+o(1))m/k^3$$

edges.

The error term given in [17] was of size $O(m^{6/7})$, although it seems likely that the truth should be $O(m^{2/3})$. The difficulty comes from the fact that we are optimizing several variables simultaneously: in a random k-partition we expect m/k^3 edges in each vertex class. If there are no large-degree vertices, then we can prove this immediately by taking a random partition and using a martingale inequality. The obstacle is that there may be vertices of large degree, and this is dealt with by first partitioning the large-degree vertices using an extremal argument and then handling the rest of the graph with probabilistic techniques.

It seems likely that a similar result should hold for r-uniform hypergraphs, and it seems surprising that it still remains open.

Conjecture 4.3 ([17]) *Let $r \geq 3$ and $k \geq 2$ be fixed integers. Then every r-uniform hypergraph with m edges has a vertex-partition into k classes, each of which contains at most*

$$\frac{m}{k^r}+o(m)$$

edges.

Perhaps an even stronger conjecture may be true.

Conjecture 4.4 ([17]) *Let $r \geq 3$. Every r-uniform hypergraph with m edges has a bipartition with at most $m/2^r$ edges in each class.*

In other words, $g^{(r)}(m) \leq m/2^r$ for $r \geq 3$. This would be very surprising if true, since it would imply the existence of a bipartition in which both sides simultaneously beat the expected number of edges in a random bipartition. Note that the conjecture would not hold for $r=2$, since we have $g(m) = m/2+\Omega(\sqrt{m})$. Even if the conjecture is not true, there may only be finitely many counterexamples.

The best current bound is the following.

Theorem 4.5 ([13]) *Let $r \geq 2$ and $k \geq 2$. Every r-uniform hypergraph with m edges has a vertex partition into k classes with at most*

$$a_r\frac{m}{k^r}+b_r m^{2r/(2r+1)}$$

edges in each class, where $a_r = O(r/\log r)$.

Finally, it seems likely that the relationship between $f^{(k)}(G)$ and $g^{(k)}(G)$ should hold, as in the graph context. We raise the following problem.

Problem 4.6 *Prove a version of Theorem 3.17 for 3-uniform hypergraphs.*

4.2 Each class meets many edges

What about hypergraph partitions in which each class *meets* many edges? Bollobás and Thomason (see [8, 15]) conjectured the following.

Conjecture 4.7 *For $r \geq 3$, every r-uniform hypergraph with m edges has a partition into r vertex classes, each of which meets at least*

$$\frac{rm}{2r-1}$$

edges.

For $r = 2$, the statement is the same as (3.1), and so is easily proved. For $r \geq 3$, Bollobás, Reed and Thomason [8] showed that there is a partition in which every class meets at least

$$\left(1 - \frac{1}{e}\right) m/3 \approx 0.21m$$

edges. It was shown in [15] that every 3-uniform hypergraph has a tripartition in which each class meets at least

$$(5m-1)/9$$

edges, which is still short of the conjectured $3m/5$; for r-uniform hypergraphs, a lower bound of

$$0.27m$$

was given. It would be a substantial step forward even to prove a lower bound of form $m/2$.

Although Conjecture 4.7 is easily proved when $r = 2$, it seems much harder to find a k-partite version of the result. The following is a version of a conjecture from [15].

Conjecture 4.8 *For every $k \geq 2$, every graph with $m \geq \binom{k}{2}$ edges has a partition into k sets, each of which meets at least*

$$\frac{2m}{2k-1}$$

edges.

The lower bound on m is necessary to avoid trivial cases such as K_{k-1}, where one vertex class may end up empty.

References

[1] N. Alon, B. Bollobás, M. Krivelevich and B. Sudakov, Maximum cuts and judicious partitions in graphs without short cycles, *J. Combin. Theory Ser. B* **88** (2003), 329–346.

[2] R. Aharoni, E.C. Milner and K. Prikry, Unfriendly partitions of a graph, *J. Combin. Theory Ser. B* **50** (1990), 1–10.

[3] N. Alon, Bipartite subgraphs, *Combinatorica* **16** (1996), 301–311.

[4] N. Alon and E. Halperin, Bipartite subgraphs of integer weighted graphs, *Discrete Math.* **181** (1998), 19–29.

[5] N. Alon, M. Krivelevich and B. Sudakov, MaxCut in H-free graphs, in press.

[6] L.D. Andersen, D.D. Grant, and N. Linial, Extremal k-colourable subgraphs, *Ars Combin* **16** (1983), 259–270.

[7] B. Bollobás, A. Gyárfás, J. Lehel and A.D. Scott, Maximum directed cuts in acyclic digraphs, submitted.

[8] B. Bollobás, B. Reed and A. Thomason, An extremal function for the achromatic number, in *Graph structure theory (Seattle, WA, 1991)* (eds. N. Robertson and P. Seymour), *Contemp. Math.*, 147, Amer. Math. Soc., Providence, RI, USA (1993), pp. 161–165.

[9] B. Bollobás and A.D. Scott, On intersections of graphs, submitted.

[10] B. Bollobás and A.D. Scott, Judicious k-partitions, submitted.

[11] B. Bollobás and A.D. Scott, Discrepancy in graphs and hypergraphs, *Combinatorica*, in press.

[12] B. Bollobás and A.D. Scott, Judicious partitions of bounded-degree graphs, *J. Graph Theory* **46** (2004), 131–143.

[13] B. Bollobás and A.D. Scott, Problems and results on judicious partitions, *Random Structures Algorithms* **21** (2002), 414–430.

[14] B. Bollobás and A.D. Scott, Better bounds for Max Cut, in *Contemporary combinatorics, Bolyai Soc. Math. Stud.*, 10, János Bolyai Math. Soc., Budapest (2002), pp. 185–246.

[15] B. Bollobás and A.D. Scott, Judicious partitions of 3-uniform hypergraphs, *European J. Combin.* **21** (2000), 289–300.

[16] B. Bollobás and A.D. Scott, Exact bounds for judicious partitions of graphs, *Combinatorica* **19** (1999), 473–486.

[17] B. Bollobás and A.D. Scott, Judicious partitions of hypergraphs, *J. Combin. Theory Ser. A* **78** (1997), 15–31.

[18] B. Bollobás and A.D. Scott, Judicious partitions of graphs, *Period. Math. Hungar.* **26** (1993), 125–137.

[19] J.A. Bondy and S.C. Locke, Largest bipartite subgraphs in triangle-free graphs with maximum degree three, *J. Graph Theory* **10** (1986), 477–504.

[20] M.M. Deza and M. Laurent, *Geometry of cuts and metrics*, Algorithms and Combinatorics, **15** Springer-Verlag, Berlin (1997).

[21] C.S. Edwards, An improved lower bound for the number of edges in a largest bipartite subgraph, in *Recent advances in graph theory (Proc. Second Czech. Sympos., Prague, 1974)* (ed. M. Fiedler), Academia, Prague (1975), pp. 167–181.

[22] C.S. Edwards, Some extremal properties of bipartite subgraphs, *Canad. J. Math.* **25** (1973), 475–485.

[23] P. Erdős, Some recent problems in Combinatorics and Graph Theory, in *Combinatorics and Computing (Proc. 26th Southeastern International Conference on Graph Theory, Boca Raton, 1995) Congressus Numerantium*, 112, (1995).

[24] P. Erdős, Problems and results in graph theory and combinatorial analysis, in *Graph theory and related topics (Proc. Conf., Univ. Waterloo, Waterloo, Ont., 1977)* Academic Press, New York–London (1979), pp. 153–163.

[25] P. Erdős, R. Faudree, J. Pach and J. Spencer, How to make a graph bipartite, *J. Comb. Theory, Ser. B* **45** (1988), 86–98.

[26] P. Erdős, A. Gyárfás and Y. Kohayakawa, The size of the largest bipartite subgraphs, *Discrete Math.* **177** (1997), 267–271.

[27] A. Hajnal and E. Szemerédi, Proof of a conjecture of P. Erdős, in *Combinatorial theory and its applications II (Proc. Colloq., Balatonfred, 1969)* North-Holland, Amsterdam (1970), pp. 601–623.

[28] T. Hofmeister and H. Lefmann, On k-partite subgraphs, *Ars Combin.* **50** (1998), 303–308.

[29] G. Hopkins and W. Staton, Extremal bipartite subgraphs of cubic triangle-free graphs, *J. Graph Theory* **6** (1982), 115–121.

[30] A. Johansson, Asymptotic choice number for triangle-free graphs, DIMACS Tech Rep., Rutgers, New Jersey, 91-5, 1996.

[31] M. Laurent, Max-Cut Problem, in *Annotated bibliographies in combinatorial optimization* (eds. M. Dell'Amico, F. Maffioli and S. Martell), *Wiley-Interscience Series in Discrete Mathematics and Optimization*, John Wiley and Sons, Ltd., Chichester (1997).

[32] J. Lehel and Zs. Tuza, Triangle-free partial graphs and edge-covering theorems, *Discrete Math.* **30** (1982), 59–63.

[33] T. Lengauer, Combinatorial algorithms for integrated circuit layout, in *Applicable Theory in Computer Science* (ed. B.G. Teubner), John Wiley and Sons, Ltd., Chichester (1990).

[34] S.C. Locke, Maximum k-colorable subgraphs, *J. Graph Theory* **6** (1982), 123–132.

[35] M. Molloy and B. Reed, *Graph colouring and the probabilistic method*, Algorithms and Combinatorics, **23** Springer-Verlag, Berlin (2002).

[36] N. Văn Ngọc and Zs. Tuza, Linear-time approximation algorithms for the max cut problem, *Combin. Probab. Comput.* **2** (1993), 201–210.

[37] S. Poljak and D. Turzık, A polynomial algorithm for constructing a large bipartite subgraph, with an application to a satisfiability problem, *Canad. J. Math.* **34** (1982), 519–524.

[38] S. Poljak and Zs. Tuza, Maximum cuts and large bipartite subgraphs, in *Combinatorial optimization (Papers from the DIMACS special year, 1992–1993)* (eds. W. Cook, L. Lovász and P. Seymour), *DIMACS Ser. Discrete Math. Theoret. Comput. Sci.*, 20, Amer. Math. Soc., Providence, RI, USA (1995), pp. 181–244.

[39] S. Poljak and Zs. Tuza, Bipartite subgraphs of triangle-free graphs, *SIAM J. Discrete Math.* **7** (1994), 307–313.

[40] T.D. Porter, Minimal partitions of a graph, *Ars Combin.* **53** (1999), 181–186.

[41] T.D. Porter, Graph partitions, *J. Combin. Math. Combin. Comput.* **15** (1994), 111–118.

[42] T.D. Porter, On a bottleneck bipartition conjecture of Erdős, *Combinatorica* **12** (1992), 317–321.

[43] T.D. Porter and Bing Yang, Graph partitions II, *J. Combin. Math. Combin. Comput.* **37** (2001), 149–158.

[44] J.B. Shearer, A note on bipartite subgraphs of triangle-free graphs, *Random Structures Algorithms* **3** (1992), 223–226.

[45] S. Shelah and E.C. Milner, Graphs with no unfriendly partitions, in *A Tribute to Paul Erdős* (eds. A. Baker, B. Bollobás & A. Thomason), CUP, Cambridge (1990), pp. 373–384.

[46] W. Staton, Edge deletions and the chromatic number, *Ars Combin.* **10** (1980), 103–106

Department of Mathematics
University College London
Gower Street
London WC1E 6BT
scott@math.ucl.ac.uk

An isoperimetric method for the small sumset problem

O. Serra

Abstract

The purpose of this paper is to survey applications of an isoperimetric method to the small sumset problem in Additive Theory. The small sumset problem asks for lower bounds of the cardinality of the sum of two sets in a group. Sample proofs are presented to illustrate the application of the method, which is based on connectivity properties of graphs. In the final part we describe some applications to several problems in number theory, group theory and combinatorics.

1 Introduction

Let G be a group written additively and let S and T be two finite subsets of G. By the *sumset* of S and T we mean the set $S+T=\{s+t:\ s\in S, t\in T\}$. We write $S+S=2S$ and, for a positive integer h, hS denotes the h-fold sum $S+\cdots+S$. The *small sumset problem* consists in finding lower bounds for the cardinality $|S+T|$ of the sumset in terms of the cardinalities of S and T. Moreover it aims at giving structural characterizations of the pairs of sets such that the cardinality of its sumset reaches (or is close to) the lower bounds. As a simple example let us consider the group of integers.

Proposition 1.1 *Let S and T be two finite nonempty sets of integers. Then*

$$|S+T|\geq |S|+|T|-1. \tag{1.1}$$

Moreover equality holds if and only if either $\min\{|S|,|T|\}=1$ *or S and T are arithmetic progressions with the same common difference.*

Proof We have $S+T\supseteq(\min(S)+T)\cup(S+\max(T))$ and these two sets intersect in the single element $\min(S)+\max(T)$. This proves (1.1).

Suppose that equality holds in (1.1) and $\min\{|S|,|T|\}\geq 2$. By deleting the largest element in T we get a set T' with $|S+T'|=|S|+|T'|-1$. Continuing in this way we eventually have $|S+\{t_1,t_2\}|=|S|+1$, where t_1 and t_2 are the two smallest elements in T. Therefore $|(S+t_1)\cap(S+t_2)|=|S\cap(S+t_2-t_1)|=|S|-1$. This means that S is an arithmetic progression of difference $d=t_2-t_1$. Similarly T is an arithmetic progression of difference $s_2-s_1=d$. □

Proposition 1.1 illustrates the two questions raised by the small sumset problem: to obtain lower bounds for the cardinality of a sumset and to characterize pairs of sets reaching these lower bounds. The problem asks for similar statements for other groups. Furthermore we would also like to extend the structural characterization to pairs of sets with 'small' sumset.

Let us start with a brief historical account of the problem. Estimating the lower bound for the cardinality $|S+T|$ of a sumset in terms of $|S|$ and $|T|$ goes back to Cauchy [7], who in 1813 found such a bound for groups of prime order. The problem was the object of interest again in the 1930's in connection to the famous $(\alpha+\beta)$ conjecture formulated by Khintchine in 1932 and solved by Mann [67] ten years later. The $(\alpha+\beta)$ Theorem of Mann can be formulated as follows. Let A and B be sets of integers with $0 \in A \cap B$. Then

$$\sigma(A+B) \geq \min\{1, \sigma(A)+\sigma(B)\},$$

where $\sigma(A) = \inf_n \frac{|A\cap[1,n]|}{n}$ is the Schnirel'man density of A. It was customary to write $\alpha = \sigma(A)$ and $\beta = \sigma(B)$, and this notation gave its name to the conjecture. In his search for the p-analogue of Mann's $(\alpha+\beta)$-Theorem, Davenport [11] rediscovered in 1935 the Cauchy inequality which is now known as the Cauchy-Davenport theorem. In 1953 Mann [68] proved an analogue of his $(\alpha+\beta)$ theorem for any finite abelian group. Kneser [60, 61] found a global inequality valid for any abelian group in 1955 and raised the question of characterizing the pairs of sets for which the cardinality of the sumset reached the lower bound. This task was accomplished by Kemperman [59] in 1960 by giving a rather intricate recursive characterization. From the early 60's Freiman [27, 28] developed a structural theory of set addition whose aim was to derive structural properties of sets from the knowledge of the cardinality of their sumsets. This theory culminated in his celebrated theorem which describes the structure of a set S whose doubling $2S$ has small cardinality. Ruzsa [74, 75] gave a different and beautiful proof of this result.

The small sumset problem has found a wide range of applications and interplay in analysis, algebra, geometry, number theory and combinatorics. Another source of motivation for the problem in the late 70's came from the study of vulnerability of networks, often modeled by vertex–transitive graphs.

Topics related to the small sumset problem are described in several surveys and books; see e.g. Freiman [27], Halberstam and Roth [33], Hamidoune [37], Mann [69] and Nathanson [70]. The volume [13] contains papers from a wide range of areas of current research and applications.

There are several methods available for approaching the small sumset problem. The classical results rely on the transforms introduced by Davenport and Dyson; see e.g. [69]. These inductive methods reduce the problem from a given pair of sets to a new pair with the same cardinality of the sumset but different cardinalities of the sets.

A second method uses trigonometric sums and was introduced by Freiman; see e.g. [25, 27]. It connects with the classical tools from analytic number theory. Freiman introduced several additional tools for studying the structure of sets under addition. Ruzsa's proof of the theorem of Freiman on the structure of sets of integers with small doubling combines different tools. They include the theorem of Menger in graphs, Fourier analysis and the theorems of Minkowski on convex sets in $\mathbf{R}^d$ and discrete lattices.

A third method is based in the so-called polynomial method. Its application to additive problems has its origin in a paper by Alon, Nathanson and Ruzsa [1]. It has also been used by Eliahou and Kervaire [18, 19] and Károlyi [55, 56].

The purpose of this paper is to survey the applications of a fourth method based on connectivity properties of graphs which may be called the *isoperimetric method*. This method has been largely developed by Hamidoune from the early 80's. It provides simple and transparent proofs for most of the classical results as well as a means of obtaining new ones. The underlying idea of the method is to translate addition theorems for groups into theorems on group actions on sets, particularly on graphs. Its combinatorial nature and its closeness to the problem makes the method suitable for exact results. The method seems most appropriate for dealing with the inverse problems of characterizing the structure of a set S with $|2S| \leq c|S|$ for constants $c < 3$. For this reason we will mainly focus on this range of applications.

The plan of the paper is the following. Section 2 introduces the isoperimetric connectivities and the basic combinatorial properties of atoms and fragments which are the main tools of the method. Section 3 describes the first applications to vertex–transitive graphs and Cayley graphs. Sections 4 and 6 are devoted to the classical addition theorems described in this introduction and include some simple proofs which illustrate the application of the isoperimetric method. An extension of the basic results from Section 3 needed for further applications is presented in Section 5. New results for abelian groups are surveyed in Section 7. In Section 8 we consider the problem for non-abelian groups. Section 9 presents a selection of applications of the small sumset problem from the viewpoint of the isoperimetric method. They include the range of diagonal forms in division rings, the Frobenius problem, a problem of Erdős and Heilbronn for subset sums, rainbow arithmetic progressions, the Cacceta and Haggkvist conjecture and the vulnerability of networks.

2 Isoperimetric connectivities

By a *graph* $X = (V, E)$ we shall always mean a locally finite directed graph, that is, all vertices of X have bounded in– and out–degree. Undirected graphs will be considered as directed graphs where each edge uv is replaced by the two arcs (u, v) and (v, u). We do not consider multiple arcs but our graphs will be assumed to have a loop at each node. This formal assumption does not modify the connectivity properties of the graph and simplifies some statements, particularly in the applications to the small sumset problem. We denote by X^{-1} the graph obtained from X by reversing the orientation of each arc. We reserve the symbol X for graphs and G will be used to denote a group.

For a subset $U \subset V$ we denote the *vertex boundary* of U by

$$\partial(U) = \{v \in V \setminus U : (u, v) \in E \text{ for some } u \in U\}.$$

We say that X is *connected* if $\partial U \neq \emptyset$ for every finite proper subset $U \subset V$. For finite graphs the usual terminology is *strongly* connected, but we shall just use the word connected. The *neighborhood* of U is

$$N(U) = U \cup \partial(U).$$

For a positive integer k, a connected graph X is *k-separable* if there is a finite set U of vertices with $|U| \geq k$ and $|V \setminus N(U)| \geq k$. A connected graph is 1-separable unless it is the complete graph. If X is k-separable, the *k–isoperimetric connectivity* $\kappa_k(X)$ is the minimum cardinality of the boundary of a finite set with at least k vertices such that the complement of its neighborhood has also at least k vertices. Let us denote by $\mathcal{F}_k$ the family of subsets of V involved in the definition of $\kappa_k(X)$,

$$\mathcal{F}_k = \{U \subset V : k \leq |U| < \infty, |V \setminus N(U)| \geq k\}.$$

The k–isoperimetric connectivity of X is

$$\kappa_k(X) = \min\{|\partial U|,\ U \in \mathcal{F}_k\}.$$

Note that $\kappa_1(X) \leq \kappa_2(X) \leq \cdots$.

A subset $F \in \mathcal{F}_k$ is a *k–fragment* of X if $|\partial F| = \kappa_k(X)$. A k–fragment of minimum cardinality is a *k–atom* of the graph. The cardinality of the k–atoms of X will be denoted by $\alpha_k(X)$.

The *k-degree* of a graph X, also known as the *k–isoperimetric number* of the graph, is defined as the minimum boundary of a set of cardinality exactly k in the graph,

$$d_k(X) = \min\{|\partial X|, X \subset V, |X| = k\}.$$

In particular $d(X) = d_1(X)$ is the minimum degree of the graph (note that the loops are not counted in vertex degrees.)

The sequence of k-degrees may have an erratic behaviour and the values of these numbers are known for very few classes of graphs. The isoperimetric connectivities supply lower bounds for them: if X is k-separable for $1 \leq k \leq s$ then $\kappa_k(X) = \min\{d_i(X), k \leq i \leq s\}$. Moreover, if a k–atom of X has size $\alpha_k(X) = k$ then $\kappa_k(X) = d_k(X)$, while $\alpha_k(X) > k$ is equivalent to $\kappa_k(X) < d_k(X)$. Thus, the connectivity of the graph equals its minimum degree if and only if the atoms have cardinality one. These relationships explain the terminology of isoperimetric connectivities.

The notion of k–isoperimetric connectivity of a connected graph X is a natural extension of the usual vertex connectivity which corresponds to the case $k = 1$. The original notions of atoms and fragments as introduced by Mader [66] correspond to 1-atoms and 1-fragments of finite undirected graphs. From now on we shall omit the reference to k when $k = 1$ and simply speak of connectivity, atoms and fragments.

The extension of the notions and properties of atoms and fragments to the directed case was considered by Hamidoune [34, 35]. Some care must be taken in this extension as it may happen that the connectivity properties of X and its inverse graph X^{-1} are different. For instance, in the infinite case we may have $\kappa(X) \neq \kappa(X^{-1})$ and in both infinite and finite graphs we may have that the size of a k–atom in X is not the same as it is in X^{-1}. This asymmetry may cause some statements about basic properties of atoms and fragments in the undirected case to be false in the directed one. However we have,

Lemma 2.1 *Let F be a k–fragment of X. Then*

$$\partial F = \partial^{-1}(V \setminus N(F)),$$

where $\partial^{-1}(U)$ denotes the vertex boundary of $U \subset V$ in the inverse graph X^{-1}.

Proof For every finite set $U \subset V$ we always have $\partial U \supseteq \partial^{-1}(V \setminus N(U))$ and, if there is a vertex $u \in \partial(F)$ not adjacent to $V \setminus N(F)$, then $|\partial(F \cup \{u\})| = |\partial(F)| - 1$ and $N(F \cup \{u\}) = N(F)$, contradicting that F is a k–fragment. □

Lemma 2.1 implies that, for finite graphs, $V \setminus N(F)$ is a k–fragment of X^{-1}. This shows that a finite graph X is k-separable if and only if X^{-1} is k-separable and both graphs have the same k-connectivity.

The following is a useful property of k–atoms proved by Hamidoune [38, 41].

Lemma 2.2 (The intersection property) *Let X be either a connected infinite locally finite graph or a finite k-separable graph with $\alpha_k(X) \le \alpha_k(X^{-1})$. Let A be a k–atom and F a k-fragment of X. Then either $A \subset F$ or $|A \cap F| \le k-1$.*

Proof The function $\mu(U) = |\partial U|$ is clearly submodular. Suppose that $|A \cap F| \ge k$ and A is not a subset of F. Then,

$$\mu(A \cup F) + \mu(A \cap F) \le \mu(A) + \mu(F) = 2\kappa_k(X). \tag{2.1}$$

Since $A \cap F$ is a proper subset of A and $|N(A \cap F)| \le |N(F)| \le |V| - k$, we have $A \cap F \in \mathcal{F}_k$. By the minimality of $|A|$, we get $\mu(A \cap F) > \kappa_k(X)$. The lemma will be proved if we show that $A \cup F \in \mathcal{F}_k$ as well since then $\mu(A \cup F) \ge \kappa_k(X)$ contradicting (2.1).

Obviously $|A \cup F| \ge k$. Let us show that $|V \setminus N(A \cup F)| \ge k$. This is clearly the case if the graph is infinite. If X is finite, denote by $F' = V \setminus N(F)$ and $A' = V \setminus N(A)$, so that we want to show $|A' \cap F'| \ge k$. By the minimality of A, $|\partial(A \cap F)| \ge |\partial A|$ which implies

$$|A \cap \partial F| \ge |(\partial(A \cap F)) \setminus \partial A| \ge |\partial A \setminus (\partial(A \cap F))| \ge |F' \cap \partial A|.$$

Therefore

$$\begin{aligned} |F'| &= |F' \cap A'| + |F' \cap \partial A| + |F' \cap A| \\ &\le |F' \cap A'| + |A \cap \partial F| + |F' \cap A| \\ &= |F' \cap A'| + |A| - |A \cap F|. \end{aligned}$$

By Lemma 2.1, F' is a k-fragment of X^{-1}. From $\alpha_k(X) \le \alpha_k(X^{-1})$ we have $|F'| \ge |A|$ which combined with the above inequality gives $|F' \cap A'| \ge |A \cap F| \ge k$ as desired. □

By Lemma 2.2, two distinct k–atoms of X intersect in at most $k-1$ points. When $k = 1$ two distinct atoms are disjoint, a fact which explains the terminology of atoms.

Note that, as seen in its proof, the condition $\alpha_k(X) \le \alpha_k(X^{-1})$ in Lemma 2.2 can be weakened to $|A| \le |V \setminus N(F)|$, a fact that may be useful in some applications.

3 Atoms of Vertex Transitive Graphs

The fact that two distinct atoms are disjoint provides the following classical application of the concept to the connectivity of vertex–transitive graphs. It was obtained by Mader [66] for undirected finite graphs and extended to the directed case by Hamidoune [34].

Theorem 3.1 *Let X be either a connected infinite locally finite graph or a finite separable graph with $\alpha(X) \le \alpha(X^{-1})$. Let G a group of automorphisms of X acting transitively on the vertices of the graph. Then the atoms of X are blocks of imprimitivity of the action of G on $V(X)$. Moreover, $\kappa(X) \ge (d+1)/2$, where $d = d(X)$ is the degree of the graph.*

Proof Each automorphism of X sends atoms to atoms. Therefore the first part of the statement is a direct consequence of the fact that distinct atoms are disjoint, so the atoms of X partition the vertex set of the graph into blocks of the action of G. For the second part, the stabiliser of an atom A of G acts vertex–transitively on the subgraph $X[A]$ of X induced by the vertices in A. In particular $X[A]$ is a finite regular graph, say of degree $d_A \leq |A| - 1$. Moreover, $(\partial A)^g = \partial(A^g)$ for any $g \in G$ so that the boundary of A is a disjoint union of atoms of X. Therefore $\kappa(X) \geq \max\{|A|, d - d_A\} \geq \max\{d_A + 1, d - d_A\} \geq (d+1)/2$. □

Theorem 3.1 implies in particular that, if G is a vertex–primitive automorphism group of X (for instance if X is a vertex–transitive graph of primer order) then the atoms of the graph are the singletons. Therefore, $\kappa(X) = d(X)$.

Note that the intersection property implies that fragments of a graph X either contain or are disjoint from each atom. If X is vertex–transitive then the vertex set is partitioned into atoms, so that fragments (and minimum cutsets) are disjoint unions of atoms. This gives a quite useful description of fragments and minimum cutsets of the graph.

Let X be as in Theorem 3.1 and denote by X/A the quotient graph which has the atoms of X as vertices and there is an arc from atom A to atom A' in X/A if there is some arc from a vertex in A to vertex in A' in X. As an additional result, Hamidoune [35] proved the following proposition.

Proposition 3.2 *Let X be as in Theorem 3.1. Then the quotient graph X/A is also vertex–transitive and $\kappa(X/A) = d(X/A)$.*

The following example shows that the equality $\kappa(X) = (d+1)/2$ in Theorem 3.1 can be reached.

Example 3.3 For two graphs X, Y the lexicographic product $X[Y]$ has set of vertices $V(X) \times V(Y)$ and $((x, y), (x', y'))$ is an arc in the product if and only if either (y, y') is an arc in Y and $x = x'$ or (x, x') is an arc in X. The lexicographic product $C_k[K_n]$ of a directed cycle of length $k \geq 3$ with a complete graph on $n \geq 2$ vertices has degree $2n - 1$ and connectivity n (the n-cliques are atoms of the graph.) By taking the infinite path in place of C_k we get an infinite locally finite example.

The lower bound on $\kappa(X)$ in Theorem 3.1 can be increased if the graph is antisymmetric, that is, at most one of the two edges $(x, y), (y, x)$ is present in the graph for each pair x, y of vertices.

Proposition 3.4 *Let X be a connected vertex–transitive antisymmetric graph with degree d. Then $\kappa(X) \geq 2d/3$.*

Proof The same proof of Theorem 3.1 applies taking into account that the degree of $X[A]$ now satisfies $d_A \leq (|A| - 1)/2$. □

The lexicographic product of a directed cycle C_k, $k \geq 3$, or an infinite directed path, with a vertex–transitive tournament provide examples reaching the lower bound for $\kappa(X)$ in Proposition 3.4.

If a connected graph is arc–transitive, that is, it has a group of automorphisms acting transitively on arcs, then the connectivity reaches its maximum value. Note that a connected arc–transitive graph is also vertex–transitive.

Proposition 3.5 *Let X be a connected arc–transitive graph with degree d. Then $\kappa(X) = d$.*

Proof Note that, if X is finite, then $d(X) = d(X^{-1})$ so we may assume that $\alpha(X) \leq \alpha(X^{-1})$. Let A be an atom of X and suppose that $|A| > 1$. The subgraph $X[A]$ is connected: if there is a proper subset $U \subset A$ whose boundary in $X[A]$ is the empty set then $|\partial U| \leq |\partial A|$, contradicting the minimality of the atom. Let (u, v) be an arc in $X[A]$ and (u', v') an arc with $u \in A$ and $v \in V \setminus A$. There is an automorphism g which sends (u, v) to (u', v'). Then $A \cap A^g \neq \emptyset$ but $A \neq A^g$, contradicting the intersection property (Lemma 2.2.) Thus $|A| = 1$ and $\kappa(X) = d(X)$. □

The lexicographic product of a directed cycle C_k, $k \geq 3$, or the infinite directed path, with a null graph (with no arcs) of order d provide examples reaching the lower bound for $\kappa(X)$ in the Proposition 3.5.

Cayley graphs provide the main class of examples of vertex–transitive graphs. Let G be a group written multiplicatively and S a finite subset of G. The (right) Cayley graph $\mathrm{Cay}(G, S)$ of G with respect to S has the elements of the group as vertices and set of arcs $\{(x, xs) : x \in G, s \in S\}$. It can be easily checked that $\mathrm{Cay}(G, S)$ is connected if and only if S generates G. If this is not the case then $\mathrm{Cay}(G, S)$ is the disjoint union of isomorphic copies of $\mathrm{Cay}(\langle S\rangle, S)$, where $\langle S\rangle$ denotes the subgroup generated by S.

The ambient group G will always be clear from the context; by abuse of notation we often write $\kappa_k(S)$ to denote the k–isoperimetric connectivity of the Cayley graph $X = \mathrm{Cay}(\langle S\rangle, S)$; we also speak of the k–isoperimetric connectivity and the k–atoms of S instead of referring to X.

We shall always assume that $1 \in S$, so that the graph has a loop at each vertex; the loops however will not be counted in the degree of the graph which we define as $d(X) = |S| - 1$. Note that the inclusion of 1 in S does not modify the connectivity properties of the graph while it allows us to simplify the statements and the proofs. With this convention, the boundary of a subset $U \subset G$ in X satisfies

$$|US| = |U| + |\partial(U)|. \tag{3.1}$$

Equality (3.1) is at the heart of the application of isoperimetric methods to the study of sets whose product has small cardinality. Thus, for example, if the connectivity of the Cayley graph satisfies $\kappa(S) = d(X) = |S| - 1$ then $|TS| \geq \min\{|G|, |T| + |S| - 1\}$ for each subset $T \subset G$.

The group G acts transitively on $X = \mathrm{Cay}(G, S)$ by left multiplication, so the graph is vertex–transitive. By Theorem 3.1, atoms of X are blocks of imprimitivity of this action of G, which means that the atom A containing 1 is a subgroup. Moreover, if $|A| > 1$, the subgraph induced by the vertices of A is connected, so that $A \cap S$ generates A. Therefore, for Cayley graphs Theorem 3.1 has the following form.

Theorem 3.6 *Let $X = \mathrm{Cay}(G, S)$ be a connected Cayley graph with $\alpha(X) \leq \alpha(X^{-1})$ and let A be an atom of X containing 1. Then A is the subgroup of G generated by $A \cap S$, and each atom is a left coset of A in G.*

Theorem 3.1 also provides the lower bound $\kappa(X) \geq |S|/2$ for the connectivity of $X = \mathrm{Cay}(G, S)$. In this setting it can be formulated as follows. The next Proposition was independently obtained by Olson [72].

Proposition 3.7 *Let S and T be finite nonempty subsets of a group G. Assume that $1 \in S$ and S generates G. Then*

$$|TS| \geq \min\{|G|, |T| + \frac{|S|}{2}\}.$$

In the following sections we give applications and extensions of Theorem 3.6 providing classical and new results for the small sumset problem.

4 Sets with small sumset in abelian groups

Throughout this section G denotes an abelian group written additively. One of the first results to be found in the literature on the small sumsets problem concerns the groups of prime order because of its connection to the Waring problem. For a positive integer n we denote by $\mathbf{Z}_n$ the cyclic group of order n. In 1935 Davenport [11] rediscovered the following result which was first stated by Cauchy [7] in 1813 and is now known as the Cauchy–Davenport theorem.

Theorem 4.1 (Cauchy–Davenport Theorem) *Let p be a prime and S, T be two nonempty subsets of $\mathbf{Z}_p$. Then*

$$|S + T| \geq \min\{p, |S| + |T| - 1\}.$$

There are many different proofs of this theorem. Davenport used the Dyson transform (see for instance [69].) A simple proof using the polynomial method can be found in Alon, Nathanson and Ruzsa [1]. Another simple proof using Theorem 3.6 is as follows.

Proof Since the cardinality of $S + T$ is invariant by translations of S and T we may assume that $0 \in S \cap T$. If $|S + T| < p$ then equality (3.1) for the Cayley graph $X = \mathrm{Cay}(\mathbf{Z}_p, S)$ gives

$$|S + T| \geq |T| + |\partial T| \geq |T| + \kappa(X).$$

Since $\mathbf{Z}_p$ has no proper subgroups, Theorem 3.6 implies that the atoms of X have cardinality 1, and thus $\kappa(X) = d(X) = |S| - 1$. □

For cyclic groups Chowla [9] obtained the following characterization of the Cauchy-Davenport inequality.

Theorem 4.2 *Let S be a subset of $\mathbf{Z}_n$ such that $0 \in S$ and $\gcd(x, n) = 1$ for each $x \in S \setminus \{0\}$. Then, for every set $T \subset \mathbf{Z}_n$,*

$$|S + T| \geq \min\{n, |S| + |T| - 1\}.$$

Proof This result can be proved along the same lines as the above proof of the Cauchy-Davenport theorem since, according to Theorem 3.6, an atom A of $\mathrm{Cay}(\mathbf{Z}_n, S)$ containing 0 is the subgroup generated by $S \cap A$ which, under the conditions of the theorem, is either trivial or the whole $\mathbf{Z}_n$. □

A general condition for the validity of the Cauchy–Davenport inequality in finite abelian groups was found by Mann [68].

Theorem 4.3 *Let S be a nonempty subset of a finite abelian group G. If there is a subset T with $S+T \neq G$ and $|S+T| < |S|+|T|-1$, then there is a finite subgroup $H < G$ with $|S+H| < |S|+|H|-1$.*

Proof Again the theorem follows from the fact that the atom containing 0 in the Cayley graph $X = \mathrm{Cay}(G, S)$ is a subgroup, where we assume without loss of generality that S generates G. In graphical terms, the inequalities $|S+T| < |S| + |T|-1 < |G|$ imply that the connectivity of X is less than the degree, so that the atom of X containing 0 is a proper subgroup H of G. Hence $|S+H|-|H| \leq |S+T|-|T|$ for each finite subset T with $S+T \neq G$. □

A global inequality valid for any abelian group was found by Kneser [60, 61]. It can be stated in the following form.

Theorem 4.4 (Kneser's Theorem) *Let S and T be finite nonempty subsets of an abelian group G. If $|S+T| < |S|+|T| \leq |G|$ then*

$$|S+T| = |S+T+H| = |S+H| + |T+H| - |H|,$$

where $H = H(S+T)$ is the largest subgroup of G such that $S+T+H = S+T$.

Note that Kneser's theorem provides yet another proof of the Cauchy–Davenport theorem since in $\mathbf{Z}_p$ we either have $H = \mathbf{Z}_p$ and $S+T$ is the whole group or $H = \{0\}$ and $|S+T| = |S|+|T|-1$.

A nonempty subset P of G is said to be *periodic* if it is a union of cosets of a nontrivial subgroup H, that is, $P = P+H$. We also say that P is H-periodic when the reference to the subgroup H is to be made explicit. Kneser's theorem can be equivalently formulated in the following way:

Theorem (Kneser's Theorem) *If $|S+T| < |S|+|T|-1$ then $S+T$ is a periodic subset of G.*

The following is a simple proof of Theorem 4.4 in the case that $S = T$ based on Theorem 3.6. It uses the following result.

Lemma 4.5 (Folklore) *Let S and T be finite subsets of a group H. If $|S|+|T| > |H|$ then $S+T = H$.*

Proof of Kneser's Theorem for $S = T$ We may assume that $0 \in S$ and S is a generating set of G. We shall show that $|S+S| < 2|S|-1$ implies that $S+S$ is a periodic subset by induction on the cardinality of S.

For $|S| = 2$ the above inequality implies $S + S = S$ so that S is a group itself. Assume that $|S| > 2$ and let A be the atom of $X = \mathrm{Cay}(G, S)$ containing 0 which, since $|S + S| - |S| < |S| - 1 = d(X)$, is a proper subgroup of G.

Let $S = \bigcup_{g \in G}(S \cap (A+g)) = S_0 \cup \ldots \cup S_{r-1}$ be the decomposition of S modulo A into nonempty subsets, where $r \geq 2$ since A is a proper subgroup and S generates G. By translating S if necessary we may assume that $0 \in S_0$ and $|S_0| = \min_{0 \leq i \leq r-1} |S_i|$. We have

$$r|A| = |S + A| = \sum_{i=0}^{r} |S_i + A| < \sum_{i=0}^{r-1} |S_i| + |A| - 1.$$

The above inequality implies $|S_i| > |A|/2$ and $|S_0|+|S_i| > |A|$ for each $i = 1, \ldots, r-1$. By Lemma 4.5 we have $|S_i + S_j| = |A|$ for all $0 \leq i, j \leq r - 1$ except possibly for $i = j = 0$. Therefore $(S+S) \setminus A$ is A-periodic. If $S_0 + S_0 = A$ then $S+S+A = S+S$ and we are done. Suppose that $S_0 + S_0 \neq A$. By Proposition 3.6 the quotient graph X/A has connectivity $\kappa(X/A) = d(X/A)$ so that we have

$$|\sigma(S) + \sigma(S)| \geq 2|\sigma(S)| - 1 = 2|\sigma(S) \setminus \{0\}| + 1,$$

where $\sigma : G \to G/A$ denotes the canonical projection. Since A is the only coset of $S + S + A$ which is not filled with $S + S$, the above inequality implies

$$|S + S| = 2|S \setminus S_0| + |S_0 + S_0|.$$

If $|S_0 + S_0| \geq 2|S_0| - 1$ then $|S+S| \geq 2|S| - 1$, a contradiction. Therefore $|S_0 + S_0| < 2|S_0| - 1$, which in particular implies $|S_0| \geq 2$. By the induction hypothesis $S_0 + S_0$ is H-periodic for some proper subgroup of A and so is $S + S$. This completes the proof. □

The theorem of Kneser also provides a minimum bound for the cardinality of $|S+T|$ which can be more precise than the one in Proposition 3.7. If $|S+T| < |S|+|T|$ then, for H the subgroup of Theorem 4.4,

$$|S + T| = \left(\frac{|S + H|}{|H|} + \frac{|T + H|}{|H|} - 1\right)|H| \geq \left(\left\lceil\frac{|S|}{|H|}\right\rceil + \left\lceil\frac{|T|}{|H|}\right\rceil - 1\right)|H|.$$

It follows that if G is a finite group of order n then

$$|S + T| \geq \min_{d|n} \left\{\left(\left\lceil\frac{|S|}{d}\right\rceil + \left\lceil\frac{|T|}{d}\right\rceil - 1\right) d\right\}.$$

Recently Eliahou, Kervaire and Plagne [20] have shown that the above bound is tight by exhibiting, for each pair of positive integers $1 \leq s, t \leq n$ a pair of sets S and T of cardinalities s and t with $|S + T|$ reaching the lower bound. This result is remarkable as it shows that the absolute minimum bound for the size of $|S + T|$ depends only on the cardinality of the group and not on its algebraic structure.

5 An extension of Theorem 3.6

Before we proceed with applications of the isoperimetric method we next discuss an extension of Theorem 3.6 which involves the 2-isoperimetric connectivity.

Let $0 \in S$ be a finite generating set of an abelian group G and $X = \text{Cay}(G, S)$. Suppose that X is 2-separable. According to Theorem 3.6, if $\kappa(X) < |S| - 1$ then the atom A of X containing 0 is a proper subgroup of G. Then A is also a 2-atom of X since it realizes the minimum boundary among all nonempty sets whose neighborhood leaves at least $|G| - |S + A| \geq |A| \geq 2$ vertices. It remains to determine the structure of the 2-atoms when $\kappa(X) = |S| - 1$. The next result shows that, to some extent, if a 2-atom has cardinality more than two then it is also a subgroup.

Theorem 5.1 *Let $0 \in S$ be a generating set of an abelian group G. Assume that $\kappa_2(S) = |S| + m$, where $m \geq -1$ if G has prime order and $-1 \leq m \leq 4$ otherwise. Let A be a 2-atom of S. If $|S| < |G| - \binom{m+4}{2}$ then either $|A| = 2$ or A is a proper subgroup of G.*

A proof of Theorem 5.1 can be found in [51, 76]. Thus, when $G = \mathbf{Z}_p$, p a prime, the 2-atoms of a set S verifying the hypothesis of Theorem 5.1 have cardinality 2. The following example shows that the condition on $|S|$ cannot be weakened.

Example 5.2 Let p be a prime number of the form $p = 3b + 1$ for some integer $b > 1$. Let $S = [0, b-1] \cup [b+1, 2b-2] \cup [2b+1, 3b-3]$ and $A = \{0, 1, b\}$. We have $|S + A| = |S| + |A| = p - 3$, so that $\kappa_2(S) \leq |S|$, and $|S| = p - 6$. Note that we cannot have $\kappa_2(S) = |S| - 1$ since this would imply $|S + T| = |S| + |T| - 1$ for some $T \subset \mathbf{Z}_p$ with $|T| \geq 2$. Then, by Vosper's theorem (see Theorem 6.1 below), S is an arithmetic progression, which can be easily checked not to be the case. Moreover, $|S + \{0, x\}| \geq |S| + 3$ for each $x \neq 0$, since otherwise S would be the union of two arithmetic progressions with difference x. This shows that A is a 2–atom of S, and A is neither a subgroup nor has cardinality 2.

When the order of G is not a prime, the inequality $m \leq 4$ in Theorem 5.1 cannot be omitted either as shown by the following example.

Example 5.3 Take $G = \mathbf{Z}_7 \times \mathbf{Z}_q$ where q is a sufficiently large prime. Consider the set $S = \{0, 1, 2, 4\} \times \{0, 1, 2, 4\}$ and $A = \{0, 1, 3\} \times \{0\}$. Then $|S + A| = |S| + |A| + 5$. Our group G has only the two proper subgroups $H_1 = \mathbf{Z}/7\mathbf{Z} \times \{0\}$ and $H_2 = \{0\} \times \mathbf{Z}/q\mathbf{Z}$, for which we have

$$|S + H_1| = |S| + |H_1| + 5 \quad \text{and} \quad |S + H_2| = 4q > |S| + |H_2| + 5.$$

On the other hand, if $B = \{(0, 0), x\}$ we can easily check that

$$|S + B| \geq |S| + 8 = |S| + |B| + 6.$$

Therefore a 2-atom of S containing $(0, 0)$ has neither cardinality 2 nor is a subgroup of G.

6 The critical pair problem

The theorem of Kneser provides information on the structure of a sumset in an abelian group when $S + T \neq G$ and $|S + T| < |S| + |T|$. The inverse problem, that is, to obtain a characterization of the structure of pairs of sets S and T verifying

$|S+T| \le |S|+|T|-1 < |G|$, was suggested by Kneser [61] and is known as the *critical pair* problem. As observed by Kemperman [59], Kneser's theorem reduces this problem to the characterization of pairs of subsets such that $S+T$ is aperiodic and thus $|S+T| = |S|+|T|-1$.

When G is a group of prime order the following theorem obtained by Vosper [81] gives such a characterization.

Theorem 6.1 *Let S and T be subsets of $\mathbf{Z}_p$, p a prime, verifying $|S| \ge 2$, $|T| \ge 2$ and*

$$|S+T| = |S|+|T|-1 \le p-1.$$

Then either $S = \mathbf{Z}_p \setminus (x-T)$ for some $x \in \mathbf{Z}_p$ or S and T are arithmetic progressions with the same difference.

For abelian groups the critical pair problem is considerably more involved. The known results give a structural description of critical pairs in terms of arithmetic progressions and subgroups. Kemperman calls a pair of nonempty subsets $\{S,T\}$ of a group G *elementary* if they satisfy one of the following conditions:

(a) Either $|S| = 1$ or $|T| = 1$.

(b) Both S and T are arithmetic progressions of difference d, where the order of d is at least $|S|+|T|-1$.

(c) Each of S and T are contained in a F-coset for some subgroup F of G and $|S|+|T| = |F|+1$.

(d) S is contained in a F-coset for some subgroup F of G and $T = g-((S+F)\setminus S)$ for some $g \in G$.

Kemperman [59, Theorem 5.1] gives the following recursive characterization of critical pairs.

Theorem 6.2 *Let S and T be two nonempty subsets of an abelian group G such that either $S+T$ is aperiodic or there is $x \in S+T$ with a unique representation in the sum. The equality*

$$|S+T| = |S|+|T|-1$$

holds if and only if there are nonempty subsets $S_1 \subset S$ and $T_1 \subset T$ and a nontrivial subgroup H of G satisfying the following conditions:

(i) Each of S_1 and T_1 is contained in an H-coset and the pair $\{S_1, T_1\}$ is elementary.

(ii) The element $\sigma(S_1)+\sigma(T_1)$ has only this representation in $\sigma(S)+\sigma(T)$, where $\sigma : G \to G/H$ is the canonical projection.

(iii) Both $S \setminus S_1$ and $T \setminus T_1$ are H-periodic.

(iv) $|\sigma(S)+\sigma(T)| = |\sigma(S)|+|\sigma(T)|-1$, where $\sigma : G \to G/H$ is the canonical projection.

Theorem 6.2 describes the structure of S and T by reducing the problem to $\sigma(S)$ and $\sigma(T)$. Therefore it provides a description of all critical pairs as long as the pairs occurring in the recursive procedure have either an aperiodic sumset or an element with a unique representation in the sum. This leaves a seemingly unanswered question. Kemperman [59] deals with this question in an informal discussion after the proof. For this reason some authors think that Kemperman's solution to the critical problem is not completely satisfactory and have tried to obtain alternative solutions. A less precise but more intuitive description of critical pairs is given by Lev [63].

An alternative approach has been considered by Hamidoune [40, 41]. He gives a simple characterization of all sets S which may appear in a critical pair provided that S satisfies the Cauchy–Davenport inequality for all subsets. Following Hamidoune, we call a finite subset S of G a *Cauchy set* if $0 \in S$, S generates G and $\kappa(S) = |S|-1$.

Note that the structure of a set S satisfying $\kappa(S) < |S| - 1$ is rather trivial: By Theorem 3.1 there is a finite proper subgroup A such that $|S + A| < |S| + |A| - 1$; such sets are all constructed from an A-periodic set $P \neq G$ by deleting an arbitrary subset $D \subset P$ with $|D| < |A| - 1$.

The characterization of Cauchy sets which may appear in a critical pair is the following. We give here a simple proof using Theorem 5.1.

Theorem 6.3 *Let S be a Cauchy set of an abelian group G. There is a finite subset $T \subset G$ with $|T| \geq 2$ verifying*

$$|S + T| = |S| + |T| - 1 < |G| - 2$$

if and only if one of the following conditions hold:

(i) S is an arithmetic progression or $G \setminus S$ is an arithmetic progression.

(ii) There is a cyclic subgroup H of G and $x \in S$ such that $S \setminus \{x\}$ is H-periodic.

(iii) There is a proper subgroup H of G with $|H| \geq 3$ such that $|S + H| = |S| + |H| - 1 < |G|$.

Proof The *if* part of the theorem can be easily checked. If d is the difference of the arithmetic progression in (i) or the generator of H in (ii) then we can take $T = \{0, d\}$, while H is a suitable choice for T in (iii).

Let us show the *only if* part. By hypothesis we know that $X = \mathrm{Cay}(G, S)$ is 2-separable and $\kappa_2(X) = |S| - 1$ and $|S| < |G| - 3$.

Let A be a 2-atom of X. By Theorem 5.1, if $|A| \geq 3$ then A is a subgroup of G and (iii) holds with $H = A$. Suppose that $A = \{0, a\}$. Let H be the cyclic subgroup of G generated by a and let $S = S_0 \cup \ldots \cup S_{r-1}$ be the decomposition of S modulo H, that is, each S_i is a nonempty intersection of S with an H-coset. Then $S + A$ is the disjoint union $\bigcup_{i=0}^{r-1}(S_i + A)$ and

$$|S + A| = \sum_{i=0}^{r-1} |S_i + A| = \sum_{i=0}^{r-1} |S_i| + 1.$$

This equality implies that $S_i + A = S_i$ for all but one of the subscripts, say $i = 0$. Therefore S_i is an H-coset for $i = 1, \ldots, r - 1$. Now let $S_0 = S_{00} \cup \ldots \cup S_{0r'}$ be the decomposition of S_0 into a minimal number of arithmetic progressions with difference a. Then $S_0 + A$ is the disjoint union $\bigcup_{i=0}^{r'}(S_{0i} + A)$ and each sumset in the union has cardinality $|S_{0i}| + 1$. Therefore $r' = 0$ and S_0 is an arithmetic progression of difference a. Hence S is the union of a (possibly empty) H-periodic subset and a nonempty arithmetic progression S_0. We may assume that $0 \in S_0$.

If H is infinite then $S = S_0$ and (i) holds. Suppose that H is finite. If $H+S=G$ then $G \setminus S = H \setminus S_0$ is an arithmetic progression and (i) holds. Finally, if $H+S \neq G$ we have $|S|+|H|-1 \leq |S+H| = |S \setminus S_0| + |H|$ which implies $|S_0| = 1$ giving (ii). This completes the proof. □

Theorem 6.3 is useful in several applications. Let us mention here that it includes Vosper's theorem as a special case: In fact, if G is a finite group and $\gcd(|G|, |S|-1) = 1$ then it can be easily checked that the cases (ii) and (iii) of the theorem cannot occur, so that S is either an arithmetic progression or the complement of an arithmetic progression. These two notions are equivalent in cyclic groups.

7 Beyond critical pairs

The complexity of the characterization of critical pairs in abelian groups described in Kemperman's Theorem 6.2 may suggest that a precise description of pairs of sets satisfying $|S+T| = |S|+|T|+m$ with $m > -1$ in abelian groups is not attainable. However quite precise results can be given for small values of m. In this section we show some results in this direction.

For groups of prime order the characterizations are simpler as illustrated by the theorem of Vosper. A direct application of Theorem 5.1 gives the following result obtained in [76].

Theorem 7.1 *Let $m \geq -1$ be an integer and let S be a subset of a group of prime order p such that $2 \leq |S| < p - \binom{m+4}{2}$. If there is a subset T such that $2 \leq |T|$ and $|T+S| \leq |T|+|S|+m \leq p-2$ then S is the union of at most $m+2$ arithmetic progressions with the same difference.*

Proof Let $A = \{0, a\}$ be a 2-atom of S. By the definition of a 2-atom, we have

$$|A+S| = |A| + |\partial A| \leq |A| + |\partial T| = |A| + |T+S| - |T| = |A| + |S| + m.$$

Take a partition $S = S_1 \cup \cdots S_r$ of S into a minimal number of arithmetic progressions of difference a. Then, by the minimality of r, $S+A$ is the disjoint union $\bigcup_{i=1}^r (S_i+A)$, and each term of the union has cardinality $|S_i+A| = |S_i|+1$, so that

$$|S| + |A| + m \geq |S+A| = \sum_{i=1}^{r} |S_i + A| = |S| + r,$$

which implies that $r \leq m+2$. □

In particular, when $m=-1$, S is an arithmetic progression as stated in Vosper's theorem. Example 5.3 shows that there are sets with $|S| \geq p - \binom{m+4}{2}$ with $\kappa_2(S) = |S|+m$ which are not the union of $m+2$ arithmetic progressions.

The fact that a set S is the union of a small number of arithmetic progressions does not imply that $S+S$ is small. If r is the minimum number of arithmetic progressions with the same difference which cover a set S with $|S| < p - \binom{r+2}{4}$, then

$$2|S| + r - 2 \leq |S+S| \leq (r+1)|S| - \binom{r+1}{2},$$

where the lower bound follows from Theorem 7.1 and the upper bound is reached when these arithmetic progressions are sufficiently spread out in $\mathbf{Z}_p$. The upper bound decreases when S can be covered by a short arithmetic progression. Let $\ell(S)$ denote the smallest length of an arithmetic progression containing S. By using trigonometric sums, Freiman [25] proved the following result.

Theorem 7.2 *Let S be a subset of $\mathbf{Z}_p$ with cardinality $|S| < p/35$ such that $|S+S| = 2|S| + m$ with $m \leq \frac{2}{5}|S| - 3$. Then $\ell(S) \leq |S| + m + 1$.*

Bilu, Lev and Ruzsa [4] mention that the statement of the above Theorem is conjectured to hold for $-1 \leq m \leq |S| - 4$ and give a proof of it for small sets. They use the so-called *rectification principle* which allows one to translate the problem to the group of integers, where an analogous theorem holds. The application of the rectification principle requires $|S| \leq cp$ where the constant depends on the ones involved in the general Freiman-Ruzsa theorem. By using recent results of Chang [8] an estimation of c is $c = (24c_1)^{-6c_1}(6c_2)^{-1}$ where $c_1 = 2^{20}(3\log 3)^2$ and $c_2 = e^{c_1}$, see Green [32].

The following precise conjecture is stated in [50].

Conjecture 7.3 *Let m be a non-negative integer and let S and T be subsets of $\mathbf{Z}_p$ such that*

$$|S+T| \leq |S| + |T| + m \leq p - (m+4).$$

If $|S| \geq m+3$ and $|T| \geq m+4$ and if $p \geq K(m)$ where $K(m)$ depends only on m, then S and T are contained in arithmetic progressions with common difference and lengths at most $|S| + m + 1$ and $|T| + m + 1$ respectively.

If true, the conditions of the above conjecture cannot be weakened. To see that one of the sets must have cardinality at least $m+4$ take $S = \{0\} \cup \{m+3, m+4, \ldots, 2m+5\}$, which verifies $|2S| = 2|S| + m$ for large enough p and is not contained in an arithmetic progression of the stated length. Now let T be the complement in $\mathbf{Z}_p$ of $x - 2S$ for some $x \notin 2S$ and p large enough. Then $|S+T| = p - |x-S| = |S| + |T| + m$ which shows that the condition $|S+T| \leq p - (m+4)$ cannot be removed.

The next theorem proves the cases $m = 0$ and $m = 1$ of Conjecture 7.3. In the following statement we denote by $\ell_r(S)$ the smallest length of an arithmetic progression of difference r containing S.

Theorem 7.4 *Let S and T be subsets of $\mathbf{Z}_p$.*

(i) Assume that $|S| \geq 3$ and $|T| \geq 4$. If

$$|S+T| \leq |S| + |T| \leq p - 4,$$

then there is $r \in (\mathbf{Z}_p)^*$ such that

$$\ell_r(S) \leq |S| + 1 \text{ and } \ell_r(T) \leq |T| + 1.$$

(ii) Assume that $|S| \geq 4$ and $|T| \geq 5$ and $p \geq 53$. If

$$|S+T| \leq |S| + |T| + 1 \leq p - 5,$$

then there is $r \in (\mathbf{Z}_p)^*$ *such that*

$$\ell_r(S) \leq |S| + 2 \text{ and } \ell_r(T) \leq |T| + 2.$$

Hamidoune and Rødseth [48] proved part (i) of Theorem 7.4 by using the transforms of Davenport and Dyson. The case (ii) is proved by Hamidoune, Zémor and the author [50] by applying the isoperimetric method.

Proof We shall only give a sketch of the proof. The first step shows that, under the conditions of the theorem, if one of the two sets S, T is contained in a short arithmetic progression then so is the other one. This "compression transfer" result enables us to reduce the proof for the pair $\{S, T\}$ to the proof for a pair of sets $\{U, V\}$ where U is a 4-atom of one of the two original sets and V is a 5-atom of U. We then use properties of atoms to show that we must have $|U| = 4$ and $|V| = 5$ so that what is left to prove is the result for a pair of sets of sizes 4 and 5. □

A similar approach can be used to give a characterization of pairs of sets $\{S, T\}$ of an abelian group for which the inequality $|S + T| \leq |S| + |T|$ holds provided that one of the sets is not larger than the smallest order of its nonzero elements. Conditions of this kind which prevent the sets to cluster around periodic subsets already appear in the following result of Kemperman [59].

Theorem 7.5 *Let* S *and* T *be subsets of an abelian group* G *with* $|S| \geq 2$ *and* $|T| \geq 2$. *If*

$$|S + T| \leq |S| + |T| - 1$$

and every nonzero element of the group has order at least $|S| + |T| + 1$ *then both* S *and* T *are arithmetic progressions with the same difference.*

Karolyi [57] obtained a proof of Theorem 7.5 by using the polynomial method. The next result obtained in [51] extends Theorem 7.5 with less restrictive hypothesis.

Theorem 7.6 *Let* $0 \in S$ *and* T *be subsets of an abelian group* G *with* $|S| \geq 2$ *and* $|T| \geq 2$. *If*

$$|S + T| \leq |S| + |T| - 1 \leq |\langle S \rangle + T| - 2,$$

and every nonzero element of S *has order at least* $|S|$, *then one of the following conditions holds:*

(i) There is $x \in T$ *such that* $T \setminus \{x\}$ *is* $\langle S \rangle$*-periodic.*

(ii) There is a subset $P \subset T$ *such that* $T \setminus P$ *is a (possibly empty)* $\langle S \rangle$*-periodic subset, and both* S *and* P *are arithmetic progressions with the same difference.*

In a similar vein the following characterization of pairs of sets S and T with $|S + T| \leq |S| + |T|$ is given in [51]. A subset U of an abelian group G is *quasi–periodic* if there is an element $x \in G \setminus U$ such that $U \cup \{x\}$ is periodic.

Theorem 7.7 *Let* $0 \in S$ *and* T *be subsets of an abelian group* G *with* $\gcd(|G|, 6) = 1$. *Assume that* $|S| \geq 4$, $|T| \geq 3$ *and* S *generates* G. *If*

$$|S+T| \leq |S| + |T| \leq |G| - 4,$$

and every nonzero element in S *has order at least* $|S| + 1$ *then either* $S \setminus \{0\}$ *is quasi-periodic or* S *and* T *are included in arithmetic progressions with the same difference and respective lengths at most* $|S| + 1$ *and* $|T| + 1$.

Deshouillers and Freiman [12] recently obtained a structural description of sets S in a cyclic group for which $|2S| \leq 2.04|S|$ and S is a small set. Roughly speaking, S is then contained in an arithmetic progression of cosets of some subgroup, and the cosets met by S are well filled.

On the other end of the spectrum there is the important case of $G = \mathbf{Z}_2^n$ where there are no nontrivial arithmetic progressions, so that sets with small sumset will be described in terms of almost periodic sets. Zémor [82] obtained the following generalization of Mann's Theorem 4.3 for binary spaces for which a proof with the isoperimetric method was given by Hamidoune [38].

Theorem 7.8 *Let* m *be a positive integer and* S *a nonempty subset of* $G = \mathbf{Z}_2^n$. *If there is a finite subset* T *with* $|T| \geq 2$ *and* $S + T \neq G$ *such that*

$$|S+T| \leq |S| + |T| + m,$$

then there is a proper subgroup H *of* G *with*

$$|S+H| \leq |S| + |H| + m.$$

Proof We may assume that $0 \in S$ and S generates G. Let A be a 2-atom of S containing 0 and let a be any element in $A \setminus \{0\}$. We have $\{0, a\} \subset A \cap (A + a)$. By the intersection property (Lemma 2.2), $A = A + a$. Therefore $A + A = A$ and A is a subgroup. We can take $H = A$. □

Thus a set S with small sumset with some other set in $\mathbf{Z}_2^n$ as in the conditions of the theorem fills in a collection of cosets of some proper subgroup H with some holes of cardinality at most $|H| + m$.

The additive group of the integers is often the source and motivation of additive problems. The analogous of Cauchy–Davenport and Vosper theorems have in $\mathbf{Z}$ simple elementary proofs which can be extended to any linearly ordered abelian group. Beyond the Cauchy–Davenport inequality, Freiman [26] obtained a characterization of sets of integers with small sumset which can be formulated as follows.

Theorem 7.9 *Let* S *and* T *be nonempty finite sets of integers with* $|S| \geq 2$ *and* $|T| \geq 2$. *Assume that* $\max(T) - \min(T) \leq \max(S) - \min(S)$ *and set* $\delta = 1$ *if there is equality and* $\delta = 0$ *otherwise. If*

$$|S+T| = |S| + |T| + m,$$

with $m \leq \min\{|S|, |T| - \delta\} - 3$ *then both* S *and* T *are contained in an arithmetic progression of the same difference and length at most* $|S| + m + 1$.

When $S = T$, Theorem 7.9 is known as the $(3k-4)$-theorem. For $k = |S|$, the upper bound $|S+S| \le 3k-4$ from the statement is best possible: if S consists of two arithmetic progressions with a sufficiently large gap between them then $|S+S| = 3k-3$ and S is not contained in a short arithmetic progression. Freiman's original proof uses elementary methods. Steinig [78] gives another version of Freiman's proof. Lev and Smeliansky [64] and Hamidoune [39] found a short proof of Theorem 7.9 which uses a reduction to finite cyclic groups and the theorem of Kneser. The idea of such a reduction is illustrated in the following simple proof of a weaker version of Theorem 7.9 for $S = T$ and $m < (|S|-3)/2$ using Theorem 3.7.

Proof of Theorem 7.9 for $S = T$ and $m < (|S|-3)/2$ Let d be the greatest common divisor of the elements in $S \setminus \{0\}$. By dividing by d and translating if necessary we may assume that S is contained in an interval $[0, s]$ with $\{0, s\} \subset S$ and $\gcd(S \setminus \{0\}) = 1$. Let $\sigma : \mathbf{Z} \to \mathbf{Z}_s$ denote the canonical projection of $\mathbf{Z}$ onto $\mathbf{Z}_s$ and write $\bar{S} = \sigma(S)$, where $|\bar{S}| = |S| - 1$. We have $2\bar{S} = \bar{S} \cup \partial(\bar{S})$ in $\mathrm{Cay}(\mathbf{Z}_s, \bar{S})$ and

$$2|S| + m = |2S| \ge |S \cup (s+S)| + |2\bar{S} \setminus \bar{S}| = 2|S| - 1 + |\partial(\bar{S})|. \tag{7.1}$$

Therefore

$$|\partial(\bar{S})| \le m + 1 < (|S|-1)/2 = (1/2)|\bar{S}|. \tag{7.2}$$

Since $\gcd(S \setminus \{0\}) = 1$, $\bar{S}$ generates $\mathbf{Z}_s$; by Theorem 3.7, we have

$$|2\bar{S}| = |\bar{S}| + |\partial(\bar{S})| \ge \min\{s, |\bar{S}| + (1/2)|\bar{S}|)\}.$$

By (7.2) we have $|\partial(\bar{S})| = s - |\bar{S}|$ and, by substitution in (7.1),

$$s \le |S| + m,$$

which proves the statement. □

Freiman [24] further extended Theorem 7.9 to the case when $|S+S| = 3|S| - 3$. Hamidoune and Plagne [47] obtained the following generalization of his $3k-3$ theorem by using the same reduction technique but applying a version of Theorem 6.3 instead of Kneser's theorem, which is no longer applicable in this case.

Theorem 7.10 *Let $S \subset [0, s]$ be a set of positive integers with $0, s \in S$ and $\gcd(S) = 1$. Let t be any integer. If*

$$|S + t \cdot S| = 3|S| - 3,$$

where $t \cdot S = \{ts,\ s \in S\}$, then one of the following holds:

(i) $|S| \ge 1 + s/2$.

(ii) $t \in \{-1, 1\}$ *and S is the union of two arithmetic progressions with the same difference.*

(iii) $|S+S| = 3|S|-3$, *s is an even number and S is of the form $S = \{0, s/2, s, x, x + s/2, 2x\}$ for some integer $0 < x < s/2$.*

8 Non-abelian groups

Throughout this section G denotes a non-abelian group written multiplicatively. For phonetic reasons we still speak of the *small sumset problem* in this case instead of the "small productset problem." The problem is less known in this context than its abelian counterpart. Note that the general bound for the size of the product of two finite sets provided by Theorem 3.7 is valid in any group.

From the results in the abelian case the problem is seemingly easier when G does not contain finite subgroups, so let us start with non-abelian torsion-free groups. In this case Theorem 3.6 shows that, for any finite nonempty subset $S \subset G$, atoms of S have cardinality 1. Therefore, the inequality

$$|ST| \geq |S| + |T| - 1$$

holds for any pair of nonempty sets S and T of a torsion–free group, a result first obtained by Kemperman [58]. The critical pair problem for torsion-free groups was solved by Brailovsky and Freiman [5]. We give below the statement and a simple proof of this characterization. A subset of the form $\{ar^i : 1 \leq i \leq m\}$ for some $a, r \in G$ and $m \in \mathbf{N}$ is said to be a *left progression* of ratio r or a left r–progression. Similarly, a set of the form $\{r^i b : 1 \leq i \leq m\}$ for some $b \in G$ is a *right r–progression*. A set which is both a left and a right r–progression is simply a *progression*.

Theorem 8.1 *Let S and T be finite subsets of a torsion free group G with* $\min\{|S|, |T|\} \geq 2$. *If*

$$|ST| = |S| + |T| - 1$$

then S is a left progression and T is a right progression.

Proof Since $|ST| = |s^{-1}STt^{-1}|$ we may assume $1 \in S \cap T$. The equality $|ST| = |S| + |T| - 1$ and $\min\{|S|, |T|\} \geq 2$ imply $\kappa_2(T) = |T| - 1$ and $\kappa_2(S^{-1}) = |S| - 1$.

To prove the statement it suffices to show that, for every finite nonempty set T with $1 \in T$ and $\kappa_2(T) = |T| - 1$, a 2-atom A of T has cardinality $|A| = 2$. For, if $A = \{1, r\}$ for some r and $T = T_1 \cup \cdots \cup T_r$ is a minimal decomposition of T into right r-progressions then

$$|A| + |T| - 1 = |AT| = \sum_{i=1}^{r} |AT_i| \geq \sum_{i=1}^{r} (|A| + |T_i| - 1) = r(|A| - 1) + |T|,$$

which implies $r = 1$. Once we know that T is a right r-progression, a similar argument shows that S is a left r-progression.

Suppose that the statement on the cardinality of the 2-atom is false and choose a counterexample U with minimum cardinality. Let A be a 2-atom of U containing 1 and denote by $X = \mathrm{Cay}(G, U)$, where we may assume that $1 \in U$ and U generates G. By the intersection property (Lemma 2.2), we have $|xA \cap A| \leq 1$ for each $x \in G$; otherwise we would have $xA = A = HA$ where H denotes the subgroup generated by x, which is not possible since A is a finite set.

If there is an element $x \in \partial A$ which can be uniquely written as a product of the form au with $a \in A$ and $u \in U$ then $|\partial(A \setminus \{a\})| \leq |\partial A \setminus \{x\} \cup \{a\}| = |\partial A|$

contradicting that A is a 2-atom. On the other hand, if there is $a \in A$ such that $aU^{-1} \cap A = \emptyset$ then $|\partial(A \setminus \{a\})| \leq |\partial A|$ again a contradiction. Hence we have

$$|xU^{-1} \cap A| \geq 2 \text{ for each } x \in A \cup \partial A. \tag{8.1}$$

Let us show that $|A| \leq |U| - 1$. By (8.1) we can define a function $f : A \to U \setminus \{1\}$ such that $x(f(x))^{-1} \in A$ for each $x \in A$. If $f(x) = f(y) = c$ for some pair x, y of distinct elements in A, then $\{x, y, xc, yc\} \subset A$ and $\{x, xc\} \subset (xy^{-1})A \cap A$ which is not possible by the intersection property. Therefore f is injective and $|A| \leq |U| - 1$.

Let A' be a 2-atom of A containing 1. By the minimality of U, $A' = \{1, r\}$ for some r and, since $\kappa_2(A) = |A| - 1$, the argument at the beginning of the proof shows that A is a left r-progression. But then $|A \cap rA| \geq 2$ contradicting the intersection property. This completes the proof. □

By using the techniques illustrated in the above proof, Hamidoune, Lladó and the author [43] obtained the following generalization of Theorem 8.1.

Theorem 8.2 *Let S be a finite generating set containing* 1 *of a non-abelian torsion free group G. Then, for each set T with $|T| \geq 4$ we have*

$$|ST| \geq |S| + |T| + 1.$$

Moreover, $|S^2| \leq 2|S|$ and $|S| = k \geq 4$ if and only if there are elements $r, x \in G$ such that $xr = rx$ and $Sx = \{1, r, \cdots, r^k\} \setminus \{c\}$ where $c \in \{1, r\}$.

The characterization of sets with small doubling in torsion free groups given in the above theorem is analogous to the one in the abelian case, and these sets are contained in a coset of an infinite cyclic group. If no translates of two finite sets S and T of a torsion free group generate an abelian subgroup, it is likely that larger lower bounds for the cardinality of the product $|ST|$ hold.

In the finite case, Zémor [83] obtained the following extension to the non-abelian case of Mann's Theorem 4.3 by using Theorem 3.6.

Theorem 8.3 *Let S be a nonempty subset of a finite group G such that, for every proper subgroup H of G,*

$$\min\{|SH|, |HS|\} \geq |S| + |H| - 1.$$

Then for every subset T such that either $|ST| \neq G$ or $|TS| \neq G$ we have

$$\min\{|ST|, |TS|\} \geq |S| + |T| - 1.$$

The following example given in [83] shows that the condition in the theorem cannot be replaced by just one of the two inequalities $|SH| \geq |S| + |H| - 1$ or $|HS| \geq |S| + |H| - 1$.

Example 8.4 Let $G = Sym(4)$ be the symmetric group on 4 elements. Take the permutations $a = (12)$ and $b = (134)$. Denote by $A = \{1, a\}$ the subgroup generated by a and define $T = (A \cup Ab)^{-1}$ and $S = G \setminus T^{-1}$. We have

$$|ST| = |S| + |T| - 2.$$

This last equality can be easily visualized in the Cayley graph $X = \mathrm{Cay}(G, T^{-1})$ where $|\partial A| = |AT^{-1}| - |A| = |T^{-1}|/2$ and, by Theorem 3.7, this cardinality reaches its minimum possible value. Hence A is an atom of X, and $ST \backslash S = \partial^{-1}(G \backslash AT^{-1}) = \partial A$. For any subgroup K such that $|KT| - |T| \leq |K| - 2$ we either have $T \subset K$ or $|K| = 2$. Since T generates G we should have $|K| = 2$ and $KT = T$. The only element of order 2 in T is a so we are led to $H = A$, but one can check that $AT \neq T$.

The formulation of an extension of Kneser's theorem to the non-abelian case has been an open question already mentioned by Diderrich [15] who conjectured a natural generalization which was however disproved by Olson [72]. Example 8.4 unveils some of the difficulties which may appear in finding the right statement. Such difficulties are easier to overcome if some normality properties are invoked. A set S in a finite non-abelian group G is *normal* if it is invariant under conjugation, that is $x^{-1}Sx = S$ for each $x \in G$. By adapting the proof of Kneser's theorem in Mann [69], Arad, Fisman and Muzychuk [2] obtain the following generalization.

Theorem 8.5 *Let S and T be finite subsets of a group G. If the product ST is a normal subset then there is a normal subgroup H of G such that*

$$|STH| = |ST| \geq |SH| + |TH| - |H|.$$

To our knowledge the applications of the above theorem to the characterization of critical pairs has not been pursued yet. The characterization of critical pairs in the finite non-abelian case has been achieved under some additional constraints. Hamidoune [38] obtained the following generalization of Vosper's theorem to any group.

Theorem 8.6 *Let S and T be nonempty subsets of a group G such that T generates G, $1 \in T$ and $\min\{|S|, |T|\} \geq 2$. If every element in $G \setminus \{1\}$ has order at least $\max\{|S|, |T|\}$ and*

$$|ST| = |S| + |T| - 1 < |G| - 1$$

then S is a left progression and T is a right progression.

Theorem 8.6 implies Vosper's Theorem 6.1, Kemperman's Theorem 7.5 and the above Theorem 8.1 of Brailovsky and Freiman.

Going back to normal sets, Arad, Fisman and Muzychuk [2] have shown by using the isoperimetric approach that, for normal subsets, Theorem 8.3 holds by replacing all subgroups in the condition by just the normal ones. With this result they give the following characterization of critical pairs in finite simple groups in which one of the subsets is normal.

Theorem 8.7 *Let S and T be subsets of a non-abelian finite simple group such that S is a normal subset and $|S| > 1$. Then*

$$|ST| \geq \min\{|G|, |S| + |T| - 1\}.$$

Moreover, if $ST \neq G$ then the equality holds if and only if either $|T| = 1$ or $T = G \setminus S^{-1}g$ for some $g \in G$.

So the only critical pairs of a non-abelian finite simple group containing a normal subset are the trivial ones. In the continuation of the above mentioned reference, Arad and Muzychuk [3] prove a strengthening of Theorem 8.7 showing that, if S is a normal set and $2 \leq |S| \leq |G|/4$ then

$$|ST| \geq |S| + |T| + 3$$

for each subset T with $|T| \geq 2$ and $|ST| \leq |G| - 2$. The results obtained by these authors are applied to give a number of significant results in the characterization of simplicity of groups and various properties related to conjugacy classes.

9 Some applications of the small sumset problem

In this section we include a selection of problems connected to the small sumset problem which have originated in number theory, group theory and combinatorics. The criterion has been to illustrate how the results described in the above sections lead to solutions or improvements of the selected problems. The selection is by no means exhaustive and a wider range of applications can be found for instance in [13]. The accounts of the contributions to the problems are even less exhaustive to keep the exposition within a reasonable size.

9.1 The range of diagonal forms

Let $a_1, \ldots, a_n$ be nonzero elements of a finite field F and let f be the diagonal form

$$f(x_1, \ldots, x_n) = a_1 x_1^k + \cdots + a_n x_n^k.$$

The estimation of the cardinality $|f(F^n)|$ of the image of f was first considered by Cauchy. For $F = \mathbf{F}_p$ the field of prime order p he gave the estimation

$$|f((\mathbf{F}_p)^n)| \geq \min\{p, n(|P| - 1) + 1\}$$

where $P = \{x^k : x \in F\}$ denotes the set of k-th powers of elements of F. As an application of Vosper's theorem, Chowla, Mann and Strauss [10] obtained the following improvement

$$\text{if } |P| \geq 3, \ |f((\mathbf{F}_p)^n)| \geq \min\{p, (2n - 1)(|P| - 1) + 1\} \tag{9.1}$$

Tietäväinen [80] generalized this last bound to finite fields of odd characteristic. The following approach allows us to easily extend inequality (9.1) to any division ring and give a simple unified proof of it.

Note that $f(F^n) = a_1P + \cdots + a_nP$, so the problem consists of estimating a lower bound for the cardinality of a sumset. Note also that $P \setminus \{0\}$ is a multiplicative subgroup of the field. The inequality (9.1) follows from the next theorem obtained by Hamidoune [41, Theorem 7.4]. We include here his short and beautiful proof.

Theorem 9.1 *Let F be a division ring and let P be a finite subset with $0 \in P$ and $|P| \geq 4$. Assume that $P \setminus \{0\}$ is a multiplicative group of R and let $a_1, \ldots, a_n$ be nonzero elements of F. Then,*

$$|a_1P + \cdots + a_nP| \geq \min\{|F_0|, (2n - 1)(|P| - 1) + 1\},$$

where F_0 is the additive group generated by P.

Proof Suppose that $P \neq F_0$. Note that the Cayley graph $X = \mathrm{Cay}(F_0, P)$ is arc-transitive: Indeed the action of F_0 by translation is transitive on the vertices of the graph and multiplication by elements in $P \setminus \{0\}$ leaves 0 fixed and acts transitively on its neighborhood; both actions are easily seen to be automorphisms of the graph. By Proposition 3.7 the connectivity of X is $\kappa(X) = |P| - 1$.

Incidentally, the fact that X is arc-transitive provides a simple proof of the fact that P is not an arithmetic progression. Otherwise, since $0 \in P$, $P = \{0, r, 2r, \ldots, (k-1)r\}$. But then,

$$|\partial(0) \cap \partial(r)| = k - 2 > |\partial(0) \cap \partial(2r)|,$$

while the $(0, r)$ and $(0, 2r)$ are arcs of the graph, thus contradicting that X is arc-transitive. A similar argument shows that P is not the complement of an arithmetic progression. Note that the cases (ii) and (iii) of Theorem 6.3 cannot occur. This is clear if F_0 is infinite and follows in the finite case from $\gcd(|F_0|, |P| - 1) = 1$. Therefore, by Theorem 6.3, for every set $T \subset F_0$ with $|T| \geq 2$ we have

$$|T + P| \geq \min\{|F_0| - 1, |T| + |P|\}.$$

The proof proceeds by induction on n, the result being trivial for $n = 1$. Dividing by a_n if necessary we may assume $a_n = 1$. By the induction hypothesis and the previous inequality,

$$|b_1 P + \cdots + b_{n-1} P + P| \geq \min\{|F_0| - 1, (2n-2)|P| + 2\}.$$

Denote by $Q = (b_1 P + \cdots + b_{n-1} P + P) \setminus \{0\}$ and $U = P \setminus \{0\}$. From $QU = Q$ we deduce that $|U|$ divides $|Q|$. Since $|U|$ also divides $|F_0| - 1$,

$$\begin{aligned} |b_1 P + \cdots + b_{n-1} P + P| &\geq \min\{|F_0|, (2n-2)(|P| - 1) + |U| + 1\} \\ &= \min\{|F_0|, (2n-1)(|P| - 1) + 1\}, \end{aligned}$$

as desired. □

9.2 The Frobenius problem

Let $A = \{a_1 < \cdots < a_n\}$ be a finite set of positive integers with $\gcd(A) = 1$. It is well known that every sufficiently large integer can be expressed as a linear combination of the elements in A with positive integer coefficients. The Frobenius problem asks for an estimation of the largest integer $G(A)$ for which the equation $G(A) = \sum_{i=1}^{k} x_i a_i$ has no solution in nonnegative integers.

The problem has an easy solution when $n = 2$ as proved by Sylvester:

$$G(\{a_1, a_2\}) = (a_1 - 1)(a_2 - 1) - 1.$$

Erdős and Graham [22] proved the inequality $G(A) \leq 2(\max(A))^2/|A|$ and conjectured that

$$G(A) \leq \frac{(\max(A))^2}{|A| - 1}.$$

Dixmier [16] proved the above conjecture and gave the exact maximum possible values of $G(A)$ when $m = \max(A)$ is 0 or 1 or 2 modulo $|A| - 1$. Dixmier's result can be stated as follows.

Theorem 9.2 *Let $A \subset [1, m]$ with $m \in A$ and* $\gcd(A) = 1$. *Then*

$$G(A) \leq (k-1)(m-r-1) - 1,$$

where $n = |A|$ and $m - 1 = k(n-1) - r$, $1 \leq r \leq n-1$.

Moreover, if $g(m, n)$ denotes the maximum value of $G(A)$ among the sets satisfying the above hypothesis, then

$$g(m,n) = \begin{cases} \frac{m(m-2)}{n-1} - m + 1, & \text{if } n-1 \text{ divides } m \text{ or } m-2 \\ \frac{(m-1)^2}{n-1} - m, & \text{if } n-1 \text{ divides } m-1 \end{cases}$$

The second part of Theorem 9.2 partially proves a conjecture of Lewin [65] which states that, for $m = \max(A)$ sufficiently large and $n = |A|$,

$$G(A) \leq \lfloor \frac{(m-2)(m-n+1)}{n-1} \rfloor - 1. \quad (9.2)$$

The proof of Dixmier uses the theorem of Kneser. Lev [62] gave an alternative proof and, by using the $(3k-3)$ theorem of Freiman, he characterized the cases of equality when $m \leq 3n$. Hamidoune [41] gives still another proof using the simpler theorem of Mann. The general idea underlying these proofs already appears in the paper by Erdős and Graham [22] and shows the role of the small sumset problem.

Let us give a hint of this connection. Following the formulation of Dixmier [16], denote by S the semigroup generated by A, so that $G(A) = \max(\mathbf{N} \setminus S)$. For a positive integer k, let $S_k = S \cap [(k-1)m+1, km]$. Let $\sigma : \mathbf{Z} \to \mathbf{Z}_m$ and, for a subset $U \subset \mathbf{Z}$, denote by $\overline{U} = \sigma(U)$. Dixmier observed that

$$\overline{S_k} + \overline{S_{k'}} \subset \overline{S_{k+k'}}, \ k, k' \geq 1.$$

In particular, $k\overline{S_1} \subset \overline{S_k}$. He also noted that S_1 generates the same semigroup as A, so that we can take $A = S_1$. Therefore we are led to determine the smallest k for which $k\overline{S_1} = \mathbf{Z}_m$.

By replacing the Theorem of Mann by a version of Theorem 6.3 Hamidoune [41] obtained an extension of the last part of Theorem 9.2 to $r \geq 3$, whenever $n \neq 5$ and $m \geq n^3 - 1$, obtaining the inequality

$$G(A) \leq \lceil \frac{m-2}{n-1}(m-n+1) \rceil - 1 < \lceil \frac{(m-2)(m-n+1)}{n-1} \rceil - 1.$$

This proves Lewin's conjecture and shows that the inequality in (9.2) is strict for $r \geq 3$. He also characterized the sets with maximal Frobenius number.

9.3 A problem of Erdős and Heilbronn on subset sums

Let G be a finite abelian group of order n and S a subset of G. The subset sum of S is denoted by

$$\Sigma(S) = \{\sum_{t \in T} T, \ T \subset S\}$$

and

$$\Sigma^*(S) = \{\sum_{t \in T} T, \ T \subset S, T \neq \emptyset\}.$$

Erdős and Heilbronn [23] proved that if $S \subset \mathbf{Z}_p \setminus \{0\}$ and $|S| \geq 3(6p)^{1/2}$ then $\Sigma^*(S) \cup \{0\} = \mathbf{Z}_p$. By taking the arithmetic progression $A = \{1, \ldots, \lfloor (k+1)/2 \rfloor\}$ with $k < 2\sqrt{p-1}$ and $S = A \cup -A$ we have $(p-1)/2 \notin \Sigma^*(S)$ which shows that the bound is almost best possible. In the same paper they conjectured that a subset S of an abelian group of order n with $|S| \geq 2n^{1/2}$ verifies $0 \in \Sigma^*(S)$. Szemerédi [79] proved the existence of an absolute constant c such that $|S| \geq cn^{1/2}$ implies $0 \in \Sigma^*(S)$. Olson showed that we can take $c = 2$ in $\mathbf{Z}_p$ [71] and $c = 3$ in a finite group [72] . Erdős [21] conjectured later that the result holds with $c = \sqrt{2}$.

If $S = \{s_1, \ldots, s_k\}$ then

$$\Sigma(S) = \{0, s_1\} + \cdots + \{0, s_k\},$$

which shows that the problem is connected to the estimation of a lower bound for a sumset. Let us summarize the method of Olson to indicate the use of the small sumset problem. For a subset $B \subseteq G$ and $x \in G$ let

$$\lambda_B(x) = |(B + x) \setminus B|.$$

Let S be a nonempty subset of $G \setminus \{0\}$ and $y \in S$. Put $B = \Sigma(S)$. Olson observed that

$$|\Sigma(S)| \geq |\Sigma(S \setminus y)| + \lambda_B(y), \tag{9.3}$$

and looked for good lower bounds for $\lambda_B(y)$. These lower bounds are obtained by an averaging technique introduced by Erdős and Heilbronn, together with lower bounds for the small sumset problem. This led Olson to obtain Theorem 3.7 for groups. Although it is not the only ingredient in the method, better lower bounds on the small sumset problem contribute to improvements of the results. Hamidoune and Zémor [52] obtained for $\mathbf{Z}_p$

$$\text{if } |S| \geq \sqrt{2p} + 5\ln p \text{ then } 0 \in \Sigma^*(S),$$

which is the current best result for the problem. For an abelian group of order n, they also obtained

$$\text{if } |S| \geq \sqrt{2n} + O(n^{1/3}\ln n) \text{ then } 0 \in \Sigma^*(S).$$

Returning to the original problem of Erdős and Heilbron, let $c(G)$ denote the minimum cardinality of a subset of a finite abelian group G such that $\Sigma^*(S) = G$ for any subset $S \subset G \setminus \{0\}$ with cardinality $|S| \leq c(G)$. Dias da Silva and Hamidoune [14] proved that, for an odd prime p, $c(\mathbf{Z}_p) \leq \sqrt{4p-7}$ which is essentially best possible.

The case of groups with composite order is of different nature and $c(G)$ has a relatively large lower bound. Let p be the smallest prime divisor of $|G| = ph$. By taking S a subgroup of order h and $p-2$ elements of a transversal of the cosets of this group, we have $\Sigma^*(S \setminus \{0\}) \neq G$ so $c(G) \geq p + h - 2$. Diderrich [15] proved that

$$p + q - 2 \leq c(\mathbf{Z}_p \oplus \mathbf{Z}_q) \leq p + q - 1,$$

where p and q are primes. The two bounds are attained. In the same paper he conjectured that, if h is composite, then

$$c(G) = p + h - 2.$$

In [46] it was proved that if $S \subset G \setminus \{0\}$ is a generating subset of G and $|S| \geq 14$ then

$$|\Sigma(S)| \geq \min\{|G| - 3, 3|S| - 3\}$$

an inequality which solves the conjecture of Diderrich for $|G| \geq 3q$ and q an odd composite number. The whole conjecture was proved by Gao and Hamidoune [29]. In [30] the critical sets of equality in the conjecture were characterized for $n = |G|$ large enough.

9.4 Rainbow arithmetic progressions

The well-known van der Waerden theorem implies that, for every sufficiently large n, every coloring of the integers in $[1, n]$ with 3 colors contains a monochromatic 3-term arithmetic progression. By the canonical version of van der Waerden theorem of Erdős and Graham, for every sufficiently large n and any arbitrary coloring of the integers in $[1, n]$ there is either a monochromatic or a rainbow 3-term arithmetic progression. Here rainbow means that no two elements have the same color. It is obviously not true that every three coloring of $[1, n]$ contains a rainbow 3-term arithmetic progression unless some conditions are set on the cardinalities of the color classes. Jungić, Licht, Mahdian, Nešetřil and Radoičić [54] have proved that in $\mathbf{Z}_p$ these conditions are very weak. They actually prove the following stronger statement by using the Hamidoune-Rødseth Theorem 7.4 (i). Let us call the *head* of an arithmetic progression $I \neq \mathbf{Z}_p$ of difference d in $\mathbf{Z}_p$ the element $x \in I$ such that $x - d \notin I$.

Theorem 9.3 *Let $a, b, c, e \in \mathbf{Z}_p$ with $abc \neq 0$. Then every 3-colouring of $\mathbf{Z}_p = R \cup S \cup T$ with $\min\{|R|, |S|, |T|\} \geq 4$ contains a rainbow solution of the linear equation*

$$ax + by + cz = e$$

unless $a = b = c$ and every colour class is an arithmetic progression with the same difference d such that the heads r, s and t of $d^{-1}R, d^{-1}S, d^{-1}T$ respectively satisfy $s + t + r = a^{-1}e + 1$ or $s + t + r = a^{-1}e + 2$.

The same authors consider the same problem in $\mathbf{Z}_n$ for arbitrary n and the equation $x + y = 2z$. They do not use results related to the small sumset problem in this case. Their results can be stated as follows.

Theorem 9.4 *Every 3-colouring of $\mathbf{Z}_n = R \cup S \cup T$ with $\min\{|R|, |S|, |T|\} > n/6$ contains a rainbow 3-term arithmetic progression. Moreover if n is odd then $n/6$ can be replaced in the above condition by n/q where q is the smallest prime divisor of n.*

The condition $\min\{|R|, |S|, |T|\} > n/6$ in the above theorem is tight for $n \equiv 0 \pmod 6$. It is also shown in [54] that, if $\mathbf{Z}_n$ does not admit rainbow free 3-colourings with no empty classes, then either $n = 2^m$ or $n = p$ is a prime for which 2 is a primitive root or $n = p \equiv 3 \pmod 4$ is a prime for which $(p - 1)/2$ is a primitive root.

9.5 The Cacceta and Haggkvist conjecture

Chartrand proposed the conjecture that a regular graph X of degree d and girth $g = g(X)$, the length of the shortest cycle in X, has at least $d(g-1)+1$ vertices. Cacceta and Haggkvist [6] proposed the same conjecture for graphs of minimum degree $d = d(X)$. The conjecture has shown to be unexpectedly hard even for $g = 3$. If true the conjecture is tight as shown by the Cayley graph $X = \text{Cay}(\mathbf{Z}_n, B)$ where $B = \{1, 2, \ldots, d\}$ is an arithmetic progression and $n = d(g-1)+1$ (here we do not include loops in X.) It is not difficult to check that the shortest cycle in this graph has length $g = 1+(n-1)/d$, so there are Cayley graphs which provide critical examples of the conjecture. The following result of Shepherdson [77] obtained in 1947 confirms the conjecture for abelian Cayley graphs. The proof of Shepherdson is short and elegant and it is based on the Davenport transform. We give here a short proof of the result based on Theorem 3.6.

Theorem 9.5 *Let B be a nonempty subset of an abelian group G with $0 \notin B$ and let k be an integer such that $k|B| \geq |G|$. Then $0 \in B \cup 2B \cup \cdots \cup kB$.*

Proof The proof is by induction on $|G|$, the result being obvious for $|G| = 2$. We may therefore assume that B generates G. Set $S = B \cup \{0\}$ and, for each positive integer j, denote by

$$B(j) = \bigcup_{1 \leq i \leq j} iB.$$

Note that

$$|B(j)| = |B(j-1)+S| \geq \min\{|G|, |B(j-1)|+\kappa(S)\} \geq \min\{|G|, |B|+(j-1)\kappa(S)\}.$$

If $\kappa(X) = |S|-1$ then $|B(k)| \geq \min\{|G|, k|B|\} = |G|$ and the result holds.

Suppose that $\kappa(S) < |S|-1$ and let A be an atom of $X = \text{Cay}(G, S)$ containing 0. Recall that A is a proper subgroup of G generated by $A \cap S = S_0$.

Let $B_0 = S_0 \setminus \{0\}$ and $B_0(j) = \cup_{1 \leq i \leq j}\, iB_0 \subset B(j)$. If $0 \in B_0(k)$ we are done. Otherwise, by the induction hypothesis, $k|B_0| < |A|$. Let $S = S_0 \cup \ldots \cup S_r$ be the decomposition of S into nonempty intersections with cosets of A. We have $|S_0| = |B_0|+1 \leq (|A|+1)/2$ which combined with $|S+A|-|S| < |A|-1$ implies

$$|S_i| > |A|/2,\ i = 1, \ldots, r. \tag{9.4}$$

Let $\sigma : G \to G/A$ be the canonical projection and $\overline{B} = \sigma(B \setminus H)$. Then $k|A||\overline{B}| \geq k(|B|-|B'|) > |G|-|A|$ so that $k|\overline{B}| \geq |G/A|$. By the induction hypothesis

$$0 \in \cup_{1 \leq i \leq k}\, i\overline{B} = \overline{B}(k),$$

so that $\overline{b_1}+\cdots+\overline{b_j} = 0$ for some $\overline{b_1}, \ldots, \overline{b_j} \in \overline{B}$ and $2 \leq j \leq k$. Therefore $b_1+\cdots+b_j = a$ for some $a \in A$, where $b_i \in \sigma^{-1}(\overline{b_i}) \cap (S \setminus S_0)$. Let $B_1 = \sigma^{-1}(\overline{b_1}) \cap (S \setminus S_0)$ and $B_2 = \sigma^{-1}(\overline{b_2}) \cap (S \setminus S_0)$, where each of B_1 and B_2 belong to $\{S_1, \ldots, S_r\}$. From inequality (9.4) we know that $|B_1|+|B_2| > |A|$ so that, by Lemma 4.5, $B_1+B_2 = b_1+b_2+A$. Therefore there are $b_1' \in B_1$ and $b_2' \in B_2$ such that $b_1'+b_2' = b_1+b_2-a$. Hence $0 = b_1'+b_2'+b_3+\cdots+b_j \in B(j)$. □

By Theorem 9.5, $X = \mathrm{Cay}(G, B)$ has girth $g \le \lceil |G|/|B| \rceil \le (|G| + |B| - 1)/|B|$ so $|G| \ge |B|(g-1)+1$. The above proof can be easily translated to the context of vertex–transitive graphs and is essentially the one used by Hamidoune [36] to prove the conjecture for this class of graphs. Better lower bounds for the isoperimetric connectivity of the graph would result in an improvement of the result as long as the characterization of the corresponding sets of the corresponding critical sets is available. By using this idea, it was shown in [43] that the class of abelian Cayley graphs which satisfy $n = d(g-1)+1$ can be fully described. Let $\mathcal{C}$ denote the class of abelian Cayley graphs of graphs of the form $\mathrm{Cay}(\mathbf{Z}_n, S)$ where either $S = \{1, 2, \ldots, d\}$ or $S = \emptyset$ for some n and d, and let

$$\mathcal{L} = \{X : X = X_1[X_2[\cdots[X_k]]\cdots],\ X_i \in \mathcal{C}, k \ge 1\}$$

be the class of lexicographic products of graphs in $\mathcal{C}$.

Theorem 9.6 *Let $X = \mathrm{Cay}(G, S)$ be a Cayley graph with girth $g = g(X) > 3$. Then $|G| \ge (d+1)(g-1) - 1$ unless $X \in \mathcal{L}$, where $d = d(X)$.*

It was conjectured in [43] that the above result can be extended to the class of vertex–transitive graphs.

9.6 Reliability of networks

A natural application of the isoperimetric connectivities is the analysis of reliability of interconnection networks. Many models proposed in the literature and used in the design of multiprocessor systems are based on Cayley graphs. The symmetry of the graphs often simplifies computational and routing algorithms, see e.g. [17, 53].

Loop networks based on Cayley graphs of a cyclic group are often used for their simplicity. It is known that abelian Cayley graphs have no good expanding properties. For example, it is shown in [49] that an abelian Cayley graph $X = \mathrm{Cay}(G, S)$ of order n and degree $d = |S| - 1 < |G| - 1$ has a cut of size at most $(8e/d)n^{1-1/d}\ln(n/2)$ which separates the graph into two equal parts. It is at least desirable that they have maximum connectivity $\kappa(X) = |S| - 1$ (we stick to our convention that $0 \in S$.) By Mann's Theorem 4.3, or equivalently by Theorem 3.1, the connectivity of a loop network $X = \mathrm{Cay}(\mathbf{Z}_n, S)$ can be written as

$$\begin{aligned} \kappa(X) &= \min\{|S+H| - |H| : H < \mathbf{Z}_n, S+H \ne \mathbf{Z}_n\} \\ &= \min\{h(|\sigma(S)| - 1) : h|n, h|\sigma(S)| \ne n\} \end{aligned}$$

where $\sigma : \mathbf{Z}_n \to \mathbf{Z}_n/H$ is the natural projection, a result rediscovered by several authors. This gives a simple criteria on $S \subset \mathbf{Z}_n$ to check for $\kappa(X) = |S| - 1$: For each divisor h of n the number of pairwise distinct residues of S modulo h must be at least $\min\{n/h, (|S| + h - 1)/h\}$.

The next property studied by network specialists is superconnectivity. In a maximally connected regular graph the set of vertices adjacent from or to a given vertex is a minimum cutset. A graph is *superconnected* when all minimum cutsets are of this form. When $\kappa_2(X) > \kappa(X)$ this is certainly the case since the neighborhood of every set of cardinality at least two whose neighborhood leaves out at least two vertices is not a minimum cutset. Thus, the characterization of abelian Cayley graphs which are superconnected is equivalent to the characterization of sets $S \subset G$ such that

$$|T+S| = |T|+|S|-1 < |G|-1 \text{ for some } T \subset G \text{ with } |T| \geq 2.$$

This characterization was carried over by Hamidoune, Lladó and the author [42] by using the theorem of Kemperman and was later simplified by Hamidoune [40] in the form of Theorem 6.3. In [42] the characterization of abelian Cayley graphs which are maximally connected and every fragment has cardinality 1 or $|G|-1$ was also given. These graphs are called vosperian and they enjoy an additional good property for network models. By the theorem of Menger, given any two sets A and B of cardinality $|A| = |B| = k \leq \kappa(X)$, there is a set of k disjoint paths from A into B in X. A graph is vosperian if and only if this property holds for every pair of sets of cardinality $\kappa(X)+1$ whenever $A \notin \{N(x), x \in V(X)\}$ and $B \notin \{N^-(x), x \in V(X)\}$, where $N(x)$ and $N^-(x)$ denote the neighborhoods of a vertex in X and X^{-1} respectively.

For non-abelian Cayley graphs general results are less known. In the context of networks some additional conditions are usually required on the generating set S. When S is a minimal generating set Godsil [31] proved that, for $X = \mathrm{Cay}(G, S \cup S^{-1})$, we have $\kappa(X) = d(X)$. This result was extended in [44] to $\kappa_2(X) = d_2(X)$ when $d(X) > 4$. More specifically, the following result was proved.

Theorem 9.7 *Let S be a minimal generating set of a finite group G and let $X = \mathrm{Cay}(G, S \cup S^{-1})$. For each $x \in G \setminus \{1\}$ denote by $S(x) = \{s \in S : s^2 = x\}$ and let $m = \max\{|S(x)|, x \in G \setminus \{1\}$. Then $\kappa_2(X) = 2(d(X)-1)$ except in the following cases.*

(i) $S = \{s,t\}$ with $s^3 = t^3 = 1$ and $(st^{-1})^2 = 1$, in which case $\kappa_2(X) = d(X) = 4$.

(ii) $m \geq 4$, in which case $\kappa_2(X) = 2d(X) - m$.

(iii) $m \leq 3$ and there are $s,t \in S$ with $s^2 = t^2 \neq 1$ and either $st = ts$ or $st = ts^{-1}$. In this case $\kappa_2(S) = 2d(X) - 4$.

(iv) S is not as in the above cases and either $m = 3$ or there is $s \in S$ with $s^3 = 1$. In this case, $\kappa_2(S) = 2d(X) - 3$.

The proof of Theorem 9.7 relies on the fact that a 2-atom of the graph is either a periodic subset or has cardinality 2.

Acknowledgements

I would like to thank Y.O. Hamidoune and G. Zémor for fruitful discussions in the development of the research surveyed in this paper. I am also very grateful to J. Nešetřil for helpful discussions in the preparation of this manuscript and to Simeon Ball for his careful reading of the preliminary version. I also thank an anonymous referee for his/her valuable comments and remarks.

References

[1] N. Alon, M. Nathanson and I. Ruzsa, A polynomial method and restricted sums of congruence classes, *J. Number Theory* **56** (1996), 404–417.

[2] Z. Arad, E. Fisman and M. Muzychuk, Order Evaluation of Products of Subsets in Finite Groups and Its Applications, I, *Transactions of the american Math. Soc.* **349** (1997), 4401–4414.

[3] Z. Arad and M. Muzychuk, Order Evaluation of Products of Subsets in Finite Groups and Its Applications, II , *J. Algebra* **182** (1996), 577–603.

[4] Y.F. Bilu, V.F. Lev and I.Z. Ruzsa, Rectification principles in Additive Number Theory, *Discrete Comput. Geom.* **19** (1998), 343–353.

[5] L.V. Brailovsky and G.A. Freiman, On a product of finite subsets in a Torsion-free group, *J. Algebra* **130** (1990), 462–476.

[6] L. Cacceta and R. Häggkvist, On minimal digraphs of given girth, in *Proceedings of the Ninth Southeastern Conference on Combinatorics, Graph Theory and Computing, Winnipeg, Utilitas Math.* (1978), pp. 181–187.

[7] A. Cauchy, Recherches sur les nombres, *J. Ecole Polytechnique* **9** (1813), 99–116.

[8] M. C. Chang, A polynomial bound in Freimans theorem, *Duke Math. J.* **113** (2002), 399-419.

[9] S. Chowla, A theorem on the addition of residue classes: applications to the number $\Gamma(k)$ in Waring's problem , *Proc. Indian Acad. Sc.* **2** (1935), 242–243.

[10] S. Chowla, H.B. Mann and E.G. Straus, Some applications of the Cauchy–Davenport theorem, *Norske Vid. Selsk. Forh. (Trondheim)* **32** (1959), 74–80.

[11] H. Davenport, On the addition of residue classes, *J. London Math. Soc.* **10** (1935), 30–32.

[12] J.M. Deshouillers and G.A. Freiman, A step beyond Kneser's theorem for abelian finite groups, *Proc. London Math. Soc.* **86** (2003), 1-28.

[13] J.M. Deshouillers, B. Landrau and A. Yudin, Eds., *Structure Theory of Set Addition*, Societé Mathématique de France, Astérisque **258** (1999).

[14] J.A. Dias da Silva and Y.O. Hamidoune, Cyclic spaces for Grassmann derivatives and additive theory, *Bull. London Math. Soc.* **26** (1994), 140–146.

[15] G.T. Diderrich, An addition theorem for abelian groups of order pq, *J. Number Theory* **7** (1975), 33–48.

[16] J. Dixmier, Proof of a conjecture of Erdős and Graham concerning the problem of Frobenius, *J. Number Theory* **34** (1990), 198–209.

[17] Z. Du and F. Hsu Eds., *Combinatorial Network Theory*, Kluwer Academic Publishers, (1996).

[18] S. Eliahou and M. Kervaire, Sumsets in vector spaces over finite fields, *J. Number Theory* **71** (1998), 12–39.

[19] S. Eliahou and M. Kervaire, Restricted sums of sets of cardinality $1 + p$ in a vector space over $\mathbf{F}_p$, *Discrete Math.* **235** (2001), 199–213.

[20] S. Eliahou, M. Kervaire and A. Plagne, Optimally small sumsets in finite Abelian groups, *J. Number Theory* **101** (2003), 338–348.

[21] P. Erdős, Problems and results in combinatorial number theory, in *A Survey of Combinatorial Number Theory* (eds. J.N. Srivastava et al.), North Holland, (1973), pp. 117–138.

[22] P. Erdős and R.L. Graham, On a linear diophantine problem of Frobenius, *Acta. Arith.* **21** (1972), 399–408.

[23] P. Erdős and H. Heilbronn, On the addition of residue classes mod p, *Acta. Arith.* **21** (1964), 149–159.

[24] G.A. Freiman, On the addition of finite sets I., *Izv. Vyssh. Uchebn. Zaved. Mat.* **13** (1959), 202–213.

[25] G.A. Freiman, Inverse problems of additive number theory. On the addition of sets of residues with respect to a prime modulus, *Soviet. Math. Doklady* **2** (1961), 1520–1522.

[26] G.A. Freiman, Inverse problems of additive number theory VI. On the addition of finite sets III. The method of trigonometric sums, *Izv. Vyssh. Uchebn. Zaved. Mat.* **28** (1962), 151–157.

[27] G.A. Freiman, *Foundations of a structural theory of set addition*, Translations of Mathematical Monographs, AMS, Vol. 37, Providence (1973).

[28] G.A. Freiman, What is the structure of K if $K + K$ is small, in *Lecture Notes in Mathematics 1240* Springer-Verlag, New York-Berlin (1987), pp. 109–134.

[29] W. Gao and Y.O. Hamidoune, On additive basis, *Acta Arith.* **88** (1999), 233–237.

[30] W. Gao, Y.O. Hamidoune, A.S. Lladó and O. Serra, Covering an abelian group by subset sums, *Combinatorica* **23** (2003), 599–611.

[31] C.D. Godsil, Connectivity of minimal Cayley graphs, *Arch. Math. (Basel)* **37** (1981), 437–476.

[32] B.J. Green, Structure Theory of Set Addition, Notes for ICMS Instructional Conference in Combinatorial Aspects of Mathematical Analysis, Edinburgh, 2002.

[33] H. Halberstam and K.F. Roth, *Sequences*, Springer-Verlag, (1982).

[34] Y.O. Hamidoune, Sur les atomes d'un graphe orienté, *C.R. Acad. Sci. Paris* **284** (1977), 1253–1256.

[35] Y.O. Hamidoune, Quelques problémes de connexité dans les graphes orientés, *J. Combin. Theory, Ser B* **30** (1981), 1–11.

[36] Y.O. Hamidoune, An application of connectivity theory in graphs to factorization of elements in groups, *European J. Combin.* **2** (1981), 349–355.

[37] Y.O. Hamidoune, Additive group theory applied to network topology, in *Combinatorial Network Theory* (eds. D. Z. Du and F. Hsu), Kluwer Academic Publishers, (1996), pp. 1–39.

[38] Y.O. Hamidoune, An isoperimetric method in additive Theory, *J. Algebra* **179** (1996), 622–630.

[39] Y.O. Hamidoune, On inverse additive problems, Rapport de Recherche EC 95/01, Institute Blaine Pascal, Paris, 1995.

[40] Y.O. Hamidoune, On subsets with a small sum in abelian groups, *Europ. J. Combinatorics* **18** (1997), 541–556.

[41] Y.O. Hamidoune, Some results in Additive Number Theory I: The critical pair Theory, *Acta Arith.* **96** (2000), 97–119.

[42] Y.O. Hamidoune, A.S. Lladó and O. Serra, Vosperian and superconnected abelian Cayley digraphs, *Graphs Combin.* **7** (1991), 143–152.

[43] Y.O. Hamidoune, A.S. Lladó and O. Serra, Minimum order of loop networks with given degree and girth, *Graphs Combin.* **11** (1995), 131–138.

[44] Y.O. Hamidoune, A.S. Lladó and O. Serra, On subsets with small product in torsion-free groups, *Combinatorica* **18** (1998), 529–540.

[45] Y.O. Hamidoune, A.S. Lladó and O. Serra, An isoperimetric problem in Cayley graphs, *Theory of Comput. Systems* **18** (1999), 507–516.

[46] Y.O. Hamidoune, A.S. Lladó and O. Serra, On sets with small subset sum, *Combin. Probab. Comput.* **8** (1999), 461–466.

[47] Y.O. Hamidoune and A. Plagne, A generalization of Freiman's $3k-3$ Theorem, *Acta Arith.* **103** (2002), 147–156.

[48] Y.O. Hamidoune and O. Rødseth, An inverse theorem modulo p, *Acta Arith.* **92** (2000), 251–262.

[49] Y.O. Hamidoune and O. Serra, On small cuts separating an abelian Cayley graph into two equal parts, *Math. Systems Theory* **29** (1996), 407–409.

[50] Y.O. Hamidoune, O. Serra and G. Zémor, On the critical pair theory in $\mathbf{Z}_p$, preprint, 2004.

[51] Y.O. Hamidoune, O. Serra and G. Zémor, On the critical pair theory in abelian groups: Beyond Chowla's theorem, preprint, 2004.

[52] Y.O. Hamidoune and G. Zémor, On zero-free subset sums, *Acta Arith.* **78** (1996), 143–152.

[53] M.C. Heydemann, Cayley graphs and interconnection networks, in *Graph symmetry* (eds. G. Hahn and G. Sabisdussi), *NATO Adv. Sci. Inst. Ser.C Math. Phys. Sci.*, 497, Kluwer Acad. Publ., Dordrecht (1997), pp. 167-224.

[54] V. Jungić, J. Licht, M. Mahdian, J. Nešetřil and R. Radoičić, Rainbow arithmetic progressions and anti-Ramsey results, *Combin. Probab. Comput.* **12** (2003), 599–620.

[55] Gy. Károlyi, The Erdős-Heilbronn problem in abelian groups, *Israel J. Math.* **139** (2004), 349-359.

[56] Gy. Károlyi, A compactness argument in the additive theory and the polynomial method, *Discrete Math.*, in press.

[57] Gy. Károlyi, An inverse theorem for the restricted set addition in abelian groups, preprint, 2004.

[58] J.H.B. Kempermann, On complexes in a semigroup, *Indag. Math.* **18** (1956), 247–254.

[59] J.H.B. Kempermann, On small sumsets in abelian groups, *Acta Math.* **103** (1960), 66–88.

[60] M. Kneser, Ein Satz über abelschen Gruppen mit Anwendungen auf die Geometrie der Zahlen, *Math. Z.* **61** (1955), 429–434.

[61] M. Kneser, Summenmengen in lokalkompakten abelesche Gruppen, *Math. Z.* **66** (1956), 88–110.

[62] V.F. Lev, Structure theorem for multiple addition and the Frobenius problem, *J. Number Theory* **58** (1996), 79–88.

[63] V.F. Lev, On small sumsets in abelian groups, in *Structure Theory of Set Addition* (eds. J.M. Deshouillers, B. Landrau and A. Yudin), *Astérisque*, 258, Societé Mathématique de France, Paris (1999), pp. 317–321.

[64] V. Lev and P. Smeliansky, On addition of two distinct sets of integers, *Acta Arith.* **70** (1995), 85–91.

[65] M. Lewin, A bound for a solution of a diophantine problem, *London Math. Soc.* **6** (1972), 61–69.

[66] W. Mader, Eine Eigenschaft der Atome endlicher Graphen, *Arch. Math.* **22** (1971), 333–336.

[67] H.B. Mann, A proof of the fundamental theorem on the density of sums of sets of positive integers, *Ann. Math.* **43** (1942), 523–527.

[68] H.B. Mann, An addition theorem for sets of elements of an abelian group, *Proc. Amer. Math. Soc.* **4** (1953), 423–427.

[69] H.B. Mann, *Addition theorems: The addition theorems of group theory and number theory*, Interscience, New York (1965).

[70] M.B. Nathanson, *Additive number theory: Inverse problems and the Geometry of sumsets*, Springer-Verlag GTM 165, (1996).

[71] J.E. Olson, An addition theorem mod p, *J. Combin. Theory* **5** (1968), 45–52.

[72] J.E. Olson, Sums of sets of group elements, *Acta Arith.* **28** (1975), 147–156.

[73] J.E. Olson, On the symmetric difference of two sets in a group, *European J. Combin.* **7** (1986), 43–54.

[74] I. Ruzsa, Arithmetic progressions in sumsets, *Acta Arith.* **60** (1991), 191–202.

[75] I. Ruzsa, Generalized arithmetic progressions and sumsets, *Acta Math. Hungar.* **65** (1994), 379–390.

[76] O. Serra and G. Zémor, On a generalization of a theorem by Vosper, *Integers* **0** (2000), 10 p., electronic only.

[77] J.C. Shepherdson, On the addition of elements of a subsequence, *J. Lond. Math. Soc.* **22** (1947), 85–88.

[78] J. Steinig, On Freiman's theorems concerning the sum of two sets of integers, *Astérisque* **258** (1999), 129–140.

[79] E. Szemerédi, On a conjecture of Erdős and Heilbronn, *Acta Arith.* **17** (1970), 227–229.

[80] A. Tietäväinen, On diagonal forms over finite fields, *Ann. Univ. Turku Ser. A* ((),1968) 1–10.

[81] G. Vosper, The critical pairs of subsets of a group of prime order, *J. London Math. Soc.* **31** (1956), 200–205.

[82] G. Zémor, Subset sums in binary spaces, *Europ. J. Combinatorics* **13** (1992), 221–230.

[83] G. Zémor, A generalisation to noncommutative groups of a theorem of Mann, *Discrete Math.* **126** (1994), 365–372.

Dept. Matemàtica Aplicada IV
Universitat Politècnica de Catunya
Jordi Girona, 1
08034 Barcelona Spain
oserra@mat.upc.es

The structure of claw-free graphs

Maria Chudnovsky and Paul Seymour

Abstract

A graph is *claw-free* if no vertex has three pairwise nonadjacent neighbours. At first sight, there seem to be a great variety of types of claw-free graphs. For instance, there are line graphs, the graph of the icosahedron, complements of triangle-free graphs, and the Schläfli graph (an amazingly highly-symmetric graph with 27 vertices), and more; for instance, if we arrange vertices in a circle, choose some intervals from the circle, and make the vertices in each interval adjacent to each other, the graph we produce is claw-free. There are several other such examples, which we regard as "basic" claw-free graphs.

Nevertheless, it is possible to prove a complete structure theorem for claw-free graphs. We have shown that every connected claw-free graph can be obtained from one of the basic claw-free graphs by simple expansion operations. In this paper we explain the precise statement of the theorem, sketch the proof, and give a few applications.

1 Introduction

A graph is *claw-free* if no vertex has three pairwise nonadjacent neighbours. (Graphs in this paper are finite and simple.) Line graphs are claw-free, and it has long been recognized that claw-free graphs are an interesting generalization of line graphs, sharing some of the same properties. For instance, Minty [16] showed in 1980 that there is a polynomial-time algorithm to find a stable set of maximum weight in a claw-free graph, generalizing the algorithm of Edmonds [9, 10] to find a maximum weight matching in a graph.

How do we construct claw-free graphs? Chvátal and Sbihi [8] and Maffray and Reed [15] studied the structure of claw-free *perfect* graphs, and indeed, it was working on an extension of their results that led us to the present project. But how can we construct the "most general" claw-free graph? This question had not been studied, as far as we know, and yet it turns out to be a very good question. There are several basic types of claw-free graphs, and we were able to show that every connected claw-free graph can be obtained starting from a graph of one of these basic types by means of "expansion" operations (or, in some restricted cases, piecing some of these basic graphs together).

The main goal of this paper is to explain this construction. We are preparing a series of about seven papers containing the results sketched here, but the titles (and the order) of these papers given in the references are provisional, and are currently being revised. In addition, some of the results quoted here are still in the form of notes and have not yet been written down formally, much less been refereed; and while we have tried hard to make sure that the theorems quoted here are true, until they are written down properly we cannot be completely sure. The reader is warned to check with the full published version of these results before relying on them completely.

Before we start to explain the structure theorem, let us introduce "antiprismatic" graphs. We say a graph G is *prismatic* if for every triangle T of G, every vertex of G

not in T has a unique neighbour in T (a *triangle* means a 3-vertex clique). Prismatic graphs have a complex structure, but it turns out that they can be completely described (see sections 6, 7). We say G is *antiprismatic* if its complement graph $\overline{G}$ is prismatic. Clearly antiprismatic graphs are claw-free, and they seem to form a subclass of claw-free graphs that is very different from the others. (At least, all our standard methods for analyzing claw-free graphs failed completely when we reached the stage of trying to analyze antiprismatic graphs, and we had to come up with totally new approaches.) Understanding antiprismatic graphs was probably the most difficult part of the project.

The Schläfli graph (a very interesting and highly symmetric graph, described later) is antiprismatic, and it and its induced subgraphs (and some other graphs derived from it by expanding vertices) form a class of antiprismatic graphs, one of (about) eight classes that we need. We showed that every antiprismatic graph either belongs to one of four of these classes, or could be constructed from a sequence of members of the other four classes by repeated hex-joins.

The most important result in this paper is of course the general structure theorem for claw-free graphs. This is explained in detail in sections 3–7, but let us give some idea of it now. We can confine ourselves to connected claw-free graphs G, and it is convenient also to assume that G admits no "homogeneous pair of cliques", which we explain in the next section. Then we find that the type of structure that G possesses depends heavily on $\alpha(G)$, the size of the largest stable set of G. When $\alpha(G) \geq 4$, it turns out that G is either a kind of generalized line graph, or a circular interval graph. When $\alpha(G) = 3$, there are several additional possibilities; for instance, that either G is a subgraph of the icosahedron, or G is expressible as a "hex-join" (explained later), or G is antiprismatic.

Unfortunately the result is rather complicated, and as a warmup we first discuss what are called "quasi-line graphs". A graph G is a *quasi-line graph* if for every vertex v, the set of neighbours of v can be expressed as the union of two cliques. (A *clique* in G is a set of pairwise adjacent vertices of G.) Note that we do not require that two neighbours of v are adjacent if and only if they both belong to one of the cliques; there may be edges between neighbours of v that do not belong to the same clique. Quasi-line graphs are clearly claw-free; they form a proper subset of the set of all claw-free graphs, and a proper superset of the set of all line graphs, and make an interesting half-way stage. We found a structure theorem for all quasi-line graphs, and for that question the answer is much prettier than for general claw-free graphs.

One reason for interest in quasi-line graphs is a problem of Ben Rebea [1]. Since Edmonds' matching algorithm generalizes to claw-free graphs, one might hope that also Edmonds' matching polytope theorem [9] extends to claw-free graphs, and several people [12, 13, 14, 17] have worked on this, although it remains open. We are asking for the list of linear inequalities defining the convex hull of the stable sets of G, where we regard a stable set of a $(0, 1)$-vector in $\Re^{V(G)}$. For general claw-free graphs G there is not even a conjecture, but Ben Rebea suggested the same problem for quasi-line graphs, and for that class there is a conjectured answer [17]. We have been able to find the desired list of inequalities for all connected quasi-line graphs that are *not* of one particular type, graphs we call "fuzzy circular interval graphs"; and in a still more recent paper, Eisenbrand, Oriolo, Stauffer and Ventura [11] solved the case that we left open. This is explained in section 2.

For a graph G, let $\chi(G), \omega(G)$ denote the chromatic number of G and the maximum size of a clique of G. Another result that we might hope to extend from line graphs to general claw-free graphs is Vizing's theorem [18], which implies that if G is a line graph (of a simple graph), then $\chi(G) \leq \omega(G) + 1$. For general claw-free graphs (or even for line graphs of non-simple graphs) this is false, although Shannon's theorem implies that $\chi(G) < \frac{3}{2}\omega(G) + 1$ for line graphs. Note also that for a n-vertex graph G with $\alpha(G) \leq 2$, even the linear bound is false; $\chi(G) \geq n/2$, and yet $\omega(G)$ may be $o(n)$. Nevertheless, for connected claw-free graphs with $\alpha(G) \geq 3$, the structure is much more controlled, and we were able to show that for any such graph, $\chi(G) \leq 2\omega(G)$ (and this is asymptotically best possible). We discuss colouring claw-free graphs in section 8.

2 Quasi-line graphs

Construct a graph G as follows. Take a circle C, and let $V(G)$ be a finite set of points of C. Take a set of intervals from C (an *interval* means a proper subset of C homeomorphic to $[0, 1]$); and say that $u, v \in V(G)$ are adjacent in G if $\{u, v\}$ is a subset of one of the intervals. We call such a graph a *circular interval graph.* All such graphs are claw-free, and indeed they are quasi-line graphs, as is easily seen. (These are a subclass of class of circular arc graphs; they are sometimes called "proper" circular arc graphs.) *Linear interval graphs* are defined in the same way, taking C to be a line instead of a circle. Every linear interval graph is also a circular interval graph.

There is another way to construct quasi-line graphs, that we explain next. A vertex $v \in V(G)$ is *simplicial* if the set of neighbours of v is a clique. A *strip* (G, a, b) consists of a claw-free graph G together with two designated simplicial vertices a, b called the *ends* of the strip. For instance, if G is a linear interval graph, with vertices $v_1, \ldots, v_n$ in order and with $n > 1$, then v_1, v_n are simplicial, and so (G, v_1, v_n) is a strip, called a *linear interval strip.*

Suppose that (G, a, b) and (G', a', b') are two strips. We compose them as follows. Let A, B be the set of vertices of $G \setminus \{a, b\}$ adjacent in G to a, b respectively, and define A', B' similarly. Take the disjoint union of $G \setminus \{a, b\}$ and $G' \setminus \{a', b'\}$; and let H be the graph obtained from this by adding all possible edges between A and A' and between B and B'. Then H is claw-free.

This method of composing two strips is symmetrical between (G, a, b) and (G', a', b'), but we do not use it in a symmetrical way. We use it as follows. Start with a graph G_0 with an even number of vertices and which is the disjoint union of complete graphs, and pair the vertices of G_0. Let the pairs be $(a_1, b_1), \ldots, (a_n, b_n)$, say. For $i = 1, \ldots, n$, let (G'_i, a'_i, b'_i) be a strip. For $i = 1, \ldots, n$, let G_i be the graph obtained by composing (G_{i-1}, a_i, b_i) and (G'_i, a'_i, b'_i); then the resulting graph G_n is called a *composition* of the strips (G'_i, a'_i, b'_i) $(1 \leq i \leq n)$. For instance, if for each of the strips (G'_i, a'_i, b'_i), G'_i is a 3-vertex path with ends a'_i, b'_i, then the effect of composing with (G'_i, a'_i, b'_i) is the identification of a_i, b_i; and so the graphs that are compositions of such 3-vertex path strips are precisely line graphs.

It is easy to check that every graph that is the composition of linear interval strips is a quasi-line graph, so this gives us a second construction for quasi-line graphs (and this includes line graphs, since the 3-vertex strip mentioned above is a linear interval strip). This is not quite the whole story for quasi-line graphs yet; we need one more concept.

A *homogeneous pair of cliques* in G is a pair (A, B) such that:

- A, B are cliques in G and $A \cap B = \emptyset$,
- no vertex of $G \setminus (A \cup B)$ has both a neighbour and a non-neighbour in A, and the same for B, and
- either $|A| \geq 2$ or $|B| \geq 2$.

Now we can state one version of our structure theorem for quasi-line graphs, the following.

2.1 *For every quasi-line graph G, if G is connected and there is no homogeneous pair of cliques in G, then either G is a circular interval graph, or G is a composition of linear interval strips.*

One might object that this is not quite a structure theorem for all quasi-line graphs. The hypothesis that G is connected is unobjectionable, because if we understand the possibilities for the connected components of a quasi-line graph, then we understand the possibilities for the entire graph; but the same is not true for the "homogeneous pair" hypothesis. Suppose that we wish to understand the structure of a connected quasi-line graph that does admit a homogeneous pair of cliques. We could delete all except one vertex from both of the cliques, and iterate (if there is still a homogeneous pair of cliques, do it again), until 2.1 can be applied; and then add back in all the homogeneous pairs we deleted. But it is not so easy to see how to describe the global structure that results. Below we give a more explicit description.

First, we need to extend the concept of a circular interval graph. We say that a graph G is a *fuzzy circular interval graph* if:

- there is a map ϕ from $V(G)$ to a circle C (not necessarily injective), and
- there is a set of intervals from C, none including another, and such that no point of C is an end of more than one of the intervals, so that
- for u, v in G, if u, v are adjacent then $\{u, v\}$ is a subset of one of the intervals, and if u, v are nonadjacent then u, v are both ends of any interval including both of them (and in particular, if $\phi(u) = \phi(v)$ then u, v are adjacent).

(If also we required ϕ to be injective, this would be equivalent to the definition of a circular interval graph.) If x, y are ends of an interval and one of the sets $\phi^{-1}(x), \phi^{-1}(y)$ has at least two members, then the pair $(\phi^{-1}(x), \phi^{-1}(y))$ is a homogeneous pair of cliques; and these turn out to be the only kinds of homogeneous pairs of cliques that we need. (Fuzzy linear interval strips are defined analogously, with the proviso that if a, b are the ends of the strip then $\phi(a), \phi(b)$ are different from $\phi(v)$ for all other vertices v of G.) The following is a more explicit version of the structure theorem for quasi-line graphs.

2.2 *For every quasi-line graph G, if G is connected, then either G is a fuzzy circular interval graph, or G is a composition of fuzzy linear interval strips.*

The current proof of 2.2 is indirect. First we apply the general structure theorem for claw-free graphs, described later; this tells us that our quasi-line graph belongs to one of several basic classes, or is an expansion of such a graph. Then we examine each of these classes separately, to figure out which quasi-line graphs it contains. The most difficult class is the one where we get least information, the class of graphs G with $\alpha(G) \leq 2$; these graphs are always claw-free, but are not necessarily quasi-line graphs, and perhaps half or more of the entire proof is spent on this case. We omit further details, which will appear in a separate paper [6].

Let us turn to the application to Ben Rebea's question, mentioned in the introduction. Let G be a graph, and for $X \subseteq V(G)$, let $\underline{X} \in \Re^{V(G)}$ be the vector where for $v \in V(G)$, $\underline{X}_v = 1$ if $v \in X$ and $\underline{X}_v = 0$ if $v \notin X$. The *stable set polytope* of G is the convex hull of all the vectors $\underline{X}$ such that $X \subseteq V(G)$ is stable. Every point p of the stable set polytope satisfies the following inequalities:

- $p_v \geq 0$ for all $v \in V(G)$
- $\sum_{v \in K} p_v \leq 1$ for every clique K of G
- for every odd list $K_1, \ldots, K_{2n+1}$ of cliques of G, if Y denotes the set of all vertices in at least two of $K_1, \ldots, K_{2n+1}$, then $\sum_{v \in Y} p_v \leq n$.

(To see this, note that since all the inequalities are linear, it suffices to check that they holds when $p = \underline{X}$ for a stable set X.) Let us call these *Edmonds' inequalities.* When G is a line graph, Edmonds' matching polytope theorem [9] asserts that a point $p \in \Re^{V(G)}$ belongs to the stable set polytope if and only if it satisfies Edmonds' inequalities. For general claw-free graphs this is not true, but the problem seems more tractable for quasi-line graphs, and in [17] there is a conjecture presenting an alternative list of inequalities that may be necessary and sufficient when G is a quasi-line graph. We were able to show the following:

2.3 *Let G be a connected quasi-line graph, that is not a fuzzy circular interval graph. Then a point* $p \in \Re^{V(G)}$ *belongs to the stable set polytope if and only if it satisfies Edmonds' inequalities.*

To prove this, we observe first that 2.2 implies that G is a composition of fuzzy linear interval strips. Each of the strips can be adequately simulated by an appropriate 5-vertex strip; if we make the corresponding composition of these simplified strips, we obtain a line graph, for which Edmonds' theorem gives the stable set polytope; and now we replace the simplified strips by the original strips, one by one, and check the effect on the stable set polytope at each step. We omit further details.

More recently, Eisenbrand, Oriolo, Stauffer and Ventura [11] have solved the same question for the case that was still open, for fuzzy circular interval graphs, and hence completed the answer to Ben Rebea's question.

3 Claw-free graphs with $\alpha(G) \geq 4$

Let us begin on the structure theorem for general claw-free graphs. In this section we consider the case of claw-free graphs G with $\alpha(G) \geq 4$. One way to make such a graph is to take the disjoint union of two claw-free graphs G_1, G_2 neither of which is complete, and we cannot hope yet to explain the structure of graphs constructed this way in any finer detail, because at this stage we know nothing about G_1, G_2 (since they might not satisfy $\alpha(G_1), \alpha(G_2) \geq 4$). Thus we had better confine ourselves to connected graphs.

Another way one might try to confound the attempt to describe all connected claw-free graphs G with $\alpha(G) \geq 4$ is the following. For $i = 1, 2$, let a_i be a simplicial vertex of a connected claw-free graph G_i; and construct G as follows. G is obtained from the disjoint union of $G_1 \setminus a_1$ and $G_2 \setminus a_2$ by making every neighbour of a_1 in G_1 adjacent to every neighbour of a_2 in G_2. (This is a version of the strip combinations of the previous section, except now we are just using one simplicial vertex instead of two.) The graph G we produce is claw-free and connected (except in trivial cases), and may well satisfy $\alpha(G) \geq 4$, even if $\alpha(G_1) \leq 3$ or $\alpha(G_2) \leq 3$ (or both). So to describe all connected claw-free graphs with $\alpha(G) \geq 4$, we will also need to describe all connected claw-free graphs G_1 with $\alpha(G_1) \leq 3$ that have a simplicial vertex. Curiously, this can be done; almost all the types of connected claw-free graphs G_1 with $\alpha(G_1) \leq 3$ cannot have simplicial vertices, and we can explicitly describe those that do. Nevertheless, to simplify the presentation here, we prefer to avoid this step.

Instead, we assume that our graph G cannot be constructed from graphs G_1, G_2 in the way just described. More precisely, let us say that G admits a *1-join* if $V(G)$ can be partitioned into four sets A_1, B_1, A_2, B_2, where $A_1 \cup A_2$ is a clique, B_1, B_2 are nonempty, and there are no edges between $A_1 \cup B_1$ and $A_2 \cup B_2$ except those between A_1, A_2. Except in trivial cases, the claw-free graphs that are expressible as compositions in the way described earlier in this section are precisely the claw-free graphs that admit 1-joins. So henceforth in this section we assume that G does not admit a 1-join. (Note that this implies that G is connected.)

For such graphs we can describe the structure completely; there is a result analogous to 2.2, except that we need two new kinds of strips, the following:

- Let G be the graph with vertex set $\{v_1, \ldots, v_{13}\}$ and with adjacency as follows. v_1-$\cdots$-v_6 is a hole in G of length 6. Next, v_7 is adjacent to v_1, v_2; v_8 is adjacent to v_4, v_5; v_9 is adjacent to v_6, v_1, v_2, v_3; v_{10} is adjacent to v_3, v_4, v_5, v_6, v_9; v_{11} is adjacent to $v_3, v_4, v_6, v_1, v_9, v_{10}$; v_{12} is adjacent to $v_2, v_3, v_5, v_6, v_9, v_{10}$; and v_{13} is adjacent to $v_1, v_2, v_4, v_5, v_7, v_8$. Let $X \subseteq \{v_{11}, v_{12}, v_{13}\}$; then the strip $(G \setminus X, v_7, v_8)$ is called an *XX-strip*.

- Let $n \geq 0$. Let $A = \{a_0, a_1, \ldots, a_n\}, B = \{b_0, b_1, \ldots, b_n\}$ and $C = \{c_1, \ldots, c_n\}$ be three cliques, pairwise disjoint. Let G be the graph with vertex set $A \cup B \cup C$ and with adjacency as follows. For $0 \leq i, j \leq n$, let a_i, b_j be adjacent if and only if $i = j > 0$, and for $1 \leq i \leq n$ and $0 \leq j \leq n$ let c_i be adjacent to a_j, b_j if and only if $i \neq j \neq 0$. Let $X \subseteq A \cup B \cup C$ with $a_0, b_0 \notin X$; then the strip $(G \setminus X, a_0, b_0)$ is called an *antihat strip*.

One version of the structure theorem in this case is the following:

3.1 *For every claw-free graph G with $\alpha(G) \geq 4$, if G does not admit a 1-join and there is no homogeneous pair of cliques in G, then either G is a circular interval graph, or G is a composition of linear interval strips, XX-strips, and antihat strips.*

The foregoing is the analogue of 2.1. There is also an analogue of 2.2, using "fuzzy" XX-strips and antihat strips, but we omit the details. The proof of 3.1 is long, about 100 pages or so, and most of it appears in [5]. We sketch some of the proof later in the paper.

4 The case $\alpha(G) = 3$

What about an analogue of 3.1 for claw-free graphs G with $\alpha(G) = 3$? There are several extra complications now. First, we have to remember the icosahedron and its induced subgraphs; they are claw-free, and not all accounted for yet. (And there are more graphs of this type to be listed.) Second, and much worse, there are the antiprismatic graphs (they require a couple of sections of their own—see sections 6, 7). Third, there is another composition operation that shows up now (and the only explanation we see for some of these graphs is that they are compositions of smaller graphs in the same class, so the structure theorem seems to need to use the new composition).

The composition is as follows. For $i = 1, 2$, let G_i be claw-free and non-null, and let A_i, B_i, C_i be disjoint cliques of G_i with union $V(G_i)$. Let G be the graph obtained from the disjoint union of G_1, G_2 by making every vertex of G_1 adjacent to every vertex of G_2 except that there are no edges between A_1 and A_2, between B_1 and B_2, and between C_1 and C_2. It is easy to see that G is claw-free. We say that G is a *hex-join* of G_1, G_2. Note that if G admits a hex-join, then the sets $A_1 \cup B_2$, $B_1 \cup C_2$ and $C_1 \cup A_2$ are three cliques with union $V(G)$, and consequently no graph G with $\alpha(G) > 3$ is expressible as a hex-join.

If $V(G)$ is not the union of three cliques (that is, if $\chi(\overline{G}) \geq 4$) then G is not expressible as a hex-join; and as for graphs G with $\chi(\overline{G}) \leq 3$, while they might admit hex-joins, the building blocks from which they are made are severely restricted. Thus there is an advantage to handling the two cases $\chi(\overline{G}) \geq 4$ and $\chi(\overline{G}) \leq 3$ separately.

For the first case, we have the following.

4.1 *For every claw-free graph G with $\alpha(G) \leq 3$ and $\chi(\overline{G}) \geq 4$, if there is no homogeneous pair of cliques in G, then either G is a circular interval graph, or G belongs to the class $\mathcal{S}_6$ (defined in the next section), or G is the graph of an XX-strip, or G is a composition of linear interval strips and antihat strips, or G is an induced subgraph of the icosahedron, or G is antiprismatic.*

We can tighten up the case when G is a composition of linear interval strips and antihat strips, but we omit those details for simplicity. There is also a "fuzzy" version of this, without the hypothesis that there is no homogeneous pair of cliques in G, but it is quite complicated and again we omit it.

If H is a graph, its line graph is denoted by $L(H)$. For the second case, we have:

4.2 *For every claw-free graph G with $\chi(\overline{G}) \leq 3$, if there is no homogeneous pair of cliques in G and G admits no hex-join, then either G is a circular interval graph, or a subgraph of $L(K_{3,n})$ for some n, or the graph of an antihat strip, or antiprismatic.*

There is more work to be done on this yet; this is still a decomposition theorem, and we need to convert it to a structure theorem. We think that every claw-free graph G with $\chi(\overline{G}) \leq 3$ can be built by a series of hex-joins, starting from graphs which are "fuzzy" versions of the graphs of 4.2; but these details are not yet worked out.

5 The decomposition theorem

To sketch the proofs of 3.1, 4.1 and 4.2, we need another definition. Suppose that V_0, V_1, V_2 are disjoint subsets with union $V(G)$, and for $i = 1, 2$ there are subsets A_i, B_i of V_i satisfying the following:

- for $i = 1, 2$, A_i, B_i are cliques, $A_i \cap B_i = \emptyset$ and A_i, B_i and $V_i \setminus (A_i \cup B_i)$ are all nonempty
- A_1 is complete to A_2, and B_1 is complete to B_2, and there are no other edges between V_1 and V_2, and
- V_0 is a clique; and for $i = 1, 2$, V_0 is complete to $A_i \cup B_i$ and anticomplete to $V_i \setminus (A_i \cup B_i)$.

In this case we say that G admits a *generalized 2-join*. Define classes $\mathcal{S}_0, \ldots, \mathcal{S}_6$ as follows.

- $\mathcal{S}_0$ is the class of all line graphs.
- $\mathcal{S}_1$ is the class of all induced subgraphs of the icosahedron.
- $\mathcal{S}_2$ is the class of all graphs of XX-strips.
- $\mathcal{S}_3$ is the class of all circular interval graphs.
- $\mathcal{S}_4$ is the class of all induced subgraphs of the graph G defined as follows. Let H be the graph with seven vertices $h_0, \ldots, h_6$, in which $h_1, \ldots, h_6$ are pairwise adjacent and h_0 is adjacent to h_1. Then G is obtained from $L(H)$ by adding one new vertex, adjacent precisely to the members of $V(L(H)) = E(H)$ that are not incident with h_1 in H.
- $\mathcal{S}_5$ is the class of graphs of antihat strips.
- $\mathcal{S}_6$ is the class of all induced subgraphs of the graph G defined as follows. Let $n \geq 0$, and let $V(G)$ be the disjoint union of sets A, B, C, D, where $A = \{a_1, \ldots, a_n\}$, $B = \{b_1, \ldots, b_n\}$, $C = \{c_1, \ldots, c_n\}$ and $D = \{d_1, d_2, d_3, d_4, d_5\}$. The edges of G are as follows.
  - A, B, C are cliques
  - a_i, b_i are adjacent for $1 \leq i \leq n$, and c_i is adjacent to a_j, b_j for $1 \leq i, j \leq n$ with $i \neq j$
  - d_1 is completely adjacent to $A \cup B \cup C$; d_2 is completely adjacent to $A \cup B \cup \{d_1\}$; d_3 is completely adjacent to $A \cup \{d_2\}$; d_4 is completely adjacent to $B \cup \{d_2, d_3\}$; and d_5 is adjacent to d_3, d_4.

The results 3.1, 4.1 and 4.2 are all consequences of the following decomposition theorem.

5.1 *Let G be claw-free and connected. Then either*

- $G \in \mathcal{S}_0 \cup \cdots \cup \mathcal{S}_6$, *or*
- *G admits either a homogeneous pair of cliques, a 1-join, a generalized 2-join, or a hex-join, or*
- *G is antiprismatic.*

This is the main theorem of [5]. The proof is very lengthy, about 100 pages. The idea of the proof is, first we choose an appropriate graph H, and assume that H is an induced subgraph of G; we analyze how the remainder of G can attach to H, and infer that either G admits one of the decompositions or falls into one of the classes. Then henceforth we can assume that G does not contain H, and we repeat for some other choice of H. But making the right sequence of choices for H is crucial, and took a great deal of experiment. The first choice should be that H is the icosahedron; then it is easy to prove that either $G = H$ or G admits one of the decompositions. The same works when H is the icosahedron with one vertex deleted. For technical reasons, it is then best to handle the case when H is the graph of an XX-strip (without the vertices called v_{11}, v_{12}, v_{13}). Now comes the first major step, that H is a line graph of a cyclically 3-connected graph, as large as possible; then as long as H is not too small, it is straightforward to analyze the structure of the remainder of G relative to H, and we can prove what we want. In particular, this works for "prisms" (two disjoint triangles joined by three disjoint paths) that are not too small, so henceforth we can assume that G contains no such prisms. Next we take H to be the largest induced cycle ("hole") in G, and assume it has length at least seven; since G contains no substantial prisms, G now tends to be a circular interval graph (unless it contains some other configurations that we have to handle), so we can assume all holes have length at most 6. And so on (and it gets worse from here)—we continue through a long series of such steps, assuming that G contains a certain subgraph H but does not contain any subgraph that we already handled. We omit further details.

Using 5.1 to deduce 3.1, 4.1 and 4.2 are rather delicate inductions, and again we omit the details.

6 Prismatic graphs—the non-orientable case

The results 3.1, 4.1 and 4.2 reduce the problem to that of understanding antiprismatic graphs. These graphs are very dense, and it seems advantageous now to work in terms of the complement graph; so we want to understand prismatic graphs. We recall that G is *prismatic* if for every triangle T of G, every vertex of G not in T has a unique neighbour in T.

If T_1, T_2 are two disjoint triangles of a prismatic graph G, then the edges between T_1 and T_2 provide a bijection from T_1 to T_2. We say G is *orientable* if there is a cyclic orientation of each triangle such that for every pair of disjoint triangles, the matching between them preserves the orientation, and *non-orientable* otherwise. It is helpful to divide the problem into two subproblems, the orientable and non-orientable cases. In this section we discuss the non-orientable case.

Before we go on, we need some more definitions. Let us say the *core* of G is the subgraph induced on the union of all triangles in G. If the core is 3-colourable then G is orientable, as is easily seen. Let G be a prismatic graph; here are some constructions that lead to more prismatic graphs. First, let T be a triangle of G, say $T = \{a_1, b_1, c\}$. We say T is a *leaf triangle at* c if a_1, b_1 both only belong to one triangle (namely, T). If this is so, let H be the graph obtained from G by adding new vertices $a_2, \dots, a_k, b_2, \dots, b_k$, where for $2 \le i \le k$, a_i has the same neighbours as a_1 in $G \setminus \{a_1, b_1, c\}$, and similarly b_i has the same neighbours as b_1, and a_i, b_i are adjacent. Then H is prismatic, and we say it is obtained from G by *multiplying a leaf triangle.*

Second, let C be the core of G, and let $V(G) \setminus V(C) = \{a_1, \dots, a_n\}$. For $1 \le i \le n$, let N_i be the set of neighbours of a_i in C. (Thus each N_i is stable.) For $1 \le i \le n$, let A_i be a set of new vertices, and let H be the graph obtained from C by adding $A_1 \cup \dots \cup A_n$, with adjacency as follows.

- For $1 \le i \le n$, A_i is stable.
- For $1 \le i \le n$, every vertex in A_i has the same neighbours in C as a_i.
- For $1 \le i < j \le n$, if $N_i \cap N_j \neq \emptyset$ then then there are no edges between A_i, A_j.
- For $1 \le i < j < k \le n$, if (N_i, N_j, N_k) is a partition of $V(C)$ into three stable sets, then there is no triangle in $A_i \cup A_j \cup A_k$.
- Otherwise, adjacency within $A_1 \cup \dots \cup A_n$ are arbitrary.

(This is not quite as wild as it might appear. For instance, if i, j, k are as in the fourth condition above, then the *only* restriction on the adjacency between A_i, A_j, A_k is that $A_i \cup A_j \cup A_k$ includes no triangle; none of these pairs of vertices are affected by any of the other conditions, so in a sense the restrictions are "separable". Note also that if C is not 3-colourable, and in particular if G is not orientable, then the fourth condition is vacuously satisfied.) Such a graph H is always prismatic, and we say it is obtained from G by *expanding around the core.*

If we hope to obtain a structure theorem for all prismatic graphs, the above two constructions will have to be accounted for in it.

Now let us begin on the non-orientable case. Here are two useful prismatic graphs, that we call P_1 and P_2.

- P_1 has nine vertices $a_1, a_2, a_3, b_1, b_2, b_3, c_1, c_2, c_3$, and edges as follows. The vertices $\{c_1, c_2, c_3\}$ are pairwise adjacent; for $1 \le i, j \le 3$, a_i is adjacent to b_j; and for $1 \le i \le 3$, c_i is adjacent to a_i, b_i.
- P_2 has ten vertices $x, y, a_1, \dots, a_8$, and adjacency as follows. For $1 \le i \le 8$, a_i is adjacent to a_{i-1}, a_{i+1} and a_{i+4} (reading subscripts modulo 8); x, y are adjacent; x is adjacent to a_1, a_3, a_5, a_7, and y is adjacent to a_2, a_4, a_6, a_8.

It is not difficult to show the following.

6.1 *Let G be prismatic. Then G is non-orientable if and only if some induced subgraph of G is isomorphic to P_1 or to P_2.*

To prove this, assign an arbitrary cyclic orientation to each triangle. Make a graph H whose vertices are the triangles of G, and whose edges correspond to disjoint pairs of triangles. Each edge of H acquires a sign, depending whether the corresponding matching between the triangles preserves or reverses the cyclic orders of the triangles. If every cycle of H has an even number of orientation-reversing edges, then G is orientable. If not, we examine the shortest cycle of H with an odd number of orientation-reversing edges; it is easy to show that it has length 3 or 4, and the corresponding three or four triangles of G induce either P_1 or P_2.

Let us see some classes of prismatic graphs.

- The *Schläfli graph* has 27 vertices, and can be described as follows. Let N be the set of all triples (i, j, k) where $1 \leq i, j, k \leq 3$. The vertex set of G is $\{a_{i,j,k} : (i, j, k) \in N\}$. For distinct $(i, j, k), (i', j', k') \in N$, let $a_{i,j,k}$ and $a_{i',j',k'}$ be adjacent if and only if either

  - $k = k'$ and either $i = i'$ or $j = j'$, or
  - $k = k' + 1 \pmod 3$ and $i \neq j'$, or
  - $k = k' + 2 \pmod 3$ and $i' \neq j$.

  This graph is much more symmetrical than is apparent from this description—see [2], for instance. (Our thanks to Adrian Bondy, who identified this mysterious graph for us.) The Schläfli graph is antiprismatic, and so all induced subgraphs of its complement are prismatic; we call them 1-*prismatic.*

- Let $k \geq 0$, and let G have vertex set the disjoint union of five sets $A = \{a_1, \ldots, a_k\}$, $B = \{b_1, \ldots, b_k\}$, $C = \{c_1, \ldots, c_k\}$, $D = \{d_1, \ldots, d_k\}$, and $\{p, q, r, s, t\}$, with adjacency as follows.

  - $\{p, q, t\}$ and $\{r, s, t\}$ are triangles, and A, B, C, D are stable
  - p is completely adjacent to $A \cup B$; q is completely adjacent to $C \cup D$; r is completely adjacent to $A \cup D$; s is completely adjacent to $B \cup C$
  - for $1 \leq i \leq k$, $a_ib_i, b_ic_i, c_id_i, d_ia_i$ are edges, and for $1 \leq i, j \leq k$ with $i \neq j$, a_ic_j and b_id_j are edges.

  We say a graph is 2-*prismatic* if it is an induced subgraph of G for some choice of k.

- Let G have vertex set the disjoint union of $\{a_0, b_0, d_1, d_2, d_3\}$ and $\{a_i, b_i, c_i\}$ ($i = 1, \ldots, k$), where $k \geq 3$. Let the edges of G be as follows:

  - a_0 is adjacent to $a_1, \ldots, a_k$ and to $c_1, \ldots, c_k$; b_0 is adjacent to $b_1, \ldots, b_k$ and to $c_1, \ldots, c_k$
  - for $1 \leq i \leq k$, c_i is adjacent to a_i, b_i
  - $1 \leq i, j \leq k$ with $i \neq j$, a_i is adjacent to b_j
  - for $i = 1, 2, 3$, d_i is adjacent to c_i and to $c_4, \ldots, c_k$

  - for $i, j \in \{1,2,3\}$ with $i \neq j$, d_i is adjacent to a_j, b_j.

  Any graph that is an induced subgraph of such a graph G is said to be *3-prismatic.*

- Let G have vertex set the disjoint union of $\{x,y,z\}$, $\{a_1,\ldots,a_m,b_1,\ldots,b_m\}$ and $\{c_1,\ldots,c_n,d_1,\ldots,d_n\}$, where $m,n \geq 2$. Let the edges of G be as follows:
  - x,y,z are pairwise adjacent
  - x is adjacent to $a_1,\ldots,a_m,b_1,\ldots,b_m$, and y is adjacent to $c_1,\ldots,c_n$, $d_1,\ldots,d_n$
  - a_i, b_i are adjacent for $1 \leq i \leq m$, and c_j, d_j are adjacent for $1 \leq j \leq n$
  - for $1 \leq i \leq m$ and $1 \leq j \leq n$, the subgraph induced on $\{a_i, b_i, c_j, d_j\}$ is a cycle of length 4.

  Any graph that is an induced subgraph of G is said to be *4-prismatic.*

We proved the following[1].

6.2 *Let G be prismatic, containing P_1 as an induced subgraph. Then G can be obtained from a 1-, 2-, or 3-prismatic graph by multiplying leaf triangles and expanding around the core.*

6.3 *Let G be prismatic, containing P_2 as an induced subgraph, and not containing P_1. Then G can be obtained from a 1- or 4-prismatic graph by multiplying leaf triangles and expanding around the core.*

These results are the main theorems of [4]. In view of 6.1, this solves our problem for the non-orientable case.

7 Prismatic graphs—the orientable case

In the orientable case, the graph may be 3-colourable, and if so then $\overline{G}$ might admit a hex-join. (Remember we are working in the complement now.) To postpone the problems with hex-joins, let us first assume that the core of G is not 3-colourable. We say that G is a *cycle of triangles* graph if for some $n \geq 5$ with $n = 2$ modulo 3, there are pairwise disjoint stable subsets $X_1, \ldots, X_{2n+1}$ of $V(G)$ with union $V(G)$, such that, reading subscripts modulo $2n$:

- for $1 \leq i \leq n$, there is a nonempty subset $\hat{X}_{2i} \subseteq X_{2i}$, and at least one of $\hat{X}_{2i}, \hat{X}_{2i+2}$ has cardinality 1;
- for $i \in \{1,\ldots,2n\}$ and all k with $2 \leq k \leq 2n-2$, let $j \in \{1,\ldots,2n\}$ with $j = i+k$ modulo $2n$:
  (1) if $k = 2$ modulo 3 and there exist $u \in X_i$ and $v \in X_j$, nonadjacent, then either i,j are odd and $k \in \{2, 2n-2\}$, or i,j are even and $u \notin \hat{X}_i$ and $v \notin \hat{X}_j$;

[1]These two theorems are not correct as stated; we have found more basic graphs that should be included. Please see [4] for the correct versions.

(2) if $k \neq 2$ modulo 3 then X_i is anticomplete to X_j;

(Note that $k = 2$ modulo 3 if and only if $2n - k = 2$ modulo 3, so these statements are symmetric between i and j.)

- for $1 \leq i \leq n+1$, X_{2i-1} is the union of three pairwise disjoint sets L_{2i-1},M_{2i-1}, R_{2i-1};
- for $1 \leq i \leq n$, X_{2i} is anticomplete to $L_{2i-1} \cup R_{2i+1}$; $X_{2i} \setminus \hat{X}_{2i}$ is anticomplete to $M_{2i-1} \cup M_{2i+1}$; and every vertex in $X_{2i} \setminus \hat{X}_{2i}$ is adjacent to exactly one end of every edge between R_{2i-1} and L_{2i+1};
- for $1 \leq i \leq n$, if $|\hat{X}_{2i}| > 1$ then
  (1) $R_{2i-1} = L_{2i+1} = \emptyset$;
  (2) if $u \in X_{2i-1}$ and $v \in X_{2i+1}$, then u, v are nonadjacent if and only if they have the same neighbour in $\hat{X}_{2i}$;
- for $1 \leq i \leq n$, if $|\hat{X}_{2i}| = 1$, then
  (1) R_{2i-1}, L_{2i+1} are matched, and every edge between $M_{2i-1} \cup R_{2i-1}$ and $L_{2i+1} \cup M_{2i+1}$ is between R_{2i-1} and L_{2i+1};
  (2) the vertex in $\hat{X}_{2i}$ is complete to $R_{2i-1} \cup M_{2i-1} \cup L_{2i+1} \cup M_{2i+1}$;
  (3) if $u \in X_{2i-1}$ and $v \in X_{2i+1}$ are nonadjacent then $u \in M_{2i-1} \cup R_{2i-1}$ and $v \in L_{2i+1} \cup M_{2i+1}$x.
  (4) $M_{2i-1}, \hat{X}_{2i-2}$ are matched and $M_{2i+1}, \hat{X}_{2i+2}$ are matched.

We proved in [3] that:

7.1 *Every cycle of triangles graph is prismatic and orientable, and its core is not 3-colourable. Conversely, if G is an orientable prismatic graph, and its core is not 3-colourable, then G is a cycle of triangles graph (with one exceptional case when G has exactly five triangles).*

Here is a sketch of the proof. The first statement is routine checking. For the second, suppose that G is orientable prismatic, and its core is not 3-colourable. Choose an orientation of each triangle as in the definition of "orientable". Every prismatic graph containing $L(K_{3,3})$ as an induced subgraph either has 3-colourable core or is not orientable; so G does not contain $L(K_{3,3})$. Since also G does not contain P_1 (since it is orientable), it follows that for every triangle T of G, some vertex $w_T \in T$ is in no other triangle. Let the other two vertices of T be u_T, v_T, where the orientation of T is the cyclic permutation $u_T \mapsto v_T \mapsto w_T \mapsto u_T$. Let H be the digraph formed by the directed edges (u_T, v_T) as T ranges over all triangles of G. Then as a graph, H has no triangles, and one can show that in every 4-edge path of H, the middle vertex is the head of exactly one of the two middle edges. It follows that the structure of H is very restricted. Moreover, H is connected (as a graph, not as a digraph), for otherwise the hypergraph formed by the triangles of G would not be connected and G would therefore have 3-colourable core, a contradiction; and H must have a directed cycle (for otherwise again the core of G would be 3-colourable); and then the theorem follows easily.

Now we turn to the case when the core is 3-colourable. Let us say that G is *triangle-connected* if the hypergraph of triangles of G is connected.

7.2 *Let G be prismatic with 3-colourable core, and not triangle-connected. Then there is a partition $(C_1, \ldots, C_k)$ of $V(G)$, and a 3-colouring (X, Y, Z) of $V(G)$, with the following properties:*

- *for $1 \leq i \leq k$, $C \cap C_i$ is nonempty (where C is the core of G), and every triangle of G is a subset of one of $C_1, \ldots, C_k$*
- *for $1 \leq i \leq k$, the hypergraph of all triangles included in C_i is connected*
- *for $1 \leq i < j \leq k$, if $u \in C_i$ and $v \in C_j$ are adjacent, then either $u \in X$ and $v \in Y$, or $u \in Y$ and $v \in Z$, or $u \in Z$ and $v \in X$*
- *for $1 \leq i < j \leq k$, if $u \in C_i$ and $v \in C_j$ are nonadjacent, and either $u \in X$ and $v \in Y$, or $u \in Y$ and $v \in Z$, or $u \in Z$ and $v \in X$, then neither of u, v belongs to C.*

This tells us that we may add edges between non-core vertices without changing the core, so that for all i with $1 \leq i < k$, the partition $C_1 \cup \cdots \cup C_i, C_{i+1} \cup \cdots \cup C_k$ corresponds to a hex-join of $\overline{G}$. Consequently, if we understand triangle-connected prismatic graphs with 3-colourable core, then we may construct all others by taking a sequence of them hooked together as in 7.2.

Let us say a prismatic graph is *substantial* if for every subset $S \subseteq V(G)$ with $|S| \leq 2$, there is a triangle disjoint from S. The prismatic graphs that are not substantial are easy to describe explicitly, but we omit the details. From now on we shall only be concerned with substantial prismatic graphs.

7.3 *Let G be prismatic, substantial and triangle-connected, with 3-colourable core, such that its core is not uniquely 3-colourable. Then the core is an induced subgraph of $L(K_{3,3})$.*

The proof is easy, but we omit it. Since in this case we understand the core completely, it is straightforward to enumerate the possible adjacencies of the non-core vertices, and we omit the details.

There remains the case when the core is uniquely 3-colourable. Let us say that G is a *path of triangles* graph if it satisfies the same axioms as a cycle of triangles graph, except we make a linear sequence rather than a circular sequence. More precisely, we say that G is a *path of triangles* graph if for some integer $n \geq 1$ there are pairwise disjoint stable subsets $X_1, \ldots, X_{2n+1}$ of $V(G)$ with union $V(G)$, such that:

- for $1 \leq i \leq n$, there is a nonempty subset $\hat{X}_{2i} \subseteq X_{2i}$; $|\hat{X}_2| = |\hat{X}_{2n}| = 1$, and for $0 < i < n$, at least one of $\hat{X}_{2i}, \hat{X}_{2i+2}$ has cardinality 1;
- for $1 \leq i < j \leq 2n+1$
  (1) if $j - i = 2$ modulo 3 and there exist $u \in X_i$ and $v \in X_j$, nonadjacent, then either i, j are odd and $j = i + 2$, or i, j are even and $u \notin \hat{X}_i$ and $v \notin \hat{X}_j$;

(2) if $j - i \neq 2$ modulo 3 then either $j = i + 1$ or X_i is anticomplete to X_j;

- for $1 \leq i \leq n+1$, X_{2i-1} is the union of three pairwise disjoint sets $L_{2i-1}, M_{2i-1}, R_{2i-1}$, where $L_1 = M_1 = M_{2n+1} = R_{2n+1} = \emptyset$;
- if $R_1 = \emptyset$ then $n \geq 2$ and $|\hat{X}_4| > 1$, and if $L_{2n+1} = \emptyset$ then $n \geq 2$ and $|\hat{X}_{2n-2}| > 1$;
- for $1 \leq i \leq n$, X_{2i} is anticomplete to $L_{2i-1} \cup R_{2i+1}$; $X_{2i} \setminus \hat{X}_{2i}$ is anticomplete to $M_{2i-1} \cup M_{2i+1}$; and every vertex in $X_{2i} \setminus \hat{X}_{2i}$ is adjacent to exactly one end of every edge between R_{2i-1} and L_{2i+1};
- for $1 < i < n$, if $|\hat{X}_{2i}| > 1$ then

  (1) $R_{2i-1} = L_{2i+1} = \emptyset$;

  (2) if $u \in X_{2i-1}$ and $v \in X_{2i+1}$, then u, v are nonadjacent if and only if they have the same neighbour in $\hat{X}_{2i}$;
- for $1 \leq i \leq n$, if $|\hat{X}_{2i}| = 1$, then

  (1) R_{2i-1}, L_{2i+1} are matched, and every edge between $M_{2i-1} \cup R_{2i-1}$ and $L_{2i+1} \cup M_{2i+1}$ is between R_{2i-1} and L_{2i+1};

  (2) the vertex in $\hat{X}_{2i}$ is complete to $R_{2i-1} \cup M_{2i-1} \cup L_{2i+1} \cup M_{2i+1}$;

  (3) if $u \in X_{2i-1}$ and $v \in X_{2i+1}$ are nonadjacent then $u \in M_{2i-1} \cup R_{2i-1}$ and $v \in L_{2i+1} \cup M_{2i+1}$

  (4) if $i > 1$ then $M_{2i-1}, \hat{X}_{2i-2}$ are matched, and if $i < n$ then $M_{2i+1}, \hat{X}_{2i+2}$ are matched.

If G is a 3-colourable graph with no triangles, we say that G is a *whirl.* We proved in [3] that:

7.4 *Every substantial path of triangles graph is prismatic and triangle-connected, with a uniquely 3-colourable core. Conversely, let G be substantial, prismatic and triangle-connected, with a non-null, uniquely 3-colourable core. Then G is 3-colourable; and either G is a path of triangles graph, or there is a partition (X, Y) of $V(G)$ so that all triangles are subsets of X, and $G|X$ is a path of triangles graph, and $G|Y$ is a whirl, and we can add edges between non-core vertices in X and Y without changing the core, so that the partition (X, Y) corresponds to a hex-join of the complement.*

This completes the statement of our structure theorem for claw-free graphs.

8 Colouring claw-free graphs

In this section we discuss some aspects of the proof of the following (proved in [7]).

8.1 *Let G be claw-free and connected, with $\alpha(G) \geq 3$. Then $\chi(G) \leq 2\omega(G)$.*

Note that the theorem is asymptotically best possible, for if G is as in the definition of "2-prismatic graph" in section 6, then H is obtained from $\overline{G}$ by deleting the vertices q, r, s, t, then H is claw-free and connected, $\alpha(H) \geq 3$, $\omega(H) = k + 1$, and $\chi(H) = 2k$. Before we discuss the proof of 8.1, we remark that there is an easy proof of the following weakening ($\Delta(G)$ denotes the maximum degree of vertices of G):

8.2 *Let G be claw-free and connected, with $\alpha(G) \geq 3$. Then $\Delta(G) \leq 4(\omega(G) - 1)$, and consequently $\chi(G) \leq 4(\omega(G) - 1)$.*

Proof The second statement follows from the first by Brook's theorem. For the first we proceed by induction on $|V(G)|$. Let v be a vertex of maximum degree $\Delta(G)$. Since $\alpha(G) \geq 3$ and G is claw-free, there is a vertex different from and nonadjacent to v; and consequently we may choose u nonadjacent to v so that $G \setminus u$ is connected. If $\alpha(G \setminus u) \geq 3$, then from the inductive hypothesis, $\Delta(G \setminus u) \leq 4(\omega(G \setminus u) - 1)$; and since $\Delta(G) = \Delta(G \setminus u)$ (because u, v are nonadjacent) and $\omega(G \setminus u) \leq \omega(G)$, the desired result would hold. We may assume then that $\alpha(G \setminus u) = 2$. Let N, M be the sets of neighbours and nonneighbours of u respectively (thus $N \cup M = V(G) \setminus \{u\}$). Since G is connected, there exists $n \in N$. If $x, y \in M$ are nonadjacent, then at least one of x, y is adjacent to n, since $\alpha(G \setminus u) = 2$; and not both are adjacent to n, since otherwise n would have three nonadjacent neighbours, which is impossible since G is claw-free. Thus n is adjacent to exactly one of x, y. It follows that the set of neighbours of n in M is a clique, and so is the set of nonneighbours of n in M, and therefore M is the union of two cliques. Since $\alpha(G) \geq 3$, there exist $m_1, m_2 \in M$, nonadjacent. Now let N_i be the set of all neighbours of m_i in N ($i = 1, 2$). We have already seen that every vertex of N is adjacent to exactly one of m_1, m_2, and so $N_1 \cup N_2 = N$ and $N_1 \cap N_2 = \emptyset$. If $x, y \in N_1$ are nonadjacent, then $\{x, y, n_2\}$ is stable, contrary to $\alpha(G \setminus u) = 2$. Hence $N_1 \cup \{u\}$ is a clique, and similarly so is $N_2 \cup \{u\}$. Thus $V(G)$ is the union of four cliques, and u is in two of them, and so $4\omega(G) \geq |V(G)| + 1 \geq \Delta(G) + 4$, as required. This proves 8.2.

However, improving the factor of 4 in 8.2 to 2 seems to be very difficult. The only proof we have of 8.1 is an application of the structure theorem for claw-free graphs; and not only is this theorem difficult itself, but also the application is difficult. Here are some of the ideas. There are two easy tricks that we used many times, the following. First, if there is a vertex whose set of neighbours is the union of two cliques, then its degree is at most $2\omega(G) - 2$, and we can delete it and use induction. And quite often there is such a vertex. For instance, if G is expressed as a generalized 2-join, then using the notation of the definition of "generalized 2-join", any vertex in A_1 has neighbour set the union of two cliques, as is easy to see. Or if G is a circular interval graph, then the neighbour set of *every* vertex is the union of two cliques. But there does not always exist such a vertex; for instance, in the Schläfli graph there is no such vertex. Every vertex has degree 16, and yet $\omega(G) = 6$.

The second trick is, that we win if $\omega(G) \geq |V(G)|/4$. For if $|V(G)|$ is even and there is a perfect matching in $\overline{G}$, then $\chi(G) \leq |V(G)|/2 \leq 2\omega(G)$; while if $|V(G)|$ is odd then $\omega(G) \geq (|V(G)| + 1)/4$, so if there is a near-perfect matching in $\overline{G}$ then $\chi(G) \leq (|V(G)| + 1)/2 \leq 2\omega(G)$ as required. So we can assume that $\overline{G}$ contains no perfect or near-perfect matching, and therefore Tutte's maximum matching theorem can be applied, and again it follows easily (but using the hypothesis that G is connected and claw-free and $\alpha(G) \geq 3$; we omit the details) that $\chi(G) \leq 2\omega(G)$. So we may assume that $V(G)$ is not the union of four cliques.

These two tricks get us a long way, and suffice to handle non-antiprismatic graphs, but for antiprismatic graphs the proof gets much more complicated and we omit the details.

We mention that an analogous statement holds in the complement, the following.

8.3 *Let G be connected and claw-free, with $\alpha(G) \geq 3$. Then $\chi(\overline{G}) \leq 2\omega(\overline{G})$.*

This is also proved in [7]. Again, it is asymptotically best possible; for if $G = L(K_n)$, then $\chi(\overline{G}) \geq n - 2$, and $\omega(\overline{G}) \leq n/2$. Its proof is not an application of the structure theorem for claw-free graphs, however; we found a direct proof. Roughly, the argument is as follows. Choose a tree T of G so that every vertex of G has a neighbour in T, with $|V(T)| \leq 2\alpha(G)$ and such that some three vertices of T are pairwise nonadjacent in G. (It is easy to see that this exists.) For every ordered pair (u, v) of adjacent vertices of T, let $N_{u,v}$ be the set of all vertices of $G \setminus (X \cup Y)$ that are adjacent to u and not v. For each (u, v), $N_{u,v} \cup \{u\}$ is a clique, and every vertex of G belongs to one of these cliques; and so $\chi(\overline{G}) \leq 2|E(T)| \leq 4\omega(\overline{G})$. In fact we can cover all the vertices of G without using all ordered pairs (u, v), and by carefully choosing the right subset of pairs (plus possibly one extra pair, not an edge of the tree) we show that $\chi(\overline{G}) \leq 2\omega(\overline{G})$.

9 Complexity issues

We have been finding an "explicit construction" for all claw-free graphs, but it is not quite clear what is meant by this. We has better mean more than just a polynomial-time recognition algorithm, for there is an obvious recognition algorithm for claw-free graphs, with running time $O(n^4)$ when the input has n vertices. So what then do we mean by an "explicit construction"? For instance, here is another attempt at an "explicit construction"—given a claw-free graph G that we have already constructed, choose a subset $X \subseteq V(G)$ such that adding a vertex with neighbour set X will not introduce a claw; and add a vertex with neighbour set X. We fervently hope that this doesn't count as solving the problem, or else we have been wasting our time; and yet why doesn't it? For one can easily test in polynomial time whether such a set X has the property we require; on what grounds do we exclude this "construction"?

It is our feeling that we should not be permitted to "guess" the set X; if we want to use this set X to construct a larger graph, we should inductively have a construction for (G, X), not just for G. With that in mind, let us go back through what we proved, and check that we pass this test. We don't, quite; for instance, we allow ourselves the operation of making a hex-join. Making a hex-join requires two smaller graphs G_1, G_2 that we already have constructed, and for each of them we need a partition of its vertex set into three cliques. Unfortunately we are currently "guessing" these cliques; if we want to use hex-joins, we should really be constructing quadruples (G, A, B, C) where A, B, C are disjoint cliques with union $V(G)$, and not just constructing G. Happily, this can be done. Such a partition (A, B, C) is just a 3-colouring of $\overline{G}$, and as we have seen, in most cases $\overline{G}$ does not admit a 3-colouring; when it does, in most cases the 3-colouring is unique; and when it is not unique, we can describe the graph explicitly and list all the 3-colourings. So that problem can be overcome.

There is a similar problem with homogeneous pairs, and that one we found more difficult. We believe it is solved, but we are still checking.

Acknowledgements

This research was performed while Maria Chudnovsky served as a Clay Mathematics Institute Research Fellow. The work of Paul Seymour was partially supported by ONR grant N00014-01-1-0608 and NSF grant DMS-0070912.

References

[1] A. Ben Rebea, Étude des stables dans les graphes quasi-adjoints, Ph.D. thesis, Univ. de Grenoble, 1981.

[2] P.J. Cameron, Strongly regular graphs, in *Topics in Algebraic Graph Theory* (eds. L.W. Beineke and R.J. Wilson), Cambridge University Press, Cambridge (2004), pp. 207–231.

[3] Maria Chudnovsky and Paul Seymour, Claw-free graphs I. Orientable prismatic graphs, manuscript, 2004.

[4] Maria Chudnovsky and Paul Seymour, Claw-free graphs II. Non-orientable prismatic graphs, manuscript, 2004.

[5] Maria Chudnovsky and Paul Seymour, Claw-free graphs IV. Sparse decomposition, manuscript, 2003.

[6] Maria Chudnovsky and Paul Seymour, Claw-free graphs VI. The structure of quasi-line graphs, manuscript, 2004.

[7] Maria Chudnovsky and Paul Seymour, Claw-free graphs VII. Colouring claw-free graphs, manuscript, 2004.

[8] V. Chvátal and N. Sbihi, Recognizing claw-free Berge graphs, *J. Combin. Theory Ser. B* **44** (1988), 154–176.

[9] J. Edmonds, Maximum matching and a polytope with 0, 1-vertices, *J. Res. Nat. Bur. Standards* **69B** (1965), 125–130.

[10] J. Edmonds, Paths, trees and flowers, *Canad. J. Math.* **17** (1965), 449–467.

[11] F. Eisenbrand, G. Oriolo, G. Stauffer, P. Ventura, Circular ones matrices and the stable set polytope of quasi-line graphs, manuscript, 2004.

[12] A. Gallucio and A. Sassano, "The rank facets of the stable set polytope for claw-free graphs", *J. Combin. Theory Ser. B* **69** (1997), 1–38.

[13] R. Giles and L. Trotter, On stable set polyhedra for $K_{1,3}$-free graphs, *J. Combin. Theory Ser. B* **31** (1981), 313–326.

[14] T.M. Liebling, G. Oriolo, B. Spille and G. Stauffer, On non-rank facets of the stable set polytope of claw-free graphs and circulant graphs, manuscript, 2004.

[15] F. Maffray and B. Reed, A description of claw-free perfect graphs, *J. Combin. Theory Ser. B* **75** (1999), 134–156.

[16] G.J. Minty, On maximal independent sets of vertices in claw-free graphs, *Journal of Combin. Theory Ser. B* **28** (1980), 284–304.

[17] G. Oriolo, Clique family inequalities for the stable set polytope for quasi-line graphs, in *Special Issue of Discrete Applied Mathematics on Stability Problems* (eds. V. Lozin and D. de Werra), in press.

[18] V.G. Vizing, On an estimate of the chromatic class of a p-graph, *Diskret. Analiz.* **3** (1964), 25–30.

Dept of Mathematics
Princeton University
Princeton NJ 08544
USA
mchudnov@math.princeton.edu
pds@math.princeton.edu

The multivariate Tutte polynomial (alias Potts model) for graphs and matroids

Alan D. Sokal

Abstract

The multivariate Tutte polynomial (known to physicists as the Potts-model partition function) can be defined on an arbitrary finite graph G, or more generally on an arbitrary matroid M, and encodes much important combinatorial information about the graph (indeed, in the matroid case it encodes the full structure of the matroid). It contains as a special case the familiar two-variable Tutte polynomial—and therefore also its one-variable specializations such as the chromatic polynomial, the flow polynomial and the reliability polynomial—but is considerably more flexible. I begin by giving an introduction to all these problems, stressing the advantages of working with the multivariate version. I then discuss some questions concerning the complex zeros of the multivariate Tutte polynomial, along with their physical interpretations in statistical mechanics (in connection with the Yang–Lee approach to phase transitions) and electrical circuit theory. Along the way I mention numerous open problems. This survey is intended to be understandable to mathematicians with no prior knowledge of physics.

1 Introduction

Let $G = (V, E)$ be a finite undirected graph with vertex set V and edge set E.[1] The *multivariate Tutte polynomial* of G is, by definition, the polynomial

$$Z_G(q, \mathbf{v}) \;=\; \sum_{A \subseteq E} q^{k(A)} \prod_{e \in A} v_e \,, \tag{1.1}$$

where q and $\mathbf{v} = \{v_e\}_{e \in E}$ are commuting indeterminates, and $k(A)$ denotes the number of connected components in the subgraph (V, A). [It is sometimes convenient to consider instead

$$\widetilde{Z}_G(q, \mathbf{v}) \;\equiv\; q^{-|V|} Z_G(q, \mathbf{v}) \;=\; \sum_{A \subseteq E} q^{k(A)-|V|} \prod_{e \in A} v_e \,, \tag{1.2}$$

which is a polynomial in q^{-1} and $\{v_e\}$.] From a combinatorial point of view, Z_G is simply the multivariate generating polynomial that enumerates the spanning subgraphs of G according to their precise edge content (with weight v_e for the edge e) and their number of connected components (with weight q for each component). As we shall see, Z_G encodes a vast amount of combinatorial information about the graph G, and contains many other well-known graph polynomials as special cases. I shall most often take an analytic point of view, and treat q and $\{v_e\}$ as real or complex variables.

[1] In this paper a "graph" is allowed to have loops and/or multiple edges unless explicitly stated otherwise.

In statistical physics, Z_G is known as the partition function of the *q-state Potts model* [102, 140, 141], for reasons to be explained in the next section.[2] The Potts model—along with its close relative, the Fortuin–Kasteleyn random-cluster model [58, 62, 76]—plays an important role in the theory of phase transitions and critical phenomena [5, 44, 96], and physicists have developed a significant amount of information (both rigorous and non-rigorous) about its properties. As will be explained in Section 5, one way of analyzing phase transitions is to study the zeros of $Z_G(q, \mathbf{v})$ when q and/or $\{v_e\}$ are treated as *complex* variables.

If we set all the edge weights v_e equal to the same value v, we obtain a two-variable polynomial $Z_G(q, v)$ that is essentially equivalent to the standard Tutte polynomial $T_G(x, y)$ [see Section 2.5 for details]. But the main message of this paper is that it is often useful to consider the multivariate polynomial $Z_G(q, \mathbf{v})$, even if one is ultimately interested in a particular two-variable or one-variable specialization. For instance, $Z_G(q, \mathbf{v})$ is *multiaffine* in the variables $\mathbf{v}$ (i.e., of degree 1 in each v_e separately); often a multiaffine polynomial in many variables is easier to handle than a general polynomial in a single variable (e.g., it may permit simple proofs by induction on the number of variables). Furthermore, many natural operations on graphs, such as the reduction of edges in series or parallel, lead out of the class of "all v_e equal". I shall illustrate the advantages of this "multivariate ideology" by showing several instances in which the multivariate extension of a single-variable result is not only vastly more powerful but also much *easier* to prove: indeed, in one case, a 20-page proof is reduced to a few lines.

All of these considerations can be extended from graphs to matroids[3]: if M is a matroid with ground set E and rank function r_M, its multivariate Tutte polynomial is defined by

$$\widetilde{Z}_M(q, \mathbf{v}) \;=\; \sum_{A \subseteq E} q^{-r_M(A)} \prod_{e \in A} v_e \,, \tag{1.3}$$

which is a polynomial in q^{-1} and $\{v_e\}$. This extends the graph definition in the sense that if G is a graph and $M(G)$ is its cycle matroid, then

$$\widetilde{Z}_{M(G)}(q, \mathbf{v}) \;=\; \widetilde{Z}_G(q, \mathbf{v}) \tag{1.4}$$

[because $r_{M(G)}(A) = |V| - k(A)$]. Since a matroid is completely determined by its rank function, $\widetilde{Z}_M$ is simply an algebraic encoding of *all* the information about the matroid M. Moreover, my earlier statement that Z_G encodes "a vast amount" of information about the graph G can now be made more precise: Z_G encodes the number of vertices $|V|$ together with all the information about G that is contained in its cycle matroid $M(G)$ [and no other information].

I am now convinced that matroids are the natural category for studying the multivariate Tutte polynomial. Of course, results that hold for graphic matroids

[2] The Potts model [102] was invented in the early 1950s by Potts' thesis advisor Domb: see [47]. The $q = 2$ case, known as the Ising model [67], was invented in 1920 by Ising's thesis advisor Lenz [84]: see [20, 78, 127] for a fascinating history. (I hasten to add that these are the only two cases I know of where the thesis advisor's invention was named after the graduate student, rather than the other way around.) The $q = 4$ case, which is a special case of the Ashkin–Teller model, was invented in 1943 by Ashkin and Teller [3]. For a review of the Potts model, see [140, 141].

[3] See Oxley [98] for an excellent introduction to matroid theory.

may or may not hold for larger classes of matroids.

2 The multivariate Tutte polynomial for graphs, and its specializations

First, a trivial but useful observation: Starting from the definition

$$Z_G(q, \mathbf{v}) \;=\; \sum_{A \subseteq E} q^{k(A)} \prod_{e \in A} v_e \tag{2.1}$$

and using the simple relation

$$k(A) \;=\; |V| - |A| + c(A) \tag{2.2}$$

where $c(A)$ is the cyclomatic number (i.e., number of linearly independent cycles) of the graph (V, A), we can rewrite Z_G in the alternate form

$$Z_G(q, \mathbf{v}) \;=\; q^{|V|} \sum_{A \subseteq E} q^{c(A)} \prod_{e \in A} \frac{v_e}{q} \,. \tag{2.3}$$

Example 2.1 (Trees) For any tree $T = (V, E)$ we have

$$Z_T(q, \mathbf{v}) \;=\; q \prod_{e \in E} (q + v_e) \,. \tag{2.4}$$

This follows immediately from (2.3), using the fact that $c(A) = 0$ for all A.

Example 2.2 (Cycles) Let $G = (V, E)$ be the cycle C_n. Then

$$Z_{C_n}(q, \mathbf{v}) \;=\; \prod_{e \in E} (q + v_e) \;+\; (q-1) \prod_{e \in E} v_e \,. \tag{2.5}$$

This follows immediately from (2.3), using the fact that $c(A) = 0$ for all $A \subsetneq E$ and $c(E) = 1$.

Let us now consider some of the polynomials that can be obtained from $Z_G(q, \mathbf{v})$ by specializing the value of q:

2.1 $q = 1$

When $q = 1$, the multivariate Tutte polynomial becomes trivial:

$$Z_G(1, \mathbf{v}) \;=\; \prod_{e \in E} (1 + v_e) \,. \tag{2.6}$$

2.2 $q = 1, 2, 3, \ldots$ (q-state Potts model) and the chromatic polynomial

Let q be a positive integer; then the *q-state Potts-model partition function* for the graph G is defined by

$$Z_G^{\rm Potts}(q, \mathbf{v}) \;=\; \sum_{\sigma\colon\, V \to \{1,2,\ldots,q\}} \; \prod_{e \in E} \Big[1 + v_e \delta(\sigma_{x_1(e)}, \sigma_{x_2(e)}) \Big] \,. \tag{2.7}$$

Here the sum runs over all maps $\sigma\colon V \to \{1, 2, \ldots, q\}$, and we sometimes write σ_x as a synonym for $\sigma(x)$; the δ is the Kronecker delta

$$\delta(a,b) \;=\; \begin{cases} 1 & \text{if } a = b \\ 0 & \text{if } a \neq b \end{cases} \tag{2.8}$$

and $x_1(e), x_2(e) \in V$ are the two endpoints of the edge e (in arbitrary order). We usually consider v_e to be a real or complex variable.

In statistical physics, this formula arises as follows: In the Potts model [102, 140, 141], an "atom" (or "spin") at the site $x \in V$ can exist in any one of q different states. The *energy* of a configuration is the sum, over all edges $e \in E$, of 0 if the spins at the two endpoints of that edge are unequal and $-J_e$ if they are equal. The *Boltzmann weight* of a configuration is then $e^{-\beta H}$, where H is the energy of the configuration and $\beta \geq 0$ is the inverse temperature. The *partition function* is the sum, over all configurations, of their Boltzmann weights. Clearly this is just a rephrasing of (2.7), with $v_e = e^{\beta J_e} - 1$. A coupling J_e (or v_e) is called *ferromagnetic* if $J_e \geq 0$ ($v_e \geq 0$), as it is then favored for adjacent spins to take the same value; *antiferromagnetic* if $-\infty \leq J_e \leq 0$ ($-1 \leq v_e \leq 0$), as it is then favored for adjacent spins to take different values; and *unphysical* if $v_e \notin [-1, \infty)$, as the weights are then no longer nonnegative.

It is far from obvious that $Z_G^{\text{Potts}}(q, \mathbf{v})$, which is defined separately for each positive integer q, is in fact the restriction to $q \in \mathbb{Z}_+$ of a *polynomial* in q. But this is in fact the case, and indeed we have:

Theorem 2.3 (Fortuin–Kasteleyn representation of the Potts model) *For integer $q \geq 1$,*

$$Z_G^{\text{Potts}}(q, \mathbf{v}) \;=\; Z_G(q, \mathbf{v}) \,. \tag{2.9}$$

That is, the Potts-model partition function is simply the specialization of the multivariate Tutte polynomial to $q \in \mathbb{Z}_+$.

Proof In (2.7), expand out the product over $e \in E$, and let $A \subseteq E$ be the set of edges for which the term $v_e \delta(\sigma_{x_1(e)}, \sigma_{x_2(e)})$ is taken. Now perform the sum over configurations $\{\sigma_x\}_{x \in V}$: in each component of the subgraph (V, A) the color σ_x must be constant, and there are no other constraints. Therefore,

$$Z_G(q, \mathbf{v}) \;=\; \sum_{A \subseteq E} q^{k(A)} \prod_{e \in A} v_e \,, \tag{2.10}$$

as was to be proved. □

The subgraph expansion (2.10) of the Potts-model partition function was discovered in the late 1960s by Fortuin and Kasteleyn [58, 76].[4] Please note that if $q \geq 0$ and $v_e \geq 0$ for all e, then the weights in (2.10) are nonnegative, and so can be interpreted probabilistically (after normalization by Z_G). The resulting probability measure on 2^E is called the *Fortuin–Kasteleyn random-cluster model* [58, 62, 76] for

[4] In the special case $v_e = -1$ it was discovered many decades earlier by Birkhoff [8] and Whitney [136]; see also Tutte [130, 131].

the graph G. Furthermore, for $q \in \mathbb{Z}_+$ a natural joint probability measure on spin states and edge occupations [i.e., a coupling of (2.7) and (2.10)] has been given by Edwards and Sokal [53]; it underlies the Swendsen–Wang [128] Monte Carlo algorithm for the ferromagnetic Potts model, and is also useful in the rigorous analysis of the Potts model.

Another important special case arises when $v_e = -1$ for all edges e: then Z_G^{Potts} gives weight 1 to each proper coloring and weight 0 to each improper coloring, and so counts the proper colorings. It follows from Theorem 2.3 that the number of proper q-colorings of G is in fact the restriction to $q \in \mathbb{Z}_+$ of a polynomial in q, called the *chromatic polynomial* $P_G(q) = Z_G(q, -1)$. The chromatic polynomial thus corresponds to the zero-temperature ($\beta \to +\infty$) limit of the antiferromagnetic ($J_e < 0$) Potts model. Many properties of the chromatic polynomial in fact extend to the entire antiferromagnetic region (i.e., $-1 \le v_e \le 0$ for all e).

2.3 $q \to 0$ limits

Let us now consider the different ways in which a meaningful $q \to 0$ limit can be taken in the q-state Potts model.

The simplest limit is to take $q \to 0$ with fixed $\mathbf{v}$. From the definition (2.1) we see that this selects out the subgraphs $A \subseteq E$ having the smallest possible number of connected components; the minimum achievable value is of course $k(G)$ itself ($=$ 1 in case G is connected). We therefore have

$$\lim_{q \to 0} q^{-k(G)} Z_G(q, \mathbf{v}) \;=\; C_G(\mathbf{v}) \,, \tag{2.11}$$

where

$$C_G(\mathbf{v}) \;=\; \sum_{\substack{A \subseteq E \\ k(A) = k(G)}} \prod_{e \in A} v_e \tag{2.12}$$

is the generating polynomial of "maximally connected spanning subgraphs" (= *connected spanning subgraphs* in case G is connected).

A different limit can be obtained by taking $q \to 0$ with fixed values of $\mathbf{w} = \mathbf{v}/q$. From (2.3) we see that this selects out the subgraphs $A \subseteq E$ having the smallest possible cyclomatic number; the minimum achievable value is of course 0. We therefore have [126, 139]

$$\lim_{q \to 0} q^{-|V|} Z_G(q, q\mathbf{w}) \;=\; F_G(\mathbf{w}) \,, \tag{2.13}$$

where

$$F_G(\mathbf{w}) \;=\; \sum_{\substack{A \subseteq E \\ c(A) = 0}} \prod_{e \in A} w_e \tag{2.14}$$

is the generating polynomial of *spanning forests*.

Finally, suppose that in $C_G(\mathbf{v})$ we replace each edge weight v_e by λv_e and then take $\lambda \to 0$. This obviously selects out, from among the maximally connected spanning subgraphs, those having the fewest edges: these are precisely the maximal spanning forests (= *spanning trees* in case G is connected), and they all have exactly $|V| - k(G)$ edges. Hence

$$\lim_{\lambda \to 0} \lambda^{k(G)-|V|} C_G(\lambda \mathbf{v}) \;=\; T_G(\mathbf{v}) \,, \tag{2.15}$$

where

$$T_G(\mathbf{v}) \;=\; \sum_{\substack{A \subseteq E \\ k(A) = k(G) \\ c(A) = 0}} \prod_{e \in A} v_e \tag{2.16}$$

is the generating polynomial of maximal spanning forests.[5] Alternatively, suppose that in $F_G(\mathbf{w})$ we replace each edge weight w_e by λw_e and then take $\lambda \to \infty$. This obviously selects out, from among the spanning forests, those having the greatest number of edges: these are once again the maximal spanning forests. Hence

$$\lim_{\lambda \to \infty} \lambda^{k(G)-|V|} F_G(\lambda \mathbf{w}) \;=\; T_G(\mathbf{w}) \,. \tag{2.17}$$

In summary, we have the following scheme for the $q \to 0$ limits of the multivariate Tutte polynomial:

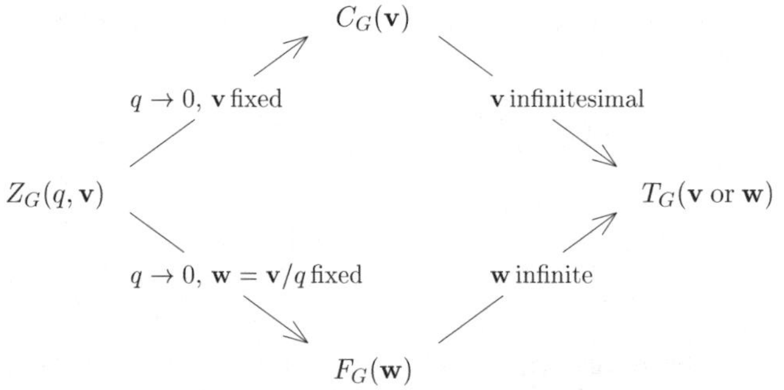

(2.18)

Finally, maximal spanning forests (= spanning trees in case G is connected) can also be obtained directly from $Z_G(q, \mathbf{v})$ by a one-step process in which the limit $q \to 0$ is taken at fixed $\mathbf{x} = \mathbf{v}/q^\alpha$, where $0 < \alpha < 1$ [58, 63, 126, 139]. Indeed, simple manipulation of (2.1) and (2.2) yields

$$Z_G(q, q^\alpha \mathbf{x}) \;=\; q^{\alpha|V|} \sum_{A \subseteq E} q^{\alpha c(A) + (1-\alpha) k(A)} \prod_{e \in A} x_e \,. \tag{2.19}$$

The quantity $\alpha c(A) + (1-\alpha) k(A)$ is minimized on (and only on) maximal spanning forests, where it takes the value $(1-\alpha) k(G)$. Hence

$$\lim_{q \to 0} q^{-\alpha|V| - (1-\alpha) k(G)} Z_G(q, q^\alpha \mathbf{x}) \;=\; T_G(\mathbf{x}) \,. \tag{2.20}$$

[5] I trust that there will be no confusion between the generating polynomial $T_G(\mathbf{v})$ and the Tutte polynomial $T_G(x, y)$. I have used here the letter T because in the most important applications the graph G is connected, so that $T_G(\mathbf{v})$ is the generating polynomial of spanning *trees*.

2.4 The multivariate flow polynomial

Let $G = (V, E)$ be a finite undirected graph, and let Γ be a finite abelian group of order $q = |\Gamma|$. Let us choose arbitrarily an orientation for each edge of G (all subsequent results will be independent of this choice). A Γ-*flow* on G is a mapping $\psi \colon E \to \Gamma$ that satisfies current conservation at every vertex. A Γ-flow ψ on G is said to be *nowhere-zero* in case $\psi(e) \neq 0$ for all $e \in E$.

Let $F_G(\Gamma)$ be the number of nowhere-zero Γ-flows on G.[6] It is a well-known (but at first sight quite surprising) fact that $F_G(\Gamma)$ depends only on the order of Γ and not on any other aspect of the structure of Γ; so we can write it as $F_G(q)$. Furthermore, it turns out that $F_G(q)$ is the restriction to $\mathbb{Z}_+$ of a polynomial in q, called the *flow polynomial* of G and also written $F_G(q)$.

In order to prove these facts (and much more), let us generalize $F_G(\Gamma)$ as follows: Assign to each edge $e \in E$ a weight u_e, and write $\mathbf{u} = \{u_e\}_{e \in E}$. Then define

$$F_G(\Gamma, \mathbf{u}) \;=\; \sum_{\Gamma\text{-flows } \psi} \; \prod_{e \in E} \Big[1 + u_e \delta(\psi(e), 0) \Big] \,, \tag{2.21}$$

where the sum runs over all Γ-flows ψ on G. If we take $u_e = -1$ for all e, this reduces to our friend $F_G(\Gamma)$. We now assert that $F_G(\Gamma, \mathbf{u})$ depends only on $q = |\Gamma|$—so we can write it as $F_G(q, \mathbf{u})$—and that $F_G(q, \mathbf{u})$ is in fact the restriction to $q \in \mathbb{Z}_+$ of a polynomial in q and $\{u_e\}$, which we shall call the *multivariate flow polynomial*. Indeed, the multivariate flow polynomial is nothing other than the multivariate Tutte polynomial in disguise:

Theorem 2.4 *Let $G = (V, E)$ be a finite undirected graph, and let Γ be a finite abelian group of order $q = |\Gamma|$. Then*

$$F_G(\Gamma, \mathbf{u}) \;=\; q^{-|V|} \left(\prod_{e \in E} u_e \right) Z_G(q, q/\mathbf{u}) \tag{2.22}$$

(where division is of course understood edge-wise) and in particular

$$F_G(\Gamma) \;=\; q^{-|V|} \, (-1)^{|E|} \, Z_G(q, -q) \,. \tag{2.23}$$

Consequently, $F_G(\Gamma)$ and $F_G(\Gamma, \mathbf{u})$ depend only on $q = |\Gamma|$ and not on any other aspect of the structure of Γ.

Proof Note first that the total number of Γ-flows on a graph G is $|\Gamma|^{c(G)}$, where $c(G)$ is the cyclomatic number of G. Moreover, the Γ-flows that vanish on a subset $A \subseteq E$ are in one-to-one correspondence with the Γ-flows on the subgraph $(V, E \setminus A)$,

[6] I trust that there will be no confusion between the flow polynomial $F_G(\Gamma)$ [and its relatives to be defined below] and the spanning-forest polynomial $F_G(\mathbf{w})$.

so their number is $|\Gamma|^{c(E\setminus A)}$. Expanding the product in (2.21), we find

$$F_G(\Gamma, \mathbf{u}) = \sum_{A \subseteq E} q^{c(E\setminus A)} \prod_{e\in A} u_e \tag{2.24a}$$

$$= \left(\prod_{e\in E} u_e\right) \sum_{A' \subseteq E} q^{c(A')} \prod_{e\in A'} \frac{1}{u_e} \tag{2.24b}$$

$$= q^{-|V|} \left(\prod_{e\in E} u_e\right) Z_G(q, q/\mathbf{u}) \tag{2.24c}$$

by comparison with (2.3). □

Theorem 2.4 is in fact a special case of a more general identity [31, 49, 113, 142] relating "spin models" taking values in an abelian group Γ to "flow models" taking values in the dual group Γ^*: here the Boltzmann weight $W_e(\sigma_x - \sigma_y)$ for an edge $e = xy$ in the spin model maps onto a weight $\widehat{W}_e(\psi(e))$ in the flow model, where $\widehat{\ }$ denotes Fourier transformation. In the special case of the Potts model, we have $W_e = \mathbf{1} + v_e\delta$, so that $\widehat{W}_e = v_e\mathbf{1} + q\delta$. Setting $u_e = q/v_e$ and relabelling Γ^* as Γ, we obtain (2.22).

2.5 Comparison to the standard Tutte polynomial

The Tutte polynomial $T_G(x,y)$ is conventionally defined by [135, p. 45] [28, pp. 124–127]

$$T_G(x,y) = \sum_{A\subseteq E} (x-1)^{r(E)-r(A)} (y-1)^{|A|-r(A)} \tag{2.25a}$$

$$= \sum_{A\subseteq E} (x-1)^{k(A)-k(E)} (y-1)^{|A|+k(A)-|V|} \tag{2.25b}$$

where $r(A) = |V| - k(A)$ is the rank of the set A in the cycle matroid $M(G)$. [Note also that $|A| - r(A)$ is the cyclomatic number $c(A)$.] Comparison with (1.1) yields

$$T_G(x,y) = (x-1)^{-k(E)} (y-1)^{-|V|} Z_G\big((x-1)(y-1),\, y-1\big) . \tag{2.26}$$

In other words, $T_G(x,y)$ and $Z_G(q,v)$ are essentially equivalent under the change of variables

$$x = 1 + q/v \tag{2.27a}$$
$$y = 1 + v \tag{2.27b}$$

$$q = (x-1)(y-1) \tag{2.27c}$$
$$v = y - 1 \tag{2.27d}$$

One advantage of the Tutte notation is that it allows a slightly smoother treatment of the $q \to 0$ limit, in which the limiting processes employed in Section 2.3 can be replaced by simple evaluation at $x = 1$ or $y = 1$. Thus, the univariate maximally-connected-spanning-subgraph and spanning-forest polynomials are given, respectively, by

$$C_G(v) = v^{|V|-k(E)}\, T_G(1, 1+v) \tag{2.28}$$

$$F_G(w) = w^{|V|-k(E)}\, T_G(1+1/w, 1) \tag{2.29}$$

A second advantage of the Tutte notation is that duality [cf. (4.7) below] takes the simple form

$$T_{G^*}(x, y) = T_G(y, x)\,. \tag{2.30}$$

But the Tutte notation also has a severe disadvantage: the use of the variables x and y conceals the fact that the particular combinations q and v play *very different roles*; q is a global variable, while v can be given separate values v_e on each edge. In particular, the Tutte notation *makes it impossible* to assign unequal weights v_e to the edges. I therefore recommend use of the notation $Z_G(q, \mathbf{v})$ whenever the "multivariate ideology" is potentially of use.

Let me conclude by observing that numerous specific evaluations of the Tutte polynomial have been given combinatorial interpretations, as counting some set of objects associated to the graph G (see e.g. [28, 135]). It would be an interesting project to seek to extend these counting problems to "counting with weights", i.e., to obtain suitably defined univariate or multivariate generating polynomials for the objects in question as specializations of $Z_G(q, v)$ or $Z_G(q, \mathbf{v})$, respectively. A few examples of this are given in Sections 2.3 and 2.4 above.

3 The multivariate Tutte polynomial for matroids

Many (but not all) of the properties of the multivariate Tutte polynomial for graphs can be carried over to the matroid version (1.3). Where the treatment is essentially identical, I shall be brief.

3.1 $q = 1$

When $q = 1$, the multivariate Tutte polynomial becomes trivial:

$$\widetilde{Z}_M(1, \mathbf{v}) = \prod_{e \in E} (1 + v_e)\,. \tag{3.1}$$

3.2 $q = 1, 2, 3, \ldots$ (generalized Potts models)

For *graphs*, we saw in Section 2.2 that the multivariate Tutte polynomial $Z_G(q, \mathbf{v})$ for integer $q \geq 1$ counts q-colorings of the vertices of G, with weights that depend on whether each edge e is properly or improperly colored [cf. (2.7)/(2.9)]. In particular, the chromatic polynomial $P_G(q) = Z_G(q, -1)$ counts proper q-colorings of G. One consequence of this is that the nonnegative integer zeros of $P_G(q)$ are consecutive integers, namely $0, 1, 2, \ldots, \chi(G) - 1$, where $\chi(G)$ is the chromatic number of G.[7]

[7] In fact, these are *all* the integer zeros of $P_G(q)$, as it is easy to show that $P_G(q)$ has no negative real zeros (when G is loopless): see e.g. [106].

Simple examples show that the latter property does not hold in general for the chromatic polynomial (also known as *characteristic polynomial*) of a matroid, $\widetilde{P}_M(q) = \widetilde{Z}_M(q, -1)$.[8] For instance, for the uniform matroid $U_{2,n}$ we have

$$\widetilde{Z}_{U_{2,n}}(q, \mathbf{v}) \;=\; q^{-2}\prod_{i=1}^{n}(1+v_i) \;+\; (1-q^{-2}) \;+\; (q^{-1}-q^{-2})\sum_{i=1}^{n} v_i \tag{3.2}$$

and hence

$$\widetilde{P}_{U_{2,n}}(q) \;=\; q^{-2}(q-1)(q+1-n)\,, \tag{3.3}$$

whose roots are $q = 1$ and $q = n - 1$; so the consecutive-root property is violated for all $n \geq 4$. It therefore seems unlikely that $\widetilde{Z}_M(q, \mathbf{v})$ can be given a "coloring" interpretation for *arbitrary* matroids M and *arbitrary* integers $q \geq 1$. Nevertheless, in some special cases such a representation can be given, as I would now like to sketch; this account is drawn from work in progress with Sergio Caracciolo and Andrea Sportiello [32].

The appropriate general context for this discussion is that of *abelian-group-valued statistical-mechanics models*, defined as follows [31]: Let V and E be finite index sets (for *variables* and *interactions*, respectively); let R be a ring with identity and let $\mathcal{M}$ be a unitary right R-module; let $B = (b_{ie})_{i\in V,\, e\in E}$ be a matrix with elements in R; let μ be Haar measure on $\mathcal{M}$, considered as an additive abelian group; and for each $e \in E$, let $W_e \colon\, \mathcal{M} \to \mathbb{C}$ be a function. We then define the partition function

$$\widetilde{Z}_{R,\mathcal{M},B}(\mathbf{W}) \;=\; \int \prod_{e\in E} W_e\Bigl(\sum_{i\in V}\sigma_i b_{ie}\Bigr) \prod_{i\in V} d\mu(\sigma_i) \tag{3.4}$$

where $\mathbf{W} = \{W_e\}_{e\in E}$. (If $\mathcal{M}$ is infinite, then the collection of functions $\mathbf{W}$ needs to satisfy suitable integrability conditions.)

[We can alternatively let $\mathcal{M}$ be an arbitrary additive abelian group $\mathcal{G}$, and let R be some subring of the ring $\mathrm{End}(\mathcal{G})$ of endomorphisms of $\mathcal{G}$. It is easy to see that the two formulations are equivalent.]

Two important special cases are:

1) $R = \mathbb{Z}$ and $\mathcal{M}$ is an arbitrary additive abelian group $\mathcal{G}$ (on which $\mathbb{Z}$ acts in the obvious way).

2) R is a field F, and $\mathcal{M}$ is a finite-dimensional vector space over F (which can be taken, without loss of generality, to be F^k for some integer $k \geq 0$).

A *generalized Potts model* is the special case of this setup in which $\mathcal{M}$ is finite, μ is normalized counting measure on $\mathcal{M}$ (i.e., each element of $\mathcal{M}$ gets weight $1/|\mathcal{M}|$), and each W_e is of the form

$$W_e(\sigma) \;=\; 1 \,+\, v_e\delta(\sigma, 0) \tag{3.5}$$

for some weights $\mathbf{v} = \{v_e\}_{e\in E}$. We then have

$$\widetilde{Z}^{\mathrm{Potts}}_{R,\mathcal{M},B}(\mathbf{v}) \;=\; |\mathcal{M}|^{-|V|} \sum_{\sigma\colon V\to\mathcal{M}} \prod_{e\in E}\Bigl[1 + v_e\delta\Bigl(\sum_{i\in V}\sigma_i b_{ie},\, 0\Bigr)\Bigr]\,. \tag{3.6}$$

---

[8] The chromatic (or characteristic) polynomial of a matroid M is ordinarily defined as $P_M(q) = q^{r(M)}\widetilde{P}_M(q)$, so that it is a polynomial in nonnegative powers of q. This is a matter of taste.

Note that if $G = (V, E)$ is a graph, B is the directed vertex-edge incidence matrix for some orientation of G (hence $R = \mathbb{Z}$), and $\mathcal{M}$ is a finite abelian group of order $q = |\mathcal{M}|$, we have $\widetilde{Z}^{\rm Potts}_{R,\mathcal{M},B}(\mathbf{v}) = \widetilde{Z}^{\rm Potts}_{G}(q, \mathbf{v}) \equiv q^{-|V|} Z^{\rm Potts}_{G}(q, \mathbf{v})$. So the models (3.6) do indeed constitute a generalization of the Potts models (2.7).

As far as I know, there does not exist, *at this level of generality*, a satisfactory analogue of the Fortuin–Kasteleyn representation (2.9). But at least in some special cases such a representation does exist [32]:

Theorem 3.1 (Fortuin–Kasteleyn representation for generalized Potts models)

(a) *Let M be a regular matroid, let B be a totally unimodular matrix of integers representing M over $\mathbb{Q}$, and let $\mathcal{G}$ be a finite abelian group of order $q = |\mathcal{G}|$. Then*

$$\widetilde{Z}^{\rm Potts}_{\mathbb{Z},\mathcal{G},B}(\mathbf{v}) \;=\; \widetilde{Z}_{M}(q, \mathbf{v})\,. \tag{3.7}$$

(b) *Let M be a matroid, let F be a finite field of order $q = |F|$ [so that q is a prime power and $F \simeq GF(q)$], let B be a matrix representing M over F, and let $\mathcal{M} = F^k$ for some integer $k \geq 0$. Then*

$$\widetilde{Z}^{\rm Potts}_{F,F^k,B}(\mathbf{v}) \;=\; \widetilde{Z}_{M}(q^k, \mathbf{v})\,. \tag{3.8}$$

We recall that an integer matrix B is called *totally unimodular* if all its subdeterminants lie in $\{0, 1, -1\}$.

Proof In (3.6), expand out the product over $e \in E$, and let $A \subseteq E$ be the set of edges for which the term $v_e \delta(\cdots)$ is taken. We then need to count the number of configurations $\boldsymbol{\sigma} = \{\sigma_i\}_{i \in V}$ satisfying $\sum_{i \in V} \sigma_i b_{ie} = 0$ for all $e \in A$. In other words, we need to count the number of solutions $\boldsymbol{\sigma} \in \mathcal{M}^V$ of the linear system $\boldsymbol{\sigma}\widehat{B} = 0$, where $\widehat{B}$ is the submatrix of B consisting of the columns from A.

Since both $R = \mathbb{Z}$ and $R = F$ are principal ideal domains, the matrix $\widehat{B}$ has a Smith normal form [2, 19, 89, 95]

$$S \;=\; P\widehat{B}Q \;=\; \begin{pmatrix} D & 0 \\ 0 & 0 \end{pmatrix} \tag{3.9}$$

where $P \in GL(m, R)$, $Q \in GL(n, R)$ and $D = \operatorname{diag}(s_1, \ldots, s_r)$; here $m = |V|$, $n = |A|$, r is the determinantal rank of $\widehat{B}$, and $s_1, \ldots, s_r \neq 0$ are the invariant factors of $\widehat{B}$. In case (a), since B is totally unimodular, we must have $s_1, \ldots, s_r \in \{1, -1\}$; and by a suitable choice of P (or Q) we can arrange to have $s_1 = \ldots = s_r = 1$. In case (b), because R is a field, we can again arrange to have $s_1 = \ldots = s_r = 1$. Therefore, a vector $\boldsymbol{\sigma} \in \mathcal{M}^V$ solves $\boldsymbol{\sigma}\widehat{B} = 0$ if and only if $\boldsymbol{\sigma}P^{-1}$ is vanishing in its first r components. The number of such solutions is thus $|\mathcal{M}|^{|V|-r}$. On the other hand, since R is an integral domain, the determinantal rank r is also equal [2, pp. 205–209] to the column rank of $\widehat{B}$ in the quotient field of R [namely, $\mathbb{Q}$ in case (a), and F itself in case (b)]; and this rank is by definition $r_M(A)$. This proves (3.7)/(3.8). □

Theorem 3.1(a) implies, in particular, the known fact [28, Exercise 6.53(e,f)] that the chromatic polynomial of a regular matroid has the consecutive-root property. See also [40, Theorem III] [27, Theorem 12.4] for results related to Theorem 3.1(a), and [41, Theorem 16.1] [144, Theorem 7.6.1] for results related to Theorem 3.1(b).

3.3 $q \to 0$ limits

The $q \to 0$ limits for the matroid multivariate Tutte polynomial (1.3) follow closely the pattern for graphs, with the obvious replacements

$$\begin{array}{rcl} \text{maximally connected spanning subgraph} & \longrightarrow & \text{spanning set} \\ \text{spanning forest} & \longrightarrow & \text{independent set} \\ \text{maximal spanning forest} & \longrightarrow & \text{basis} \end{array}$$

Thus, the limit $q \to 0$ with fixed $\mathbf{v}$ selects out the subsets $A \subseteq E$ of maximum rank, i.e. the spanning sets. We therefore have

$$\lim_{q\to 0} q^{r_M(E)} \widetilde{Z}_M(q, \mathbf{v}) \;=\; S_M(\mathbf{v}) \,, \tag{3.10}$$

where $r_M(E)$ is the rank of M, and

$$S_M(\mathbf{v}) \;=\; \sum_{\substack{A \subseteq E \\ r_M(A) = r_M(E)}} \prod_{e \in A} v_e \tag{3.11}$$

is the generating polynomial of *spanning sets* in M.

Similarly, the limit $q \to 0$ with fixed $\mathbf{w} = \mathbf{v}/q$ selects out the subsets $A \subseteq E$ having the smallest value of $|A| - r_M(A)$; the minimum achievable value is 0, and the sets attaining it are the independent sets. We therefore have

$$\lim_{q\to 0} \widetilde{Z}_M(q, q\mathbf{w}) \;=\; I_M(\mathbf{w}) \,, \tag{3.12}$$

where

$$I_M(\mathbf{w}) \;=\; \sum_{\substack{A \subseteq E \\ r_M(A) = |A|}} \prod_{e \in A} w_e \tag{3.13}$$

is the generating polynomial of *independent sets* in M.

Finally, we have

$$\lim_{\lambda\to 0} \lambda^{-r_M(E)} S_M(\lambda \mathbf{v}) \;=\; B_M(\mathbf{v}) \tag{3.14a}$$

$$\lim_{\lambda\to \infty} \lambda^{-r_M(E)} I_M(\lambda \mathbf{w}) \;=\; B_M(\mathbf{w}) \tag{3.14b}$$

where

$$B_M(\mathbf{v}) \;=\; \sum_{\substack{A \subseteq E \\ |A| = r_M(A) = r_M(E)}} \prod_{e \in A} v_e \tag{3.15}$$

is the generating polynomial of *bases* in M.

In summary, we have the following scheme for the $q \to 0$ limits of the multivariate Tutte polynomial for matroids:

$$\begin{array}{ccccc} & & S_M(\mathbf{v}) & & \\ & \nearrow & & \searrow & \\ & q \to 0,\ \mathbf{v}\ \text{fixed} & & \mathbf{v}\ \text{infinitesimal} & \\ \widetilde{Z}_M(q,\mathbf{v}) & & & & B_M(\mathbf{v}\ \text{or}\ \mathbf{w}) \\ & \searrow & & \nearrow & \\ & q \to 0,\ \mathbf{w} = \mathbf{v}/q\ \text{fixed} & & \mathbf{w}\ \text{infinite} & \\ & & I_M(\mathbf{w}) & & \end{array} \qquad (3.16)$$

3.4 A final remark

Several alternative definitions of multivariate (or "weighted") Tutte polynomials for graphs and/or matroids can be found in the literature. Most of these are essentially equivalent to $\widetilde{Z}_M(q, \mathbf{v})$ after suitable changes of variables. Exceptions can, however, be found in the work of Zaslavsky [145] and Bollobás–Riordan [9]; their results are generalized and clarified in an illuminating recent paper of Ellis-Monaghan and Traldi [54].

4 Elementary identities

I now wish to prove some elementary identities for the multivariate Tutte polynomial. There are two alternative approaches to proving such identities: one is to prove the identity directly for indeterminate (or complex) q, using the subgraph expansion (1.1) or its generalization (1.3) to matroids; the other is to prove the identity first for *positive integer* q, using the spin/coloring representation (2.7), and then to extend it to general q by arguing that two polynomials (or rational functions) that coincide at infinitely many points must be equal. The latter approach is perhaps less elegant, but it is often simpler or more intuitive. However, only the former approach extends to matroids.

One way to guess (albeit not to prove) an identity for matroids is to prove it first for graphs, and then translate it from Z_G to $\widetilde{Z}_G = q^{-|V|} Z_G$; usually the latter identity carries over verbatim to matroids, *mutatis mutandis*.

4.1 Disjoint unions and direct sums

If G is the disjoint union of G_1 and G_2, then trivially

$$Z_G(q, \mathbf{v}) \;=\; Z_{G_1}(q, \mathbf{v})\, Z_{G_2}(q, \mathbf{v}) \,. \qquad (4.1)$$

That is, Z_G "factorizes over components".

A slightly less trivial situation arises when G consists of subgraphs G_1 and G_2 joined at a single cut vertex x; in this case

$$Z_G(q, \mathbf{v}) \;=\; \frac{Z_{G_1}(q, \mathbf{v})\, Z_{G_2}(q, \mathbf{v})}{q} \,. \qquad (4.2)$$

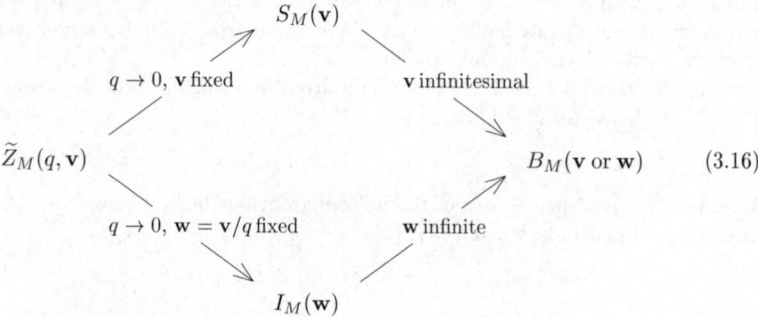

This is easily seen from the subgraph expansion in the form (2.3). It is also easily seen from the coloring representation (2.7), by first fixing the color σ_x at the cut vertex and then summing over it; from this viewpoint, (4.2) reflects the S_q permutation symmetry of the q-state Potts model.[9] We summarize (4.2) by saying that Z_G "factorizes over blocks" modulo a factor q.

The identities (4.1) and (4.2) can be written in a unified form, by using $\widetilde{Z}_G = q^{-|V|} Z_G$: in both cases we have

$$\widetilde{Z}_G(q, \mathbf{v}) \;=\; \widetilde{Z}_{G_1}(q, \mathbf{v})\, \widetilde{Z}_{G_2}(q, \mathbf{v}) \,. \tag{4.3}$$

This, in turn, is a special case of the following obvious fact: if a matroid M is the direct sum of matroids M_1 and M_2, then

$$\widetilde{Z}_M(q, \mathbf{v}) \;=\; \widetilde{Z}_{M_1}(q, \mathbf{v})\, \widetilde{Z}_{M_2}(q, \mathbf{v}) \,. \tag{4.4}$$

4.2 Duality

Suppose first that $G = (V, E)$ is a connected *planar* graph. Consider any plane embedding of G, and let $G^* = (V^*, E^*)$ be the corresponding dual graph. There is a natural bijection between E and E^* (namely, an edge $e \in E$ is identified with the unique edge $e^* \in E^*$ that it crosses), so we shall henceforth identify E^* with E. Of course, the vertex set V^* can be identified with the faces in the given embedding of G, so by Euler's relation we have

$$|V| - |E| + |V^*| \;=\; 2 \,. \tag{4.5}$$

Consider now any subset $A \subseteq E$, and draw in G^* the *complementary* set of edges $(E \setminus A)$. Simple topological arguments then yield the relations

$$k_G(A) \;=\; c_{G^*}(E \setminus A) + 1 \tag{4.6a}$$

$$k_{G^*}(E \setminus A) \;=\; c_G(A) + 1 \tag{4.6b}$$

where k = components and c = cyclomatic number. [Note that (4.6a) and (4.6b) are equivalent, as a consequence of (4.5).] Substituting (4.6) into (2.1)–(2.3), we deduce the *duality relation* [140]

$$Z_{G^*}(q, \mathbf{v}) \;=\; q^{1-|V|} \left(\prod_{e \in E} v_e \right) Z_G(q, q/\mathbf{v}) \,. \tag{4.7}$$

In brief, duality takes $v_e \mapsto q/v_e$ (and inserts some prefactors). Note that two applications of (4.7) lead us back to where we started, thanks to (4.5). In terms of $\widetilde{Z}_G = q^{-|V|} Z_G$, we have

$$\widetilde{Z}_{G^*}(q, \mathbf{v}) \;=\; q^{1-|V^*|} \left(\prod_{e \in E} v_e \right) \widetilde{Z}_G(q, q/\mathbf{v}) \tag{4.8a}$$

$$\;=\; q^{|V|-1} \left(\prod_{e \in E} \frac{v_e}{q} \right) \widetilde{Z}_G(q, q/\mathbf{v}) \,. \tag{4.8b}$$

[9] More precisely, it reflects the symmetry of the spin model under a global transformation $\sigma_y \mapsto g\sigma_y$ (simultaneously for all $y \in V$) that acts *transitively* on each single-spin space.

In the $q \to 0$ limit, we obtain the following special cases of (4.7):

$$C_{G^*}(\mathbf{v}) \;=\; \left(\prod_{e\in E} v_e\right) F_G(1/\mathbf{v}) \tag{4.9}$$

$$F_{G^*}(\mathbf{w}) \;=\; \left(\prod_{e\in E} w_e\right) C_G(1/\mathbf{w}) \tag{4.10}$$

$$T_{G^*}(\mathbf{v}) \;=\; \left(\prod_{e\in E} v_e\right) T_G(1/\mathbf{v}) \tag{4.11}$$

We can also relate the multivariate Tutte polynomial on G^* to the multivariate flow polynomial on G: by (2.22) we have

$$Z_{G^*}(q,\mathbf{v}) \;=\; qF_G(q,\mathbf{v})\,. \tag{4.12}$$

In particular, the chromatic polynomial on G^* is essentially identical to the flow polynomial on G:

$$P_{G^*}(q) \;=\; qF_G(q)\,. \tag{4.13}$$

The chromatic polynomial and the flow polynomial are thus "dual" objects.

Among graphs, only *planar* graphs have duals with good properties [43, section 4.6] [98, section 5.2]; one major advantage of considering matroids is that *every* matroid has a dual. The duality formula for the matroid multivariate Tutte polynomial is, not surprisingly, identical in form to (4.8):

$$\widetilde{Z}_{M^*}(q,\mathbf{v}) \;=\; q^{-r_{M^*}(E)} \left(\prod_{e\in E} v_e\right) \widetilde{Z}_M(q,q/\mathbf{v}) \tag{4.14a}$$

$$\;=\; q^{r_M(E)} \left(\prod_{e\in E} \frac{v_e}{q}\right) \widetilde{Z}_M(q,q/\mathbf{v})\,. \tag{4.14b}$$

[Here $r_M(E)$ is the rank of M and $r_{M^*}(E)$ is the rank of M^*; their sum is $|E|$.] Indeed, (4.14) is an easy consequence of the definition (1.3) together with the formula for the rank function of a dual:

$$r_{M^*}(A) \;=\; |A| + r_M(E\setminus A) - r_M(E)\,. \tag{4.15}$$

4.3 Deletion-contraction identity

If $e \in E$, let $G \setminus e$ denote the graph obtained from G by deleting the edge e, and let G/e denote the graph obtained from $G \setminus e$ by contracting the two endpoints of e into a single vertex (please note that we retain in G/e any loops or multiple edges that may be formed as a result of the contraction). Then, for any $e \in E$, we have the identity

$$Z_G(q,\mathbf{v}) \;=\; Z_{G\setminus e}(q,\mathbf{v}_{\neq e}) + v_e Z_{G/e}(q,\mathbf{v}_{\neq e}) \tag{4.16}$$

where $\mathbf{v}_{\neq e} = \{v_f\}_{f\in E\setminus e}$. This is easily seen either from the coloring representation (2.7) or the subgraph expansion (1.1). Please note that the deletion-contraction identity (4.16) takes the same form regardless of whether e is a normal edge, a loop, or a bridge (in contrast to the situation for the usual Tutte polynomial T_G). Of course, if e is a loop, then $G/e = G \setminus e$, so we can also write $Z_G = (1+v_e)Z_{G\setminus e} = (1+v_e)Z_{G/e}$. Similarly, if e is a bridge, then $G \setminus e$ is the disjoint union of two subgraphs G_1 and G_2 while G/e is obtained by joining G_1 and G_2 at a cut vertex, so that $Z_{G/e} = Z_{G\setminus e}/q$ and hence $Z_G = (1+v_e/q)Z_{G\setminus e} = (q+v_e)Z_{G/e}$.

In terms of $\widetilde{Z}_G = q^{-|V|}Z_G$, the deletion-contraction identity takes the form

$$\widetilde{Z}_G \;=\; \widetilde{Z}_{G\setminus e} + \frac{v_e}{q}\widetilde{Z}_{G/e} \qquad \text{if } e \text{ is not a loop} \tag{4.17a}$$

$$\begin{aligned} \widetilde{Z}_G \;&=\; \widetilde{Z}_{G\setminus e} + v_e\widetilde{Z}_{G/e} \\ &=\; (1+v_e)\widetilde{Z}_{G\setminus e} \\ &=\; (1+v_e)\widetilde{Z}_{G/e} \qquad \text{if } e \text{ is a loop} \end{aligned} \tag{4.17b}$$

as easily follows from (4.16) together with the counting of vertices in $G\setminus e$ and G/e.

Not surprisingly, the deletion-contraction formula for matroids is identical in form to (4.17):

$$\widetilde{Z}_M \;=\; \widetilde{Z}_{M\setminus e} + \frac{v_e}{q}\widetilde{Z}_{M/e} \qquad \text{if } e \text{ is not a loop} \tag{4.18a}$$

$$\begin{aligned} \widetilde{Z}_M \;&=\; \widetilde{Z}_{M\setminus e} + v_e\widetilde{Z}_{M/e} \\ &=\; (1+v_e)\widetilde{Z}_{M\setminus e} \\ &=\; (1+v_e)\widetilde{Z}_{M/e} \qquad \text{if } e \text{ is a loop} \end{aligned} \tag{4.18b}$$

This easily follows from the formulae for the rank function of a deletion or contraction: if $A \subseteq E \setminus e$, then

$$r_{M\setminus e}(A) \;=\; r_M(A) \tag{4.19a}$$

$$r_{M/e}(A) \;=\; \begin{cases} r_M(A\cup e) - 1 & \text{if } e \text{ is not a loop} \\ r_M(A\cup e) & \text{if } e \text{ is a loop} \end{cases} \tag{4.19b}$$

Remark Many treatments in the literature *define* the Tutte polynomial by the deletion-contraction identity (4.16) together with the initial condition $Z_G(q,\mathbf{v}) = q^{|V|}$ for an edgeless graph (or the analogous thing for the standard Tutte polynomial T_G). But this approach has the disadvantage that one must prove that this Z_G is well-defined, i.e. that the result does not depend on the order in which one applies (4.16) to the various edges. A much cleaner approach, it seems to me, is to define $Z_G(q,\mathbf{v})$ by the explicit formula (1.1), and then deduce the deletion-contraction identity as an immediate property.

4.4 Parallel-reduction identity

If G contains edges e_1, e_2 connecting the same pair of vertices x, y, they can be replaced, without changing the value of Z, by a single edge $e = xy$ with weight

$$v_e \;=\; (1+v_{e_1})(1+v_{e_2}) \;-\; 1 \;=\; v_{e_1} + v_{e_2} + v_{e_1}v_{e_2}\,. \tag{4.20}$$

This is easily seen either from the coloring representation (2.7) or the subgraph expansion (1.1). More formally, we can identify the new edge e with (for instance) the old edge e_1 after deletion of e_2, and thus write

$$Z_G(q, \mathbf{v}_{\neq e_1,e_2}, v_{e_1}, v_{e_2}) \;=\; Z_{G\setminus e_2}(q, \mathbf{v}_{\neq e_1,e_2}, v_{e_1} + v_{e_2} + v_{e_1}v_{e_2})\,. \tag{4.21}$$

The parallel-reduction rule $(v_1, v_2) \mapsto v_{\text{eff}}$ with $1 + v_{\text{eff}} = (1 + v_1)(1 + v_2)$ can be remembered by the mnemonic "1+v multiplies". We write $v_1 \| v_2 \equiv (1+v_1)(1+v_2)-1$.

A virtually identical formula holds for matroids: if e_1 and e_2 are parallel elements in a matroid M (i.e., form a two-element circuit), then

$$\widetilde{Z}_M(q, \mathbf{v}_{\neq e_1,e_2}, v_{e_1}, v_{e_2}) \;=\; \widetilde{Z}_{M\setminus e_2}(q, \mathbf{v}_{\neq e_1,e_2}, v_{e_1} + v_{e_2} + v_{e_1}v_{e_2})\,. \tag{4.22}$$

The formula (4.22) also holds trivially if e_1 and e_2 are both loops.

4.5 Series-reduction identity

We say that edges $e_1, e_2 \in E$ are *in series (in the narrow sense)* if there exist vertices $x, y, z \in V$ with $x \neq y$ and $y \neq z$ such that e_1 connects x and y, e_2 connects y and z, and y has degree 2 in G. In this case the pair of edges e_1, e_2 can be replaced, without changing the value of Z, by a single edge $e = xz$ with weight

$$v_e \;=\; \frac{v_{e_1}v_{e_2}}{q + v_{e_1} + v_{e_2}} \tag{4.23}$$

provided that we then multiply Z by the prefactor $q + v_{e_1} + v_{e_2}$. More formally, we can identify the new edge e with (for instance) the old edge e_1 after contraction of e_2, and thus write

$$Z_G(q, \mathbf{v}_{\neq e_1,e_2}, v_{e_1}, v_{e_2}) \;=\; (q + v_{e_1} + v_{e_2})\, Z_{G/e_2}(q, \mathbf{v}_{\neq e_1,e_2}, v_{e_1}v_{e_2}/(q + v_{e_1} + v_{e_2}))\,. \tag{4.24}$$

This identity can be derived from the coloring representation (2.7) by noting that

$$\begin{aligned} &\sum_{\sigma_y=1}^{q} [1 + v_{e_1}\delta(\sigma_x, \sigma_y)]\, [1 + v_{e_2}\delta(\sigma_y, \sigma_z)] \\ &\quad = q + v_{e_1} + v_{e_2} + v_{e_1}v_{e_2}\delta(\sigma_x, \sigma_z) && \text{(4.25a)} \\ &\quad = (q + v_{e_1} + v_{e_2}) \left[1 + \frac{v_{e_1}v_{e_2}}{q + v_{e_1} + v_{e_2}}\delta(\sigma_x, \sigma_z)\right]. && \text{(4.25b)} \end{aligned}$$

Alternatively, it can be derived from the subgraph expansion (1.1) by considering the four possibilities for the edges e_1 and e_2 to be occupied or empty and analyzing the number of connected components thereby created. The series-reduction rule $(v_1, v_2) \mapsto v_{\text{eff}} \equiv v_1v_2/(q + v_1 + v_2)$ can be remembered by the mnemonic "$1 + q/v$ multiplies": namely,

$$1 + \frac{q}{v_{\text{eff}}} \;=\; \left(1 + \frac{q}{v_1}\right)\left(1 + \frac{q}{v_2}\right). \tag{4.26}$$

We write $v_1 \bowtie_q v_2 \equiv v_1v_2/(q + v_1 + v_2)$.

Consider now the more general situation in which $\{e_1, e_2\}$ is a two-edge cut of G (not necessarily the cut associated with a degree-2 vertex y); we then say that e_1, e_2 are *in series (in the wide sense)*. It turns out that the identity (4.24) still holds. To see this, let us prove the generalization of this identity to matroids. Let e_1 and e_2 be series elements in a matroid M, i.e., suppose that $\{e_1, e_2\}$ is a cocircuit. Then, for any $A \subseteq E \setminus \{e_1, e_2\}$, we have

$$r_M(A \cup e_1) \;=\; r_M(A \cup e_2) \;=\; r_M(A) + 1 \tag{4.27}$$

(since the complement of a cocircuit is a hyperplane). A short calculation using (4.19b) with $e = e_2$ then yields

$$\widetilde{Z}_M(q, \mathbf{v}_{\neq e_1,e_2}, v_{e_1}, v_{e_2}) \;=\; \frac{q + v_{e_1} + v_{e_2}}{q}\, \widetilde{Z}_{M/e_2}(q, \mathbf{v}_{\neq e_1,e_2}, v_{e_1}v_{e_2}/(q + v_{e_1} + v_{e_2}))\,. \tag{4.28}$$

The formula (4.28) also holds trivially if e_1 and e_2 are both coloops.

Please note that duality $v \mapsto q/v$ interchanges the parallel-reduction rule ("$1+v$ multiplies") with the series-reduction rule ("$1+q/v$ multiplies"). This is no accident, since we now see that parallel-reduction and series-reduction (in the wide sense) are indeed duals of each other: $\{e_1, e_2\}$ is a circuit (resp. cocircuit) in M if and only if it is a cocircuit (resp. circuit) in M^*.

Remark Using the series-reduction formula (4.24) together with the multivariate approach, one can give a simple proof of (a generalization of) the Brown–Hickman [17] theorem on chromatic roots of large subdivisions [122, Appendix A]. Likewise, using the parallel- and series-reduction formulae (4.21)/(4.24), one can give a simple proof of (a slight generalization of) Thomassen's [129] construction of 2-degenerate graphs with arbitrarily large real chromatic roots [122, Appendix B].

4.6 Reduction formulae for 2-rooted subgraphs

Let $G = (V, E)$ be a finite graph, and let x, y be distinct vertices of G. We define G/xy to be the graph in which x and y are contracted to a single vertex. (N.B.: If G contains one or more edges xy, then these edges are *not* deleted, but become loops in G/xy.) There is a canonical one-to-one correspondence between the edges of G and the edges of G/xy; for simplicity (though by slight abuse of notation) we denote an edge of G and the corresponding edge of G/xy by the same letter. In particular, we can apply a given set of edge weights $\{v_e\}_{e\in E}$ to both G and G/xy.

Let us now define

$$Z_G^{(x\leftrightarrow y)}(q, \mathbf{v}) \;=\; \sum_{\substack{A \subseteq E \\ A \text{ connects } x \text{ to } y}} q^{k(A)} \prod_{e\in A} v_e \tag{4.29}$$

$$Z_G^{(x\not\leftrightarrow y)}(q, \mathbf{v}) \;=\; \sum_{\substack{A \subseteq E \\ A \text{ does not connect } x \text{ to } y}} q^{k(A)} \prod_{e\in A} v_e \tag{4.30}$$

From (1.1) we have trivially

$$Z_G(q, \mathbf{v}) \;=\; Z_G^{(x\leftrightarrow y)}(q, \mathbf{v}) \,+\, Z_G^{(x\not\leftrightarrow y)}(q, \mathbf{v}) \tag{4.31}$$

and almost as trivially

$$Z_{G/xy}(q,\mathbf{v}) \;=\; Z_G^{(x\leftrightarrow y)}(q,\mathbf{v}) \,+\, q^{-1} Z_G^{(x\not\leftrightarrow y)}(q,\mathbf{v})\,. \tag{4.32}$$

Let now q be an integer ≥ 1, and define the restricted Potts-model partition function

$$Z_{G,x,y}^{\rm Potts}(q,\mathbf{v};\sigma_x,\sigma_y) \;=\; \sum_{\sigma\colon\, V\setminus\{x,y\}\to\{1,\ldots,q\}} \; \prod_{e\in E} \Big[1 + v_e\delta(\sigma_{x1(e)},\sigma_{x2(e)})\Big] \tag{4.33}$$

where $\sigma_x,\sigma_y \in \{1,\ldots,q\}$. We then have the following refinement of the Fortuin–Kasteleyn identity (2.9):

Proposition 4.1

$$Z_{G,x,y}^{\rm Potts}(q,\mathbf{v};\sigma_x,\sigma_y) \;=\; A_{G,x,y}(q,\mathbf{v}) \,+\, B_{G,x,y}(q,\mathbf{v})\,\delta(\sigma_x,\sigma_y) \tag{4.34}$$

where

$$A_{G,x,y}(q,\mathbf{v}) \;=\; q^{-2} Z_G^{(x\not\leftrightarrow y)}(q,\mathbf{v}) \tag{4.35a}$$

$$B_{G,x,y}(q,\mathbf{v}) \;=\; q^{-1} Z_G^{(x\leftrightarrow y)}(q,\mathbf{v}) \tag{4.35b}$$

are polynomials in q and $\{v_e\}$, whose degrees in q are

$$\deg A \;=\; |V| - 2 \tag{4.36a}$$

$$\deg B \;\leq\; |V| - 1 - {\rm dist}(x,y) \tag{4.36b}$$

where ${\rm dist}(x,y)$ *is the length of the shortest path from x to y using edges having $v_e \neq 0$ (if no such path exists, then $B = 0$).*

First Proof In (4.33), expand out the product over $e \in E$, and let $A \subseteq E$ be the set of edges for which the term $v_e\delta(\sigma_{x1(e)},\sigma_{x2(e)})$ is taken. Now perform the sum over configurations $\{\sigma_z\}_{z\in V\setminus\{x,y\}}$: in each connected component of the subgraph (V,A) the spin value σ_z must be constant. In particular, in each component containing x and/or y, the spins must all equal the specified value σ_x and/or σ_y; in all other components, the spin value is free. Therefore,

$$Z_{G,x,y}^{\rm Potts}(q,\mathbf{v};\sigma_x,\sigma_y) \;=\; \begin{cases} q^{-2} Z_G^{(x\not\leftrightarrow y)}(q,\mathbf{v}) & \text{if } \sigma_x \neq \sigma_y \\ q^{-2} Z_G^{(x\not\leftrightarrow y)}(q,\mathbf{v}) + q^{-1} Z_G^{(x\leftrightarrow y)}(q,\mathbf{v}) & \text{if } \sigma_x = \sigma_y \end{cases} \tag{4.37}$$

The claims about the degrees are easily verified. □

Second Proof For each positive integer q, the S_q permutation symmetry of the Potts model implies that

$$Z_{G,x,y}^{\rm Potts}(q,\mathbf{v};\sigma_x,\sigma_y) \;=\; A \,+\, B\delta(\sigma_x,\sigma_y) \tag{4.38}$$

for some numbers A and B depending on G, x, y, q and $\mathbf{v}$. But then, summing over σ_x, σ_y without and with the constraint $\sigma_x = \sigma_y$, we get

$$Z_G \;=\; q^2 A \,+\, qB \tag{4.39a}$$

$$Z_{G/xy} \;=\; qA \,+\, qB \tag{4.39b}$$

Hence

$$A = \frac{Z_G - Z_{G/xy}}{q(q-1)} = q^{-2} Z_G^{(x \not\leftrightarrow y)} \tag{4.40a}$$

$$B = \frac{qZ_{G/xy} - Z_G}{q(q-1)} = q^{-1} Z_G^{(x \leftrightarrow y)} \tag{4.40b}$$

by virtue of (4.31) and (4.32). □

Let us now consider inserting the 2-rooted graph (G, x, y) in place of an edge e_* in some other graph H. From (4.34)/(4.35) we see that (G, x, y) then acts as a single edge with effective weight

$$v_{\rm eff} \equiv v_{{\rm eff},G,x,y}(q, \mathbf{v}) = \frac{B_{G,x,y}(q, \mathbf{v})}{A_{G,x,y}(q, \mathbf{v})} = \frac{qZ_G^{(x \leftrightarrow y)}(q, \mathbf{v})}{Z_G^{(x \not\leftrightarrow y)}(q, \mathbf{v})}, \tag{4.41}$$

which is a rational function of q and $\{v_e\}$; in addition, the partition function is multiplied by an overall prefactor $A_{G,x,y}(q, \mathbf{v})$. This follows from Proposition 4.1 whenever q is an integer ≥ 1; and the corresponding identity then holds for all q, because both sides are rational functions of q that agree at infinitely many points. It is also worth noting that the "transmissivity" $t_{\rm eff} \equiv v_{\rm eff}/(q + v_{\rm eff})$ is given by the simple formula

$$t_{\rm eff} = \frac{Z_G^{(x \leftrightarrow y)}(q, \mathbf{v})}{Z_G(q, \mathbf{v})}. \tag{4.42}$$

The most general version of this construction appears to be the following [138]: Let $H = (V, E)$ be a finite undirected graph, and let $\vec{H}$ be a directed graph obtained by assigning an orientation to each edge of H. For each edge $e \in E$, let $G_e = (V_e, E_e, x_e, y_e)$ be a 2-rooted finite undirected graph (so that $x_e, y_e \in V_e$ with $x_e \neq y_e$) equipped with edge weights $\{v_{\tilde{e}}\}_{\tilde{e} \in E_e}$. We denote by $\mathbf{G}$ the family $\{G_e\}_{e \in E}$, and we denote by $\vec{H}^{\mathbf{G}}$ the undirected graph obtained from H by replacing each edge $e \in E$ with a copy of the corresponding graph G_e, attaching x_e to the tail of e and y_e to the head. Its edge set is thus $\mathbf{E} = \biguplus_{e \in E} E_e$ (disjoint union). We then have:

Proposition 4.2 *Let $H = (V, E)$ and $\{G_e\}_{e \in E}$ be as above. Suppose that*

$$\begin{aligned} & Z_{G_e, x_e, y_e}(q, \{v_{\tilde{e}}\}_{\tilde{e} \in E_e}; \sigma_{x_e}, \sigma_{y_e}) \\ & \quad = A_{G_e, x_e, y_e}(q, \{v_{\tilde{e}}\}_{\tilde{e} \in E_e}) + B_{G_e, x_e, y_e}(q, \{v_{\tilde{e}}\}_{\tilde{e} \in E_e}) \, \delta(\sigma_{x_e}, \sigma_{y_e}), \end{aligned} \tag{4.43}$$

and define

$$v_{{\rm eff},e} \equiv \frac{B_{G_e, x_e, y_e}(q, \{v_{\tilde{e}}\}_{\tilde{e} \in E_e})}{A_{G_e, x_e, y_e}(q, \{v_{\tilde{e}}\}_{\tilde{e} \in E_e})}. \tag{4.44}$$

Then

$$Z_{\vec{H}^{\mathbf{G}}}(q, \{v_{\tilde{e}}\}_{\tilde{e} \in \mathbf{E}}) = \left(\prod_{e \in E} A_{G_e, x_e, y_e}(q, \{v_{\tilde{e}}\}_{\tilde{e} \in E_e}) \right) \times Z_H(q, \{v_{{\rm eff},e}\}_{e \in E}). \tag{4.45}$$

In particular, $Z_{\vec{H}^{\mathbf{G}}}$ does not depend on the orientations of the edges of H.

4.7 Analogy with electrical circuit theory

The identities discussed in Sections 4.4–4.6 are strongly reminiscent of analogous identities in electrical circuit theory: there are elementary formulae for the reduction of linear circuit elements placed in series or parallel; and more generally, any 2-terminal subnetwork consisting of linear passive circuit elements is equivalent (at a fixed frequency ω) to some single "effective admittance". But the relation of the multivariate Tutte polynomial to electrical circuit theory goes beyond mere analogy. For, as was discovered by Kirchhoff [77] in 1847 and will be reviewed in Section 6, linear electrical circuits are intimately related to spanning trees; and as was seen in (2.18), the spanning-tree polynomial $T_G(\mathbf{v})$ arises from the double limit of the multivariate Tutte polynomial in which $q \to 0$ and $\mathbf{v}$ is infinitesimal.

Look back, for instance, at the Tutte–Potts parallel law $v_1 \| v_2 \equiv v_1 + v_2 + v_1 v_2$; it is a nonlinear generalization of the familiar law $v_1 \| v_2 \equiv v_1 + v_2$ for combining electrical conductances in parallel, and reduces to it when $\mathbf{v}$ is infinitesimal. Likewise, the Tutte–Potts series law $v_1 \bowtie_q v_2 \equiv v_1 v_2/(q + v_1 + v_2)$ is a nonlinear generalization of the familiar law $v_1 \bowtie v_2 \equiv v_1 v_2/(v_1 + v_2)$ for combining electrical conductances in series, and reduces to it when $q \to 0$.

It thus makes sense to ask which results of electrical circuit theory can be generalized from the spanning-tree polynomial $T_G(\mathbf{v})$ to the connected-spanning-subgraph polynomial $C_G(\mathbf{v})$, the spanning-forest polynomial $F_G(\mathbf{w})$, or even the full multivariate Tutte polynomial $Z_G(q, \mathbf{v})$.

5 Complex zeros of Z_G: Why should we care?

Since $Z_G(q, \mathbf{v})$ is a polynomial in q and $\{v_e\}$, it is mathematically natural to inquire about the zeros of $Z_G(q, \mathbf{v})$ as a function of q or $\{v_e\}$ or both. Here we can treat q and $\{v_e\}$ as either real or complex variables.

The real zeros of the chromatic polynomial have been extensively studied, and to a lesser extent this is true also for the flow polynomial; see [70] for an excellent recent survey. Many of these results for the chromatic polynomial extend to the multivariate Tutte polynomial $Z_G(q, \mathbf{v})$ in the antiferromagnetic regime (i.e., $-1 \le v_e \le 0$ for all e) [74].[10]

The complex zeros of the partition function play a very important role in statistical mechanics, as was shown by Yang and Lee [143] in 1952. This arises as follows: Statistical physicists are interested in *phase transitions*, namely in points where one or more physical quantities (e.g. the energy or the magnetization) depend nonanalytically (in many cases even discontinuously) on one or more control parameters (e.g. the temperature or the magnetic field). Now, such nonanalyticity is manifestly impossible in (1.1)/(2.7)—or in any other reasonable statistical-mechanical model, for that matter—for any finite graph G. Rather, phase transitions arise only in the *infinite-volume limit*. That is, we consider some countably infinite graph $G_\infty = (V_\infty, E_\infty)$—usually a regular lattice, such as $\mathbb{Z}^d$ with nearest-neighbor edges—and an increasing sequence of finite subgraphs $G_n = (V_n, E_n)$. It can then be shown

---

[10] Indeed, some of the proofs become *simpler* when the multivariate generalizations are considered: see [74] for details. In particular, the multivariate approach clarifies the meaning of the mysterious number $q = 32/27$ [52,69].

(under modest hypotheses on the G_n) that the *(limiting) free energy per unit volume*

$$f_{G_\infty}(q,v) \;=\; \lim_{n\to\infty} |V_n|^{-1} \log Z_{G_n}(q,v) \tag{5.1}$$

exists for all nondegenerate physical values of the parameters[11], namely either

(a) q integer ≥ 1 and $-1 < v < \infty$ (using (2.7), see [68, Section I.2])

or (b) q real > 0 and $0 \leq v < \infty$ (using (1.1), see [61, Theorem 4.1] and [60, 116])

This limit $f_{G_\infty}(q,v)$ is in general a continuous function of v [and of q in case (b)]; but it can fail to be a real-analytic function of v and/or q, because complex singularities of $\log Z_{G_n}(q,v)$—namely, complex zeros of $Z_{G_n}(q,v)$—can approach the real axis in the limit $n \to \infty$. Therefore, the possible points of physical phase transitions are precisely the real limit points of such complex zeros. As a result, theorems that constrain the possible location of complex zeros of the partition function are of great interest.[12]

In particular, if it can be proven that the partition function Z_{G_n} is nonvanishing in some complex domain D *that does not vary with the size of the graph* G_n, then (under mild hypotheses) the infinite-volume free energy f_{G_∞} will be analytic in D—that is, in physical terms, there will be no phase transitions in D. Study of the complex zeros of the partition function is thus one way of proving the absence of phase transitions in specified regions of parameter space.

It is useful to distinguish between "soft" and "hard" theorems of this type. The "soft" theorems assert the nonvanishing of the partition function Z_G for some specified statistical-mechanical model in some "natural" large domain of multidimensional complex space (e.g. the unit polydisc or the product of right half-planes) for arbitrary graphs G. The "hard" theorems, by contrast, assert the nonvanishing of Z_G in a domain that depends quantitatively on some parameter associated to G, such as its maximum degree. ("Hard" theorems are of physical interest only when the parameter in question is "local", so that the domain obtained is uniform in the G_n. E.g. maximum degree is OK, but total number of vertices is not.)

The "soft" theorems are particularly interesting from a mathematical point of view: they assert that some combinatorially natural generating polynomial is non-

---

[11] Here "physical" means that the weights are nonnegative, so that the model has a probabilistic interpretation; and "nondegenerate" means that we exclude the limiting cases $v = -1$ in (a) and $q = 0$ in (b), which cause difficulties due to the existence of configurations having zero weight.

[12] I don't want to give the impression that the partition function is the *only* quantity of interest to statistical physicists. Indeed, what is really of physical significance is not the partition function, but rather the *Boltzmann–Gibbs probability measure*, which gives the probabilities for different configurations of the system: in the Potts (resp. random-cluster) model this measure is given by the summand of (2.7) [resp. (2.10)] divided by the partition function $Z_G(q, \mathbf{v})$. From this point of view, the partition function is merely a normalization factor, of no intrinsic physical interest whatsoever! Nevertheless, *derivatives* of the partition function with respect to the parameters occurring in it yield correlation functions (i.e., expectations of particular random variables with respect to the Boltzmann–Gibbs measure), which *are* of direct physical interest. Furthermore, the complex zeros of the partition function are of interest for the reason just explained; the Yang–Lee picture thus provides *one* method for investigating phase transitions, to be used in conjunction with other approaches.

vanishing in some large domain of $\mathbb{C}^n$. I am aware of only three classes of such theorems (but I urge readers to try to discover new ones!):

1) The Lee–Yang theorem for the ferromagnetic Ising model at complex magnetic field [83] and its generalizations.[13]

2) The Heilmann–Lieb theorem for the matching polynomial [64].[14]

3) The half-plane theorem for the spanning-tree polynomial and its generalizations [36].

The Lee–Yang and Heilmann–Lieb theorems can, in fact, be given a unified combinatorial formulation: Let us equip the graph $G = (V, E)$ with edge weights $\boldsymbol{\lambda} = \{\lambda_e\}_{e\in E}$ and vertex weights $\mathbf{t} = \{t_i\}_{i\in V}$; and let us form the generating polynomial

$$P_G(\mathbf{t}, \boldsymbol{\lambda}) = \sum_{A\subseteq E} \left(\prod_{e\in A} \lambda_e\right)\left(\prod_{i\in V} w_i(A)\right) \tag{5.2}$$

where

$$\text{Lee–Yang:} \qquad w_i(A) = \begin{cases} 1 & \text{if } \deg_A(i) \text{ is even} \\ t_i & \text{if } \deg_A(i) \text{ is odd} \end{cases} \tag{5.3}$$

$$\text{Heilmann–Lieb:} \qquad w_i(A) = \begin{cases} 1 & \text{if } \deg_A(i) = 0 \\ t_i & \text{if } \deg_A(i) = 1 \\ 0 & \text{if } \deg_A(i) > 1 \end{cases} \tag{5.4}$$

Then the Lee–Yang and Heilmann–Lieb theorems assert that $P_G(\mathbf{t}, \boldsymbol{\lambda}) \neq 0$ whenever $\lambda_e \geq 0$ for all e and $\operatorname{Re} t_i > 0$ for all i (or $\operatorname{Re} t_i < 0$ for all i). Indeed, the Lee–Yang and Heilmann–Lieb theorems can be given a unified proof.[15] The half-plane theorem for the spanning-tree polynomial will be discussed in Section 6 below.

For a prototypical example of a "hard" theorem, consider the multivariate generating polynomial for independent sets of vertices in a graph $G = (V, E)$:[16]

$$Z_G(\mathbf{w}) = \sum_{\substack{A \subseteq V \\ A \text{ independent}}} \prod_{i\in A} w_i \,, \tag{5.5}$$

---

[13] A partial bibliography (through 1980) of generalizations of the Lee–Yang theorem can be found in [85].

[14] This theorem is most often quoted in the form of its univariate corollary, namely that the roots of the univariate matching polynomial are all (negative) real. But it is the multivariate theorem that is truly fundamental. Two simple proofs of the multivariate Heilmann–Lieb theorem can be found in [36, Theorem 10.1 and Remark 4 following it].

[15] See [36, Sections 4.7 and 4.8, and Remark 4 in Section 10.1]. See also [110, 111] for related results, proven by a different method.

[16] We recall that a set $A \subseteq V$ is called *independent* (or *stable*) if it does not contain any pair of adjacent vertices. This is totally unrelated to the concept of independent sets of elements in a matroid; it is unfortunate that the same word is used.

where $\mathbf{w} = \{w_i\}_{i \in V}$ are (complex) vertex weights. [In statistical mechanics, (5.5) is called the grand partition function of the *hard-core lattice gas* on G; the weights w_i are called *activities* or *fugacities*.] It can then be shown [114] that if G is a graph of maximum degree Δ, and $|w_i| \le (\Delta-1)^{\Delta-1}/\Delta^\Delta$ for all $i \in V$, then $Z_G(\mathbf{w}) \neq 0$. Indeed, this is a corollary of a more general result [114] that provides a sufficient condition on a set of radii $\mathbf{R} = \{R_i\}_{i \in V}$ for Z_G to be nonvanishing in the closed polydisc $|w_i| \le R_i$.[17]

These theorems belong, in fact, to a long line of results in the mathematical-physics literature asserting that the grand partition function of a repulsive lattice gas is nonvanishing in some polydisc at small fugacity. (Physically this corresponds to the absence of phase transition for a gas at low density.) Two main methods of proof are used:

1) The coefficients in the Taylor expansion of $\log Z_G(\mathbf{w})$ are interpreted combinatorially: this is the so-called *Mayer expansion* [132]. Then, by direct estimation involving some nontrivial combinatorics, the Mayer expansion is proven to be convergent in a suitable polydisc $|w_i| < R_i$ [12, 21, 24–26, 29, 80, 91, 101, 103, 104, 115, 120]. It immediately follows that Z_G is nonvanishing in this same polydisc.

2) The nonvanishing of Z_G in the closed polydisc $|w_i| \le R_i$ is proven directly, by induction on the number of vertices in G [45,46,114,121]. It then immediately follows that the Taylor series for $\log Z_G$ is convergent in the interior of this polydisc.

The foregoing results have a wider application than one might at first think, because the hard-core lattice gas is not merely *one* interesting statistical-mechanical model; it is, in fact, the *universal* statistical-mechanical model in the sense that any statistical-mechanical model living on a vertex set V_0 can be mapped onto a gas of nonoverlapping "polymers" on V_0, i.e. a hard-core lattice gas on the intersection graph of V_0 [120, Section 5.7].[18] This construction, which is termed the "polymer expansion" or "cluster expansion", is an important tool in mathematical statistical mechanics [11, 21, 25, 59, 115]; it is widely employed to prove the absence of phase transition at high temperature, low temperature, large magnetic field, low density, or weak nonlinear coupling. In the antiferromagnetic Potts model it can be used to prove the nonvanishing of $Z_G(q, \mathbf{v})$ in a region $|q| > R$ (see Section 9).

6 $q = 0$: Spanning trees, electric circuits, and all that

As explained in Section 2.3, the spanning-tree polynomial $T_G(\mathbf{v})$ can be obtained from the multivariate Tutte polynomial $Z_G(q, \mathbf{v})$ via the double limit in which $q \to 0$ and $\mathbf{v}$ is infinitesimal. In this section we shall study some remarkable properties of the spanning-tree polynomial and its generalization to matroids.

[17] These results are, in turn, closely related to the Lovász local lemma [55, 56, 124, 125] in probabilistic combinatorics; see [114] for a detailed development of this connection.

[18] The *intersection graph* of a finite set S is the graph whose vertices are the nonempty subsets of S, and whose edges are the pairs with nonempty intersection.

6.1 The matrix-tree theorem

Let $G = (V, E)$ be a connected graph, and let $L_G(\mathbf{x})$ be the Laplacian matrix for G with edge weights $\mathbf{x} = \{x_e\}_{e \in E}$:

$$L_G(\mathbf{x}) = \begin{cases} -\sum_{e \sim ij} x_e & \text{if } i \neq j \\ \sum_{k \neq i} \sum_{e \sim ik} x_e & \text{if } i = j \end{cases} \tag{6.1}$$

where $e \sim ij$ denotes that e connects i to j. [Equivalently, $L_G(\mathbf{x}) = BXB^{\mathrm{T}}$ where B is the directed vertex-edge incidence matrix for any orientation of G, and $X = \mathrm{diag}(\mathbf{x})$.] By construction, the row (and column) sums of $L_G(\mathbf{x})$ are all zero, so its determinant is zero. Now let i be any vertex of G, and let $L_G(\mathbf{x})_{\setminus i}$ be the matrix obtained from $L_G(\mathbf{x})$ by deleting the ith row and column. Kirchhoff [77] proved in 1847 the following striking result (see also [15, 34, 35, 92, 94, 146]):

Proposition 6.1 (matrix-tree theorem) $\det L_G(\mathbf{x})_{\setminus i}$ *is independent of i and equals $T_G(\mathbf{x})$, the generating polynomial of spanning trees in G.*

Many different proofs of the matrix-tree theorem are now available; one simple proof is based on the Cauchy–Binet theorem in matrix theory (see e.g. [92]).

More generally, it turns out that *each* minor of $L_G(\mathbf{x})$ enumerates a suitable class of rooted spanning forests [1, 33, 93]. To formulate this result, suppose that $I, J \subseteq V$, and let us denote by $L_G(\mathbf{x})_{\setminus I,J}$ the matrix obtained from $L_G(\mathbf{x})$ by deleting the columns I and the rows J; when $I = J$, we write simply $L_G(\mathbf{x})_{\setminus I}$. Then the "principal-minors matrix-tree theorem" states that

$$\det L_G(\mathbf{x})_{\setminus \{i_1,\ldots,i_r\}} = \sum_{F \in \mathcal{F}(i_1,\ldots,i_r)} \prod_{e \in F} x_e \,, \tag{6.2}$$

where the sum runs over all spanning forests F in G composed of r disjoint trees, each of which contains exactly one of the "root" vertices $i_1, \ldots, i_r$. This theorem can easily be derived by applying Proposition 6.1 to the graph in which the vertices $i_1, \ldots, i_r$ are contracted to a single vertex. Furthermore, the "all-minors matrix-tree theorem" (whose proof is a bit more intricate, see [1, 33, 93]) states that for any subsets I, J of the same cardinality r, we have

$$\det L_G(\mathbf{x})_{\setminus I,J} = \sum_{F \in \mathcal{F}(I|J)} \epsilon(F, I, J) \prod_{e \in F} x_e \,, \tag{6.3}$$

where the sum runs over all spanning forests F in G composed of r disjoint trees, each of which contains exactly one vertex from I and exactly one vertex (possibly the same one) from J; here $\epsilon(F, I, J) = \pm 1$ are signs whose precise definition is not needed here.

The virtue of the matrix-tree theorem is that enumerative questions about spanning trees (and *rooted* spanning forests) can be reduced to linear algebra.

6.2 Electric circuits and the half-plane property for graphs

Now let us consider the graph G as an electrical network: to each edge e we associate a complex number x_e, called its *conductance* (or *admittance*).[19] [The *resistance* (or *impedance*) is $1/x_e$.] Suppose that we inject currents $\mathbf{J} = \{J_i\}_{i \in V}$ into the vertices. What node voltages $\boldsymbol{\varphi} = \{\varphi_i\}_{i \in V}$ will be produced? Applying Kirchhoff's law of current conservation at each vertex and Ohm's law on each edge, it is not hard to see that the node voltages and current inflows satisfy the linear system

$$L_G(\mathbf{x})\, \boldsymbol{\varphi} \;=\; \mathbf{J}\,. \tag{6.4}$$

It is then natural to ask: Under what conditions does this system have a (unique) solution? Two obvious constraints arise from the fact that the row and column sums of $L_G(\mathbf{x})$ are zero: firstly, the current vector must satisfy $\sum_{i \in V} J_i = 0$ ("conservation of total current"), or else no solution will exist; and secondly, if $\boldsymbol{\varphi}$ is any solution, then so is $\boldsymbol{\varphi} + c\mathbf{1}$ for any c ("only voltage *differences* are physically observable"). So let us assume that $\sum_{i \in V} J_i = 0$; and let us break the redundancy in the solution by fixing the voltage to be zero at some chosen reference node $i_0 \in V$ ("ground"). Does the modified system

$$L_G(\mathbf{x})_{\setminus i_0}\, \boldsymbol{\varphi}_{\setminus i_0} \;=\; \mathbf{J}_{\setminus i_0} \tag{6.5}$$

then have a unique solution? This will be so if and only if $\det L_G(\mathbf{x})_{\setminus i_0}$ is nonzero—which, by virtue of the matrix-tree theorem, is equivalent to $T_G(\mathbf{x})$ being nonzero.

Simple counterexamples show that this is not always the case. Suppose, for instance, that G consists of a pair of vertices connected by two edges e, f in parallel. With $x_e = 1$ and $x_f = -1$, it is easy to see that no solution exists (except when $\mathbf{J} = 0$). Of course, the reader will object that negative resistances are physically unrealizable! Fine: consider instead $x_e = i$ and $x_f = -i$; once again there is no solution (unless $\mathbf{J} = 0$). This example corresponds physically to a perfectly lossless capacitor together with a perfectly lossless inductor, exactly at their resonant frequency. But once again, the reader will rightly object that *perfectly* lossless components are physically unrealizable; every component in the real world exhibits *some* dissipation. This reasoning leads us to conjecture, on physical grounds, that if $\operatorname{Re} x_e > 0$ for all e (i.e. every branch is strictly dissipative), then the network is uniquely solvable once we fix the voltage at a single reference node $i_0 \in V$. This conjecture turns out to be true, and we have:

Theorem 6.2 *Let G be a connected graph. Then the spanning-tree polynomial T_G has the "half-plane property" (HPP), i.e. $\operatorname{Re} x_e > 0$ for all e implies $T_G(\mathbf{x}) \neq 0$.*

The proof of Theorem 6.2 is not difficult: Consider any nonzero complex vector $\boldsymbol{\varphi} = \{\varphi_i\}_{i \in V}$ satisfying $\varphi_{i_0} = 0$. Because G is connected, we have $B^{\mathrm{T}}\boldsymbol{\varphi} \neq 0$. Therefore, the quantity

$$\boldsymbol{\varphi}^* L_G(\mathbf{x})\boldsymbol{\varphi} \;=\; \boldsymbol{\varphi}^* B X B^{\mathrm{T}} \boldsymbol{\varphi} \;=\; \sum_{e \in E} |(B^{\mathrm{T}}\boldsymbol{\varphi})_e|^2\, x_e \tag{6.6}$$

---

[19] Complex admittances are relevant, for instance, in alternating-current circuits. Indeed, the reader is invited to imagine that we are considering an alternating-current circuit at some fixed frequency ω.

has strictly positive real part whenever $\operatorname{Re} x_e > 0$ for all e; so in particular $(BXB^{\mathrm{T}}\varphi)_i \neq 0$ for some $i \neq i_0$. It follows that the submatrix of $L_G(\mathbf{x})$ obtained by suppressing the i_0th row and column is nonsingular, and so has a nonzero determinant. Theorem 6.2 then follows from the matrix-tree theorem.[20]

It cannot be overemphasized how remarkable this theorem is. Suppose, for instance, that G is the complete graph on n vertices; then $T_G(\mathbf{x})$ is a homogeneous polynomial of degree $n-1$ in $n(n-1)/2$ variables, containing n^{n-2} monomials. It seems utterly miraculous, at first sight, that a polynomial of such complexity should be nonvanishing whenever $\operatorname{Re} x_e > 0$ for all e. The fact that it is so clearly expresses a deep property arising from the underlying combinatorial structure.

An immediate corollary of Theorem 6.2 is that the complementary spanning-tree polynomial

$$\widetilde{T}_G(\mathbf{x}) \;=\; \left(\prod_{e \in E} x_e\right) T_G(1/\mathbf{x}) \tag{6.7}$$

also has the half-plane property, since the map $x_e \mapsto 1/x_e$ takes the right half-plane onto itself.

6.3 The half-plane property for matroids

From a combinatorial point of view, the noteworthy fact is that the spanning trees of G constitute the bases of the graphic matroid $M(G)$, and their complements constitute the bases of the cographic matroid $M^*(G)$. So T_G and $\widetilde{T}_G$ are the multivariate basis generating polynomials for $M(G)$ and $M^*(G)$, respectively. This naturally suggests generalizing Theorem 6.2 to more general matroids and, perhaps, to more general set systems. Before posing these questions precisely, we need to fix some notation and terminology.

A *set system* (or *hypergraph*) $\mathcal{S}$ on the (finite) ground set E is simply a collection $\mathcal{S}$ of subsets of E. Given any set system $\mathcal{S}$ on E, we define its *(multivariate) generating polynomial* to be

$$P_{\mathcal{S}}(x) \;=\; \sum_{S \in \mathcal{S}} \prod_{e \in S} x_e \,. \tag{6.8}$$

The *rank* of a set system is the maximum cardinality of its members (by convention we set rank $= -\infty$ if $\mathcal{S} = \varnothing$); equivalently, it is the degree of the generating polynomial $P_{\mathcal{S}}$. A set system $\mathcal{S}$ is *r-uniform* if $|S| = r$ for all $S \in \mathcal{S}$, or equivalently if its generating polynomial $P_{\mathcal{S}}$ is homogeneous of degree r. We shall be particularly interested in the *basis generating polynomial* of a matroid M,

$$P_{\mathcal{B}(M)}(x) \;=\; B_M(x) \;=\; \sum_{B \in \mathcal{B}(M)} \prod_{e \in B} x_e \,. \tag{6.9}$$

We can now pose the following questions concerning possible extensions of Theorem 6.2:

[20] This proof is well known in the circuit-theory literature: see e.g. [35, Section 2.7] as well as the related results in [42, pp. 398–401, 430–431 and 850–851] [100, pp. 52–53 and 67–69]. It has, moreover, a natural physical interpretation: if $\varphi = \{\varphi_i\}_{i \in V}$ are the node voltages, then the real part of the quadratic form (6.6) is the total power dissipated in the circuit.

Question 6.3 *For which matroids M does the basis generating polynomial $P_{\mathcal{B}(M)}$ have the half-plane property?*

More generally:

Question 6.4 *For which r-uniform set systems $\mathcal{S}$ does the generating polynomial $P_{\mathcal{S}}$ have the half-plane property?*

These questions were recently studied in a long paper by Choe, Oxley, Wagner and myself [36]. Our original conjecture was that all matroids (and no non-matroidal set systems) have the half-plane property. That would be nice and neat, but it turns out to be false; and the truth is considerably more interesting and subtle. Our conjecture is half right: an r-uniform set system with the half-plane property is necessarily the set of bases of a matroid [36, Theorem 7.1]. But not every matroid has the half-plane property, and we do not yet have a complete characterization of those that do. Nevertheless, we can find large classes of matroids with the half-plane property:

(a) Every complex unimodular matroid—or what is equivalent, sixth-root-of-unity matroid [137]—has the half-plane property [36, Theorem 8.1 and Corollary 8.2].[21] This class properly includes the regular matroids, which in turn properly includes the graphic and cographic matroids. The proof of the half-plane property for complex unimodular matroids is, in fact, a direct generalization of the proof just given for Theorem 6.2.

(b) Every uniform matroid has the half-plane property. (Indeed, every loopless uniform matroid has the Brown–Colbourn property, which is stronger than the half-plane property: see Section 7 below.)

(c) A significant subclass of transversal matroids—those we call "nice"—have the half-plane property [36, Section 10].[22]

(d) All matroids of rank or corank at most 2 have the half-plane property [36, Corollary 5.5], as do all matroids on a ground set of at most 6 elements [36, Proposition 10.4].

(e) The class of matroids with the half-plane property is closed under minors, duality, direct sums, 2-sums, series and parallel connection, full-rank matroid union, and some special cases of principal truncation, principal extension, principal cotruncation and principal coextension [36, Section 4].

Furthermore, we show that:

---

[21] It is proven in [36, Theorem 8.9] that a matroid is complex unimodular if and only if it is sixth-root-of-unity.

[22] In [36, Conjecture 13.16 and Question 13.17] it was conjectured that all rank-3 transversal matroids have the half-plane property, and the question was raised whether it might even be true that *all* transversal matroids have the half-plane property. Both conjectures are in fact false. Choe and Wagner [37] have exhibited a rank-4 transversal matroid that lacks the half-plane property (and indeed lacks the weaker Rayleigh property, see Section 6.4 below). I have found a class of rank-3 transversal matroids that lack the half-plane property, and plan to publish them elsewhere [123].

(f) A binary matroid has the half-plane property if and only if it is regular [36, Corollary 8.16].

(g) A ternary matroid has the half-plane property if and only if it is a sixth-root-of-unity matroid [36, Corollary 8.17].

Finally, we can show that certain matroids do *not* have the half-plane property: among these are the Fano matroid F_7, the non-Fano matroid F_7^-, their relaxations F_7^{--}, F_7^{-3} and $M(K_4)+e$, the matroids P_8, P_8' and P_8'', the Pappus and non-Pappus matroids, the free extension (non-Pappus $\setminus 9)+e$, and all their duals [36, Section 11]. The first six of these examples are minor-minimal, and we conjecture that the others are as well; but we strongly suspect that this list is incomplete, and indeed we consider it likely that the set of minor-minimal non-half-plane-property matroids is infinite.

More generally, we consider homogeneous multiaffine polynomials

$$P(\mathbf{x}) \;=\; \sum_{S \subseteq E, |S|=r} a_S \prod_{e \in S} x_e \qquad (6.10)$$

with arbitrary complex coefficients a_S (not necessarily 0 or 1). We prove two *necessary* conditions for $P \not\equiv 0$ to have the half-plane property:

(a) P must have the "same-phase property", i.e. all the nonzero coefficients a_S must have the same phase [36, Theorem 6.1]. So without loss of generality we can assume that all the a_S are real and nonnegative.

(b) The *support* $\operatorname{supp}(P) = \{S \subseteq E\colon\ a_S \neq 0\}$ must be the collection of bases of a matroid [36, Theorem 7.1].

This latter fact is particularly striking: it shows that matroids arise *naturally* from a consideration of homogeneous multiaffine polynomials with the half-plane property. We do not know whether the converse of this result is true, i.e. whether for every matroid M there exists a homogeneous multiaffine polynomial P with the half-plane property such that $\operatorname{supp}(P) = \mathcal{B}(M)$. But it is true, at least, for all matroids representable over $\mathbb{C}$ [36, Corollary 8.2].

We also also give two *sufficient* conditions for a homogeneous multiaffine polynomial P to have the half-plane property (or be identically zero):

(a) *Determinant condition* [36, Theorem 8.1]: $a_S = |\det(A \upharpoonright S)|^2$ for some $r \times n$ complex matrix A [here $n = |E|$, and $A \upharpoonright S$ denotes the square submatrix of A using the columns indexed by the set S]. This corresponds to $P(\mathbf{x}) = \det(AXA^*)$ where $X = \operatorname{diag}(\mathbf{x})$ and * denotes Hermitian conjugate.

(b) *Permanent condition* [36, Theorem 10.2]: $a_S = \operatorname{per}(\Lambda \upharpoonright S)$ for some $r \times n$ nonnegative matrix Λ. This corresponds to $P(\mathbf{x}) = \operatorname{per}(\Lambda X)$.

Unfortunately, the relationship between these sufficient conditions and the half-plane property looks complicated. Neither family of polynomials contains the other; their intersection is nonempty; and their union is a proper subset of the set of all homogeneous multiaffine polynomials with the half-plane property.

In any case, the matroids with the half-plane property form a very natural class, which deserves further study. A long list of open questions can be found in [36, Section 13].

6.4 The Rayleigh property for graphs and matroids

Consider once again the graph G as an electrical network, with conductances $\{x_e\}_{e\in E}$ on the edges, and select a pair of distinct vertices $i,j \in V$. Then Kirchhoff [77] showed in 1847 that the effective conductance between nodes i and j is

$$\mathcal{Y}_{ij}(\mathbf{x}) \;=\; \frac{T_G(\mathbf{x})}{T_{G/ij}(\mathbf{x})}\,, \tag{6.11}$$

where G/ij is the graph obtained from G by contracting i and j to a single vertex.[23]

Suppose now (in contrast to the preceding subsections) that all the conductances x_e are real and positive. Physical intuition suggests that increasing one or more of the branch conductances x_e cannot cause the effective conductance between any pair of nodes to *decrease*; in other words, we conjecture that $\mathcal{Y}_{ij}(\mathbf{x})$ is a nondecreasing function of all the x_e on $[0,\infty)^E$. Let us rephrase this statement by considering the graph H obtained from G by adjoining a new edge f connecting i to j; then $G = H\setminus f$ and $G/ij = H/f$. Using

$$T_{H\setminus f}(\mathbf{x}) \;=\; T_{H\setminus\{e,f\}}(\mathbf{x}) + x_e T_{(H\setminus f)/e}(\mathbf{x}) \tag{6.12a}$$

$$T_{H/f}(\mathbf{x}) \;=\; T_{(H/f)\setminus e}(\mathbf{x}) + x_e T_{H/\{e,f\}}(\mathbf{x}) \tag{6.12b}$$

we find

$$\frac{\partial}{\partial x_e}\mathcal{Y}_{ij}(\mathbf{x}) \;=\; \frac{T_{(H/e)\setminus f}(\mathbf{x})\, T_{(H/f)\setminus e}(\mathbf{x}) \;-\; T_{H/\{e,f\}}(\mathbf{x})\, T_{H\setminus\{e,f\}}(\mathbf{x})}{T_{H/f}(\mathbf{x})^2} \tag{6.13a}$$

$$\;=\; \frac{T_{H/e}(\mathbf{x})\, T_{H/f}(\mathbf{x}) \;-\; T_{H/\{e,f\}}(\mathbf{x})\, T_H(\mathbf{x})}{T_{H/f}(\mathbf{x})^2}\,. \tag{6.13b}$$

The conjecture then becomes:

Theorem 6.5 (Rayleigh property for graphs) *Let $H = (V,E)$ be a finite undirected graph, and let $e,f \in E$ with $e \neq f$. Then*

$$T_{H/e}(\mathbf{x})\, T_{H/f}(\mathbf{x}) \;-\; T_{H/\{e,f\}}(\mathbf{x})\, T_H(\mathbf{x}) \;\ge\; 0 \tag{6.14}$$

for all $\mathbf{x} \ge 0$.

Please note that (6.14) has a beautiful probabilistic interpretation: Consider the probability measure on spanning trees $T \subseteq H$ giving weight $\big(\prod_{e\in T} x_e\big)/T_H(\mathbf{x})$ to the tree T. Then (6.14) asserts that the events $e \in T$ and $f \in T$ are *negatively correlated*.

---

[23] PROOF: By definition, $1/\mathcal{Y}_{ij}(\mathbf{x})$ is the voltage difference induced between nodes i and j if we inject 1 ampere of current into i and extract 1 ampere from j. So let node j be "ground", and apply (6.5) with $i_0 = j$ and $\mathbf{J} = \delta_i - \delta_j$. We then have

$$1/\mathcal{Y}_{ij}(\mathbf{x}) \;=\; [L_G(\mathbf{x})_{\setminus j}]^{-1}_{ii} \;=\; \frac{\det L_G(\mathbf{x})_{\setminus\{i,j\}}}{\det L_G(\mathbf{x})_{\setminus j}}$$

by Cramer's rule. The principal-minors matrix-tree theorem (6.2) then tells us that $\det L_G(\mathbf{x})_{\setminus j} = T_G(\mathbf{x})$ and $\det L_G(\mathbf{x})_{\setminus\{i,j\}} = T_{G/ij}(\mathbf{x})$.

Theorem 6.5 has both algebraic and variational/probabilistic proofs: see e.g. [4, Section 3.8] [57, Theorem 2.1] [37, Theorems 5.2 and 5.6] for the former and [48, 87, 88, 99] for the latter.

In an important recent paper, Choe and Wagner [37] have investigated the generalization of this problem to matroids. Let M be a matroid with ground set E, and let $B_M(\mathbf{x}) = P_{\mathcal{B}(M)}(\mathbf{x})$ be its (multivariate) basis generating polynomial. We say that M is a *Rayleigh matroid* (or has the *Rayleigh property*) in case

$$B_{M/e}(\mathbf{x})\, B_{M/f}(\mathbf{x}) \;-\; B_{M/\{e,f\}}(\mathbf{x})\, B_M(\mathbf{x}) \;\geq\; 0 \qquad (6.15)$$

for all $e, f \in E$ with $e \neq f$ and all $\mathbf{x} \geq 0$. Choe and Wagner [37] have shown that:

(a) The class of Rayleigh matroids is closed under minors, duality, direct sums, and 2-sums.

(b) A binary matroid is Rayleigh if and only if it does not contain S_8 as a minor. Equivalently, a binary matroid is Rayleigh if and only if it can be constructed from regular matroids, F_7, F_7^* and $AG(3,2)$ by taking direct sums and 2-sums.

(c) Every matroid with the half-plane property is Rayleigh (but the converse is false).

(d) Every matroid on a ground set of at most 7 elements is Rayleigh.

More recently, Wagner [134] has also shown that

(e) Every matroid of rank or corank at most 3 is Rayleigh.

On the other hand, Choe and Wagner [37] have exhibited a rank-4 transversal matroid that is not Rayleigh.

The most striking of these results, in my opinion, are (b) and (c). For binary matroids, (b) provides a pair of beautiful characterizations of the Rayleigh property: one by excluded minors and the other by internal structure. It is at least conceivable that analogous results could be obtained for ternary matroids, $GF(4)$-representable matroids, or even arbitrary matroids. As for (c), one might have thought, *a priori*, that the half-plane property and the Rayleigh property are distinct but unrelated properties of electric circuits (one for complex edge conductances, the other for positive real edge conductances), which thus hold for all graphic matroids but extend to distinct and unrelated families of non-graphic matroids. Item (c) shows that the truth is quite different: the half-plane property is strictly stronger than the Rayleigh property. Indeed, in some vague sense it seems that the half-plane property is *quite a bit* stronger than the Rayleigh property: non-HPP matroids seem to be plentiful, while non-Rayleigh matroids seem to be fairly rare. In any case, the Rayleigh matroids form a very natural class and deserve further study; see [37] for a list of open problems.

7 $q = 0$ again: The reliability polynomial

7.1 The Brown–Colbourn property for graphs

Once again let $G = (V, E)$ be a connected graph, considered now as a communications network with unreliable communication channels: the edge e is assumed to be operational with probability p_e and failed with probability $1 - p_e$, independently for each edge. Let $R_G(\mathbf{p})$ be the probability that every node is capable of communicating with every other node (this is the so-called *all-terminal reliability*). Clearly we have

$$R_G(\mathbf{p}) \;=\; \sum_{\substack{A \subseteq E \\ (V,A)\,\text{connected}}} \prod_{e\in A} p_e \prod_{e\in E\setminus A} (1-p_e)\,, \tag{7.1}$$

where the sum runs over all connected spanning subgraphs of G, and we have written $\mathbf{p} = \{p_e\}_{e\in E}$. We call $R_G(\mathbf{p})$ the (multivariate) *reliability polynomial* [38] for the graph G; it is a multiaffine polynomial, i.e. of degree at most 1 in each variable separately. If the edge probabilities p_e are all set to the same value p, we write the corresponding univariate polynomial as $R_G(p)$, and call it the univariate reliability polynomial. We are interested in studying the zeros of these polynomials when the variables p_e (or p) are taken to be *complex* numbers.

The reliability polynomial $R_G(\mathbf{p})$ is, of course, just the connected-spanning-subgraph polynomial $C_G(\mathbf{v})$ in disguise:

$$R_G(\mathbf{p}) \;=\; \left[\prod_{e\in E}(1-p_e)\right] C_G\!\left(\frac{\mathbf{p}}{1-\mathbf{p}}\right) \tag{7.2}$$

$$C_G(\mathbf{v}) \;=\; \left[\prod_{e\in E}(1+v_e)\right] R_G\!\left(\frac{\mathbf{v}}{1+\mathbf{v}}\right) \tag{7.3}$$

So we can equally well work with $C_G(\mathbf{v})$, making the change of variables $v_e = p_e/(1-p_e)$.

Brown and Colbourn [16] studied a number of examples and made the following conjecture:

> **Univariate Brown–Colbourn conjecture.** For any connected graph G, the zeros of the univariate reliability polynomial $R_G(p)$ all lie in the closed disc $|p-1| \le 1$. In other words, if $|p-1| > 1$, then $R_G(p) \neq 0$.

Subsequently, I proposed [121] a multivariate extension of the Brown–Colbourn conjecture:

> **Multivariate Brown–Colbourn conjecture.** For any connected graph G, if $|p_e - 1| > 1$ for all edges e, then $R_G(\mathbf{p}) \neq 0$.

In terms of C_G, the multivariate Brown–Colbourn conjecture states that if G is a *loopless* connected graph and $|1 + v_e| < 1$ for all edges e, then $C_G(\mathbf{v}) \neq 0$. Loops must be excluded because a loop e multiplies C_G by a factor $1 + v_e$ but leaves R_G unaffected.

A few years ago, Wagner [133] proved, using an ingenious and complicated construction, that the univariate Brown–Colbourn conjecture holds for all series-parallel graphs.[24] As an illustration of the power of the "multivariate ideology", let us now

[24] Unfortunately, there seems to be no completely standard definition of "series-parallel graph"; a plethora of slightly different definitions can be found in the literature [13, 38, 50, 97, 98]. So let

prove the stronger result [121] that the multivariate Brown–Colbourn conjecture holds for all series-parallel graphs—and let us moreover do it in two lines:[25]

Theorem 7.1 *Let G be a loopless connected series-parallel graph. If $|1+v_e| < 1$ for all edges e, then $C_G(\mathbf{v}) \neq 0$.*

Proof How should we prove a theorem for series-parallel graphs? The answer is obvious: just show that series and parallel reduction preserve the hypothesis $|1+v_e| < 1$. If edges e_1, e_2 (with weights v_1, v_2) are in parallel, they can be replaced by a single edge with weight v_* satisfying $1 + v_* = (1+v_1)(1+v_2)$; obviously $|1+v_i| < 1$ for $i = 1, 2$ implies that $|1+v_*| < 1$. Likewise, if edges e_1, e_2 (with weights v_1, v_2) are in series, they can be replaced by a single edge with weight v_* satisfying $1/v_* = 1/v_1 + 1/v_2$, provided that we multiply C_G by the prefactor $v_1 + v_2$ [cf. (4.23)/(4.24) specialized to $q = 0$]. Now $|1+v| < 1$ is equivalent to $\mathrm{Re}(1/v) < -1/2$, so $\mathrm{Re}(1/v_i) < -1/2$ for $i = 1, 2$ implies that $\mathrm{Re}(1/v_*) < -1 < -1/2$; furthermore, the prefactor $v_1 + v_2$ in (4.24) is nonzero. Since the multivariate Brown–Colbourn conjecture manifestly holds for the base case of trees, it necessarily holds for all loopless connected series-parallel graphs. □

Does the multivariate Brown–Colbourn conjecture hold for all graphs? For several years I would have bet that it does (though I was unable to find a proof); but in 2002, Gordon Royle sent me the surprising news that the multivariate Brown–Colbourn conjecture fails already for the simplest non-series-parallel graph, namely the complete graph K_4. The construction turns out to be very simple [108]. Since the *uni*variate Brown–Colbourn conjecture holds for K_4, let us try the next simplest situation, namely the *bi*variate one in which the six edges receive two different weights a and b. There are five cases:

(a) One edge receives weight a and the other five receive weight b:

$$C_{K_4}(a,b) = (8b^3 + 5b^4 + b^5) + (8b^2 + 10b^3 + 5b^4 + b^5)a \qquad (7.4)$$

(b) A pair of nonintersecting edges receive weight a and the other four edges receive weight b:

$$C_{K_4}(a,b) = (4b^3 + b^4) + (8b^2 + 8b^3 + 2b^4)a + (4b + 6b^2 + 4b^3 + b^4)a^2 \quad (7.5)$$

(c) A pair of intersecting edges receive weight a and the other four edges receive weight b:

$$C_{K_4}(a,b) = (3b^3 + b^4) + (10b^2 + 8b^3 + 2b^4)a + (3b + 6b^2 + 4b^3 + b^4)a^2 \quad (7.6)$$

me be completely precise about my own usage: I shall call a loopless graph *series-parallel* if it can be obtained from a forest by a finite sequence of series and parallel extensions of edges (i.e. replacing an edge by two edges in series or two edges in parallel). I shall call a general graph (allowing loops) series-parallel if its underlying loopless graph is series-parallel. Some authors write "obtained from a tree", "obtained from K_2" or "obtained from C_2" in place of "obtained from a forest"; in my terminology these definitions yield, respectively, all *connected* series-parallel graphs, all connected series-parallel graphs whose blocks form a path, or all *2-connected* series-parallel graphs. See [13, Section 11.2] for a more extensive bibliography.

[25] This proof can be found in [121, Remark 3 in Section 4.1] or in [108, Theorem 5.6(c) $\Longrightarrow$ (a)].

(d) A 3-star receives weight a and the complementary triangle receives weight b:

$$C_{K_4}(a,b) \;=\; (9b^2+3b^3)a + (6b+9b^2+3b^3)a^2 + (1+3b+3b^2+b^3)a^3 \quad (7.7)$$

(e) A three-edge path receives weight a and the complementary three-edge path receives weight b:

$$C_{K_4}(a,b) \;=\; b^3+(7b^2+3b^3)a+(7b+9b^2+3b^3)a^2+(1+3b+3b^2+b^3)a^3 \quad (7.8)$$

Now plot the roots a when b traces out the circle $|1+b|=1$, or vice versa. In cases (b) and (d) it turns out that the roots can enter the "forbidden discs" $|1+a|<1$ and $|1+b|<1$ (see [108, Figures 1 and 2]). In fact, this is quite easy to understand analytically: if we solve the equation $C_{K_4}(a,b)=0$ for a in terms of b, expanding in power series for b near 0, in cases (b) and (d) we obtain

$$a \;=\; \delta_1 b + \delta_2 b^{3/2} + O(b^2) \quad (7.9)$$

with $\delta_1 < 0$ and $\delta_2 \neq 0$. If we now set $b = -1+e^{i\theta}$, then one of the roots will have $|1+a|<1$ for small $\theta \neq 0$ (arising from the $\delta_2 b^{3/2}$ term). A small perturbation (so that $|1+b|<1$) then yields a counterexample to the multivariate Brown–Colbourn conjecture.

In fact, with a little more work Royle and I are able to prove the following [108, Theorem 5.6]:

Theorem 7.2 *A loopless connected graph G has the multivariate Brown–Colbourn property* if and only if *it is series-parallel.*

Moreover, as a corollary of the failure of the multivariate Brown–Colbourn conjecture for the complete graph K_4, Royle and I are able to show [108] that the univariate conjecture is false as well: counterexamples include a 4-vertex, 16-edge planar graph that can be obtained from K_4 by adding parallel edges, and a 1512-vertex, 3016-edge simple planar graph that can be obtained from K_4 by adding parallel edges and then subdividing edges. These counterexamples to the univariate conjecture are fairly easy to find once one has in hand the K_4 counterexample to the multivariate conjecture, but would probably not otherwise be guessed. This illustrates once again the advantages of considering the multivariate problem, even if one is ultimately interested in a particular univariate specialization.

7.2 The Brown–Colbourn property for matroids

The foregoing considerations can be extended to matroids. Let M be a matroid with ground set E, and let $S_M(\mathbf{v})$ be its spanning-set polynomial. We then say that M has the (multivariate) *Brown–Colbourn property* in case $S_M(\mathbf{v}) \neq 0$ whenever $|1+v_e|<1$ for all $e \in E$.[26]

Gordon Royle and I are currently investigating the Brown–Colbourn property for matroids [109]. Here are some of our results so far:

[26]Unfortunately, in [36] we used the dual definition (i.e., a matroid M has the Brown–Colbourn property in the sense of [36] if and only if M^* has the Brown–Colbourn property as defined here). I think the present definition is more natural in view of the relation to graphs (namely, G has the Brown–Colbourn property as defined here if and only if its cycle matroid $M(G)$ does). I therefore pledge to stick to this definition in the future.

(a) The class of matroids with the Brown–Colbourn property is closed under deletion, direct sums, 2-sums, series connection and parallel connection.[27]

(b) If M has the Brown–Colbourn property, then it has the half-plane property as well [36]. But the converse does not hold, as the example of $M(K_4)$ shows.

(c) If M has the Brown–Colbourn property, it must be loopless.

(d) Every loopless uniform matroid (i.e., every uniform matroid $U_{r,n}$ with $1 \le r \le n$) has the Brown–Colbourn property [36, 133].

(e) A binary matroid has the Brown–Colbourn property if and only if it is the cycle matroid of a loopless series-parallel graph.

Most of these results are fairly easy to prove.

8 $q = 0$ yet again: Spanning forests

As discussed in Section 6.1, Kirchhoff's matrix-tree theorem [15,34,77,92,94,146] and its generalizations to arbitrary minors [1,33,93] express the generating polynomials of spanning trees and rooted spanning forests in a graph G as determinants associated to the graph's Laplacian matrix. Here I would like to discuss a recently discovered extension of the matrix-tree theorem that provides a compact representation also for the generating polynomial of *unrooted* spanning forests, $F_G(\mathbf{w})$.

Recall first that one useful formalism for manipulating determinants is Grassmann algebra; indeed, any determinant can be written as a Gaussian "integral" over Grassmann variables.[28] In particular, it follows from the matrix-tree theorem that the generating polynomials of spanning trees and rooted spanning forests can be written as Gaussian Grassmann integrals.

Very recently, Caracciolo, Jacobsen, Saleur, Sportiello and I [30] have proven a generalization of the matrix-tree theorem in which a large class of combinatorial objects are represented by suitable *non-Gaussian* Grassmann integrals. As a special case, we show that the generating polynomial of unrooted spanning forests, which arises as a $q \to 0$ limit of the multivariate Tutte polynomial [cf. (2.18)], can be represented by a Grassmann integral involving a Gaussian term and a particular quartic term. Although this representation has not yet led to any new rigorous results concerning $F_G(\mathbf{w})$, it has led to important non-rigorous insights into the behavior of $F_G(\mathbf{w})$ for large subgraphs of a regular two-dimensional lattice [30, 75]—insights that may yet be translatable into theorems by exploiting the rigorous renormalization-group methods developed in recent decades by mathematical physicists (see e.g. [107]).

As in Section 6.1, let $G = (V, E)$ be a finite undirected graph, and let $L = L_G(\mathbf{w})$ be the Laplacian matrix for G with edge weights $\mathbf{w} = \{w_e\}_{e \in E}$. Let us now

---

[27] The class of matroids with the B–C property fails to be closed under contraction, at least for a trivial reason: namely, contraction of one of a set of parallel elements will produce one or more loops, which are incompatible with the B–C property [see item (c) below]. But we do not know whether the contraction of a *simple* matroid with the B–C property can fail to have the B–C property.

[28] For introductions to Grassmann algebra and Grassmann–Berezin integration, see e.g. [1, 147].

introduce, at each vertex $i \in V$, a pair of Grassmann variables ψ_i, $\bar{\psi}_i$. All of these variables are nilpotent of order 2 (i.e., $\psi_i^2 = \bar{\psi}_i^2 = 0$), anticommute, and obey the usual rules for Grassmann integration [1, 147]. Writing $\mathcal{D}(\psi, \bar{\psi}) = \prod_{i \in V} d\psi_i \, d\bar{\psi}_i$, we have, for any matrix A,

$$\int \mathcal{D}(\psi, \bar{\psi}) \; e^{\bar{\psi} A \psi} \;=\; \det A \tag{8.1}$$

and more generally

$$\int \mathcal{D}(\psi, \bar{\psi}) \; \bar{\psi}_{i_1} \psi_{j_1} \cdots \bar{\psi}_{i_r} \psi_{j_r} \; e^{\bar{\psi} A \psi} \\ = \epsilon(i_1, \ldots, i_r | j_1, \ldots, j_r) \; \det A_{\setminus \{i_1, \ldots, i_r\}, \{j_1, \ldots, j_r\}} \,, \tag{8.2}$$

where the sign $\epsilon(i_1, \ldots, i_r | j_1, \ldots, j_r) = \pm 1$ depends on how the vertices are ordered but is always $+1$ when $(i_1, \ldots, i_r) = (j_1, \ldots, j_r)$. These formulae allow us to rewrite the matrix-tree theorems in Grassmann form; for instance, (6.2) becomes

$$\int \mathcal{D}(\psi, \bar{\psi}) \left(\prod_{\alpha=1}^{r} \bar{\psi}_{i_\alpha} \psi_{i_\alpha} \right) e^{\bar{\psi} L \psi} \;=\; \sum_{F \in \mathcal{F}(i_1, \ldots, i_r)} \; \prod_{e \in F} w_e \,. \tag{8.3}$$

Let us next introduce, for each connected (not necessarily spanning) subgraph $\Gamma = (V_\Gamma, E_\Gamma)$ of G, the object

$$Q_\Gamma \;=\; \left(\prod_{e \in E_\Gamma} w_e \right) \left(\prod_{i \in V_\Gamma} \bar{\psi}_i \psi_i \right) . \tag{8.4}$$

(Note that each Q_Γ is even and hence commutes with the entire Grassmann algebra.) Now consider an unordered family $\mathbf{\Gamma} = \{\Gamma_1, \ldots, \Gamma_l\}$ with $l \ge 0$, and let us try to evaluate an expression of the form

$$\int \mathcal{D}(\psi, \bar{\psi}) \; Q_{\Gamma_1} \cdots Q_{\Gamma_l} \; e^{\bar{\psi} L \psi} \,. \tag{8.5}$$

If the subgraphs $\Gamma_1, \ldots, \Gamma_l$ have one or more vertices in common, then this integral vanishes on account of the nilpotency of the Grassmann variables. If, by contrast, the $\Gamma_1, \ldots, \Gamma_l$ are vertex-disjoint, then (8.3) expresses $\int \mathcal{D}(\psi, \bar{\psi}) \left(\prod_{k=1}^{l} \prod_{i \in V_{\Gamma_k}} \bar{\psi}_i \psi_i \right) e^{\bar{\psi} L \psi}$ as a sum over forests rooted at the vertices of $V_{\mathbf{\Gamma}} = \bigcup_{k=1}^{l} V_{\Gamma_k}$. In particular, all the edges of $E_{\mathbf{\Gamma}} = \bigcup_{k=1}^{l} E_{\Gamma_k}$ must be absent from these forests, since otherwise two or more of the root vertices would lie in the same component (or one of the root vertices would be connected to itself by a loop edge). On the other hand, by adjoining the edges of $E_{\mathbf{\Gamma}}$, these forests can be put into one-to-one correspondence with what we shall call $\mathbf{\Gamma}$-*forests*, namely, spanning subgraphs H in G whose edge set contains $E_{\mathbf{\Gamma}}$ and which, after deletion of the edges in $E_{\mathbf{\Gamma}}$, leaves a forest in which each tree component contains exactly one vertex from $V_{\mathbf{\Gamma}}$. (Equivalently, a $\mathbf{\Gamma}$-forest is a subgraph H with l connected components in which each component contains exactly one Γ_i, and which does not contain any cycles other than those lying entirely within the Γ_i. Note, in particular, that a $\mathbf{\Gamma}$-forest is a forest if and only if all the

Γ_i are trees.) Furthermore, adjoining the edges of $E_{\mathbf{\Gamma}}$ provides precisely the factor $\prod_{e \in E_{\mathbf{\Gamma}}} w_e$. Therefore

$$\int \mathcal{D}(\psi, \bar{\psi})\, Q_{\Gamma_1} \cdots Q_{\Gamma_l}\, e^{\bar{\psi} L \psi} \;=\; \sum_{H \in \mathcal{F}_{\mathbf{\Gamma}}} \prod_{e \in H} w_e \tag{8.6}$$

where the sum runs over all $\mathbf{\Gamma}$-forests H.

We can now combine all the formulae (8.6) into a single generating function, by introducing a variable t_Γ for each connected subgraph Γ of G. Since $1 + t_\Gamma Q_\Gamma = e^{t_\Gamma Q_\Gamma}$, we have

$$\int \mathcal{D}(\psi, \bar{\psi})\, e^{\bar{\psi} L \psi + \sum\limits_{\Gamma} t_\Gamma Q_\Gamma} \;=\; \sum_{\substack{\mathbf{\Gamma} \text{ vertex-} \\ \text{disjoint}}} \left(\prod_{\Gamma \in \mathbf{\Gamma}} t_\Gamma \right) \sum_{H \in \mathcal{F}_{\mathbf{\Gamma}}} \prod_{e \in H} w_e\,. \tag{8.7}$$

We can express this in another way by interchanging the summations over $\mathbf{\Gamma}$ and H. Consider an arbitrary spanning subgraph H with connected components $H_1, \ldots, H_l$; let us say that Γ *marks* H_i (denoted $\Gamma \prec H_i$) in case H_i contains Γ and contains no cycles other than those lying entirely within Γ. Define the weight

$$W(H_i) \;=\; \sum_{\Gamma \prec H_i} t_\Gamma\,. \tag{8.8}$$

Then saying that H is a $\mathbf{\Gamma}$-forest is equivalent to saying that each of its components is marked by exactly one of the Γ_i; summing over the possible families $\mathbf{\Gamma}$, we obtain

$$\int \mathcal{D}(\psi, \bar{\psi})\, e^{\bar{\psi} L \psi + \sum\limits_{\Gamma} t_\Gamma Q_\Gamma} \;=\; \sum_{\substack{H \text{ spanning} \subseteq G \\ H = (H_1, \ldots, H_l)}} \left(\prod_{i=1}^{l} W(H_i) \right) \prod_{e \in H} w_e\,. \tag{8.9}$$

This is our general combinatorial formula. Extensions allowing prefactors $\bar{\psi}_{i_1} \psi_{j_1} \cdots \bar{\psi}_{i_r} \psi_{j_r}$ are also easily derived.

Now consider the special case in which $t_\Gamma = t$ whenever Γ consists of a single vertex with no edges, $t_\Gamma = u$ whenever Γ consists of two vertices linked by a single edge, and $t_\Gamma = 0$ otherwise. We have

$$\begin{aligned} &\int \mathcal{D}(\psi, \bar{\psi}) \exp\!\left[\bar{\psi} L \psi + t \sum_i \bar{\psi}_i \psi_i + u \sum_{\langle ij \rangle} w_{ij} \bar{\psi}_i \psi_i \bar{\psi}_j \psi_j \right] \\ &\qquad = \sum_{\substack{F \in \mathcal{F} \\ F = (F_1, \ldots, F_l)}} \left(\prod_{i=1}^{l} \left(t|V_{F_i}| + u|E_{F_i}| \right) \right) \prod_{e \in F} w_e \end{aligned} \tag{8.10}$$

where the sum runs over spanning forests F in G with components $F_1, \ldots, F_l$; here $|V_{F_i}|$ and $|E_{F_i}|$ are, respectively, the numbers of vertices and edges in the tree F_i. [The four-fermion term $u \sum_{\langle ij \rangle} w_{ij} \bar{\psi}_i \psi_i \bar{\psi}_j \psi_j$ can equivalently be written, using nilpotency of the Grassmann variables, as $-(u/2) \sum_{i,j} \bar{\psi}_i \psi_i L_{ij} \bar{\psi}_j \psi_j$.] If $u = 0$, this formula represents vertex-weighted spanning forests as a determinant ("massive fermionic free field") [6, 51]. More interestingly, since $|V_{F_i}| - |E_{F_i}| = 1$ for each tree

F_i, we can take $u = -t$ and obtain the generating function of *unrooted* spanning forests with a weight t for each component. This is furthermore equivalent to giving each edge e a weight w_e/t, and then multiplying by an overall prefactor $t^{|V|}$. We have therefore proven:

Proposition 8.1 *Let $G = (V, E)$ be a finite undirected graph, let L be the Laplacian matrix for G with edge weights $\mathbf{w} = \{w_e\}_{e \in E}$, and let F_G be the generating polynomial of spanning forests in G. Then*

$$\int \mathcal{D}(\psi, \bar{\psi}) \exp\Big[\bar{\psi} L \psi + t \sum_i \bar{\psi}_i \psi_i + \frac{t}{2} \sum_{i,j} \bar{\psi}_i \psi_i L_{ij} \bar{\psi}_j \psi_j\Big] = t^{|V|} F_G(\mathbf{w}/t) \,. \quad (8.11)$$

This representation of unrooted spanning forests is the translation to generating functions and Grassmann variables of a little-known but important paper by Liu and Chow [86].

It is an open question whether there exists some analogue of Proposition 8.1 for the object dual to $F_G(\mathbf{w})$, namely the generating polynomial $C_G(\mathbf{v})$ of connected spanning subgraphs, when the graph G is non-planar.

It would also be interesting to know whether the foregoing identities are in any way related to the forest-root formula of Brydges and Imbrie [22, 23, 65, 66].

9 Absence of zeros at large $|q|$

Combinatorialists have long been interested in the real or complex zeros of the chromatic polynomial $P_G(q)$. A fair number of interesting theorems have by now been proven, notably concerning the real zeros, but even here a vast number of open problems remain; moreover, very little is known rigorously concerning the complex zeros. (See [70] for an excellent recent survey treating both real and complex zeros.)

In this section I would like to discuss one aspect of the complex-zero problem where some progress has recently been made by exploiting methods from mathematical statistical mechanics: namely, the absence of zeros at large enough $|q|$.

9.1 Bounds in terms of maximum degree and its relatives

The bounds I want to discuss [121] apply, in fact, not only to the chromatic polynomial but to the multivariate Tutte polynomial $Z_G(q, \mathbf{v})$ throughout the "complex antiferromagnetic regime" $|1 + v_e| \le 1$. These bounds come in several variants, but here is a typical one:

Theorem 9.1 *Let $G = (V, E)$ be a loopless finite undirected graph of maximum degree Δ. Suppose that the edge weights $\mathbf{v} = \{v_e\}_{e \in E}$ satisfy $|1 + v_e| \le 1$ for all e. Let $v_{\max} = \max_{e \in E} |v_e|$. Then all the zeros of $Z_G(q, \mathbf{v})$ lie in the disc $|q| < 7.963907 v_{\max} \Delta$.*

For the chromatic polynomial ($v_e = -1$ for all e) one obtains the immediate corollary:

Corollary 9.2 *Let $G = (V, E)$ be a loopless finite undirected graph of maximum degree Δ. Then all the zeros of the chromatic polynomial $P_G(q)$ lie in the disc $|q| < 7.963907\Delta$.*

Corollary 9.2 answers in the affirmative a question posed by Brenti, Royle and Wagner [14, Question 6.1], generalizing an earlier conjecture of Biggs, Damerell and Sands [7] limited to regular graphs.

Of course, the constant 7.963907 is an artifact of the proof; it is presumably far from sharp. The linear dependence on Δ is, however, best possible, since the complete graph $K_{\Delta+1}$ has chromatic roots $0, 1, 2, \ldots, \Delta$. Furthermore (and perhaps surprisingly), the complete graph $K_{\Delta+1}$ is *not* the extremal graph for this problem, and a bound $|q| \leq \Delta$ is *not* possible. In fact, a non-rigorous (but probably rigorizable) asymptotic analysis, confirmed by numerical calculations, shows [112] that the complete bipartite graph $K_{\Delta,\Delta}$ has a chromatic root $\alpha\Delta + o(\Delta)$, where $\alpha = -2/W(-2/e) \approx 0.678345 + 1.447937i$; here W denotes the principal branch of the Lambert W function (the inverse function of $w \mapsto we^w$) [39]. So the constant in Corollary 9.2 cannot be better than $|\alpha| \approx 1.598960$.

A complete proof of Theorem 9.1 and its variants can be found in [121] (see also [11]); here I would like simply to sketch the method, which I hope will be of wider use.

It is immediately apparent that Theorem 9.1 and Corollary 9.2 are "hard" results in the sense of Section 5: the zero-free region depends on the maximum degree Δ. The proof of these results follows the plan sketched there: namely, the multivariate Tutte polynomial on the graph $G = (V, E)$ is first mapped onto a gas of nonoverlapping "polymers" on V, i.e. a hard-core lattice gas on the intersection graph of V; then this hard-core lattice gas is controlled by means of the theorems mentioned at the end of Section 5.

The mapping to a polymer gas is really quite simple; indeed, it is nearly trivial. The definition (1.2) writes the multivariate Tutte polynomial $\widetilde{Z}_G(q, \mathbf{v})$ as a sum over spanning subgraphs (V, A). Let us perform this sum in two stages: First we sum over partitions of V into disjoint nonempty subsets $V_1, \ldots, V_k$ ($k \geq 1$ arbitrary); these subsets will correspond to the vertex sets of the connected components of (V, A). Then we sum over the ways of choosing edges within each component so as to make it connected. From (1.2) we see that a component $V_i = S$ will get a weight

$$w(S) \;=\; q^{1-|S|} \sum_{\substack{B \subseteq E \\ (S,B)\,\text{connected}}} \prod_{e \in B} v_e \tag{9.1a}$$

$$\;=\; q^{1-|S|}\, C_{G|S}(\mathbf{v}) \tag{9.1b}$$

where $G|S$ denotes the induced subgraph. Note, in particular, that (provided G is loopless) any set S of cardinality 1 gets weight $w(S) = 1$. So we need not consider such sets explicitly: it suffices to define the "polymers" to be the sets V_i of cardinality 2 or more; we then know that any vertex not covered by a polymer must be covered by a set V_i of cardinality 1. The polymers are therefore an arbitrary family (possibly empty) of disjoint sets $S_i \subseteq V$ of cardinality 2 or more. We have thus written $\widetilde{Z}_G(q, \mathbf{v})$ as an independent-set polynomial (= hard-core lattice gas) of the form (5.5) for a graph $\widehat{G} = (\widehat{V}, \widehat{E})$ whose vertices are the subsets of V of cardinality 2 or more, and whose edges are the pairs with nonempty intersection. The fugacities are given by (9.1).

The usefulness of this representation for proving the absence of zeros at large $|q|$ comes from the fact that the fugacities $w(S)$ are all suppressed by powers of q^{-1}

(since $|S| \geq 2$), hence are small for large $|q|$. Furthermore, this suppression operates more strongly for larger polymers. This raises the hope that, when $|q|$ is sufficiently large, the model lies in a "low-fugacity" regime where the methods mentioned in Section 5 can be brought to bear. This is in fact the case.

Of course, I have sloughed over one crucial point: the sum over B in (9.1a). It is here, not surprisingly, that the maximum degree Δ enters (along with $v_{\max}$). What one needs, it turns out, is a bound on the number of connected subgraphs $H \subseteq G$ containing a specified vertex x and having a specified number of vertices $m = |S|$. What is easier to prove, however, is a bound on the number of connected subgraphs $H \subseteq G$ containing a specified vertex x and having a specified number of *edges* [121, Section 4.2] [72]. The relation between the two is far from obvious, but it turns out that in the complex antiferromagnetic region $|1+v_e| \leq 1$ (and only there!), a beautiful inequality due to Oliver Penrose [101] allows connected subgraphs to be bounded by *trees*.[29] The upshot is that the sum over B in (9.1a) can be controlled, with the result that $w(S)$ satisfies a uniform bound that is exponentially decaying in $|S|$, provided that $|q|$ is large enough compared to Δ. This suffices to ensure the convergence of the "polymer expansion", and hence the nonvanishing of $\widetilde{Z}_G(q, \mathbf{v})$.

I believe it should be possible to extend this result beyond the complex antiferromagnetic regime, by bounding directly the number of connected subgraphs having a specified number of vertices [73]. But the resulting bound on $|q|$ will no longer be linear in Δ and $v_{\max}$; rather, it will behave roughly like $v_{\max}^{\Delta/2}$. And this is not simply an artifact of the method of proof: in the q-state Potts *ferromagnet* ($v > 0$, $q > 0$) on the the simple hypercubic lattice $\mathbb{Z}^d$ with nearest-neighbor edges (hence $\Delta = 2d$), there is a first-order phase transition at $v_t(q) = q^{1/d}[1 + O(1/q)]$ for all sufficiently large q [10, 79, 81, 82, 90]. As discussed in Section 5, this means that for v near $v_t(q)$ there will be complex zeros of $Z_G(q, v)$ for large subgraphs $G \subset \mathbb{Z}^d$.

By an *ad hoc* (and aesthetically unsatisfying) method, Theorem 9.1 and Corollary 9.2 can be strengthened so as to bound the roots, not in terms of the maximum degree Δ, but in terms of the *second-largest* degree Δ_2 [121, Section 6]. For simplicity let me state only the result for chromatic polynomials:

Corollary 9.3 *Let $G = (V, E)$ be a loopless finite undirected graph of second-largest degree Δ_2. Then all the zeros of the chromatic polynomial $P_G(q)$ lie in the union of the discs $|q| < 7.963907\Delta_2$ and $|q - 1| < 7.963907\Delta_2$. In particular, they all lie in the disc $|q| < 7.963907\Delta_2 + 1$.*

It should be stressed that "second-largest" in Corollary 9.3 *cannot* be replaced by "third-largest". Indeed, the generalized theta graphs $\Theta^{(s,p)}$ [18, 122] constitute a family of planar (in fact, series-parallel) graphs in which all but two vertices have degree 2, yet their chromatic roots are dense in the region $\{q \in \mathbb{C}\colon\ |q - 1| \geq 1\}$ [122].[30] So, with *one* large-degree vertex, the chromatic roots remain bounded; but with *two*, all hell can break loose.

[29] See [121, Section 4.1] and [114, Section 2.2] for further discussion.

[30] The graph $\Theta^{(s,p)}$ consists of a pair of endvertices connected by p internally disjoint paths each of length s.

9.2 Bounds in terms of maxmaxflow?

Alas, all of the foregoing bounds have a severe defect: they are *unnatural*, because the multivariate Tutte polynomial $Z_G(q, \mathbf{v})$ factorizes over blocks [cf. (4.2)], while Δ and Δ_2 can grow arbitrarily large when blocks are glued together at a cut vertex. As a consequence, the bounds obtained from Theorem 9.1 and its variants are sometimes very far from sharp: for instance, a tree can have arbitrarily large Δ and Δ_2, but its chromatic roots are only 0 and 1! Of course, this defect can be cured by the *ad hoc* technique of applying Theorem 9.1 ff. to each block of G rather than directly to G. But the deeper significance of this remark is that the quantities Δ and Δ_2 are *unnatural* for studying the multivariate Tutte polynomial because they are *not matroidal*.

One would like, therefore, to find a graph invariant that is smaller than Δ and Δ_2, that "trivializes over blocks" (and ideally generalizes to matroids), and that is strong enough to bound the roots of $P_G(q)$ and $Z_G(q, \mathbf{v})$. I have a candidate: it is the *maxmaxflow* [72, 119, 121], defined as follows: For $x, y \in V$ with $x \neq y$, let $\lambda_G(x, y)$ be the max flow from x to y:

$$\begin{aligned} \lambda_G(x,y) &= \text{max \# of edge-disjoint paths from } x \text{ to } y && (9.2a) \\ &= \text{min \# of edges separating } x \text{ from } y && (9.2b) \end{aligned}$$

The *maxmaxflow* $\Lambda(G)$ is then defined by

$$\Lambda(G) \;=\; \max_{\substack{x, y \in V \\ x \neq y}} \lambda_G(x, y)\,. \qquad (9.3)$$

[Note the contrast with the edge-connectivity, which is the *minimum* of $\lambda_G(x, y)$ over $x \neq y$.] Clearly $\lambda_G(x, y) \leq \min[d_G(x), d_G(y)]$, so that

$$\Lambda(G) \;\leq\; \Delta_2(G)\,. \qquad (9.4)$$

Furthermore, it is easy to see that maxmaxflow "trivializes over blocks" in the sense that $\Lambda(G) = \max_{1 \leq i \leq b} \Lambda(G_i)$ where $G_1, \ldots, G_b$ are the blocks of G.

An apparently very different quantity can be defined via cocycle bases. For X, Y disjoint subsets of V, let $E(X, Y)$ denote the set of edges in G between X and Y. A *cocycle* of G is a set $E(X, X^c)$ where $X \subseteq V$ and $X^c \equiv V \setminus X$. It is well-known that the cocycles of G form a vector space over $GF(2)$ with respect to symmetric difference; this is called the *cocycle space* of G. Let $\widetilde{\Lambda}(G)$ be the minmax cardinality of the cocycles in a basis, i.e.

$$\widetilde{\Lambda}(G) \;=\; \min_{\mathcal{B}} \max_{C \in \mathcal{B}} |C|\,, \qquad (9.5)$$

where the min runs over all bases $\mathcal{B}$ of the cocycle space of G, and the max runs over all the cocycles C in the basis $\mathcal{B}$. Since one special class of cocycle bases consists of taking the stars $C(x) = E(\{x\}, \{x\}^c)$ for all but one of the vertices in each component of G, we clearly have

$$\widetilde{\Lambda}(G) \;\leq\; \Delta_2(G)\,. \qquad (9.6)$$

The relationship, if any, between maxmaxflow and cocycle bases is perhaps not obvious at first sight. But Bill Jackson and I have proven [72] that

$$\Lambda(G) \;=\; \widetilde{\Lambda}(G)\,. \tag{9.7}$$

The two definitions thus give dual approaches to the same quantity.

All this theory, extends in fact, to the more general situation of a finite undirected graph G equipped with nonnegative real edge weights $\mathbf{w} = \{w_e\}_{e\in E}$. (In the application to the multivariate Tutte polynomial, one might want to take $w_e = |v_e|$ or something similar.)

The ultimate goal of this work is to prove (or disprove!) strengthenings of Theorem 9.1 ff. in which maximum degree or second-largest degree are replaced by maxmaxflow. For instance, for chromatic polynomials one would like to prove the following:

Conjecture 9.4 *There exist universal constants $C(\Lambda) < \infty$ such that all the chromatic roots (real or complex) of all loopless graphs of maxmaxflow Λ lie in the disc $|q| \le C(\Lambda)$. Indeed, I conjecture that $C(\Lambda)$ can be taken to be linear in Λ.*

It is natural to try to prove Conjecture 9.4 by modifying the arguments of [121] so as to decompose a spanning subgraph of G into its *2-connected* (rather than merely *connected*) components. Such an argument will require, at a minimum, a good bound on the number of 2-connected subgraphs $H \subseteq G$ containing a specified edge e and having a specified number of vertices or edges. Bill Jackson and I have found some bounds of this type [72]. But this is only a first step, and the other obstacles in generalizing this proof—such as controlling the interaction between the 2-connected components—have yet to be overcome.

It is worth mentioning that $\widetilde{\Lambda}$ can be defined also for binary matroids, generalizing the definition for graphs. But we do not yet know whether it is strong enough to give a good (i.e., singly exponential) bound on the number of 2-connected deletion minors containing a specified element e and having a specified size. Nor do we know whether it is useful in bounding the zeros of the matroid chromatic polynomial $\widetilde{Z}_M(q,-1)$.

9.3 Some further questions

There is no compelling reason, *a priori*, to limit attention in the complex q-plane to discs centered at the origin; many other types of regions are of interest. For instance, the following conjecture might conceivably be true:

Conjecture 9.5 *All the chromatic roots (real or complex) of all loopless graphs of maxmaxflow Λ lie in the half-plane $\operatorname{Re} q \le \Lambda$.*

By factoring $P_G(q)/q$ and using the fact that its roots are either real or occur in complex-conjugate pairs, one easily deduces that the truth of Conjecture 9.5 would imply:

Conjecture 9.6 *If G has maxmaxflow Λ, then $P_G(q)/q$ and all its derivatives are nonnegative at $q = \Lambda$. [The same then holds also for $P_G(q)$ and all its derivatives.]*

Since the truth of Conjecture 9.6 would imply that $P_G(q)$ has all nonnegative Taylor coefficients at $q = \Lambda$, and since $P_G(q)$ is not identically zero (when G is loopless), Conjecture 9.6 would imply:

Conjecture 9.7 *If G is a loopless graph of maxmaxflow Λ, then $P_G(q) > 0$ for $q > \Lambda$. In particular, $P_G(q)$ has no roots in the real interval (Λ, ∞).*

Even the weaker versions of these conjectures with Λ replaced by Δ_2 or Δ are open! So one has the 3×3 matrix of conjectures

$$\begin{array}{ccccc} 9.5 & \Longrightarrow & 9.5' & \Longrightarrow & 9.5'' \\ \Downarrow & & \Downarrow & & \Downarrow \\ 9.6 & \Longrightarrow & 9.6' & \Longrightarrow & 9.6'' \\ \Downarrow & & \Downarrow & & \Downarrow \\ 9.7 & \Longrightarrow & 9.7' & \Longrightarrow & 9.7'' \end{array}$$

in which the strongest might well be true and the weakest is still an open question.[31] It might be useful to seek counterexamples, either by systematic calculation on small graphs up to ≈ 20 vertices or, perhaps more fruitfully, by constructing suitable infinite families of graphs. It goes without saying that I have no idea how to prove any of these conjectures, or even any compelling reason to believe they are true (other than my inability thus far to find counterexamples).

We can pose these questions most generally as follows: Let $\mathcal{G}$ be a class of finite graphs, and let $\mathcal{V}$ be a subset of the complex plane. Then we can ask about the sets

$$S_1(\mathcal{G}, \mathcal{V}) \;=\; \bigcup_{G \in \mathcal{G}} \bigcup_{v \in \mathcal{V}} \{q \in \mathbb{C}\colon\; Z_G(q, v) = 0\} \tag{9.8}$$

$$S_2(\mathcal{G}, \mathcal{V}) \;=\; \bigcup_{G \in \mathcal{G}} \bigcup_{\mathbf{v}\colon\, v_e \in \mathcal{V}\ \forall e} \{q \in \mathbb{C}\colon\; Z_G(q, \mathbf{v}) = 0\} \tag{9.9}$$

Among the interesting cases are the chromatic polynomials $\mathcal{V} = \{-1\}$, the antiferromagnetic regime $\mathcal{V} = [-1, 0]$, and the complex antiferromagnetic regime $\mathcal{V} = \mathcal{A} \equiv \{v \in \mathbb{C}\colon\; |1 + v| \le 1\}$.

The theorems discussed in this section show that the sets $S_2(\mathcal{G}_r, \mathcal{A})$ and $S_2(\mathcal{G}'_r, \mathcal{A})$ are *bounded*, where $\mathcal{G}_r$ (resp. $\mathcal{G}'_r$) is the set of all graphs whose maximum degree (resp. second-largest degree) is less than or equal to r. But boundedness is a rather weak statement about a set; one would like to learn more about its location in the complex plane.

In the other direction, I have recently shown [122] that if $\mathcal{G}$ is the family of all generalized theta graphs $\Theta^{(s,p)}$ [18, 122], then $S_1(\mathcal{G}, \{-1\})$ is dense in the region $\{q \in \mathbb{C}\colon\; |q - 1| \ge 1\}$. Moreover, if $\mathcal{G}'$ denotes the family of joins of $\Theta^{(s,p)}$ with the complete graph K_2, then $S_1(\mathcal{G} \cup \mathcal{G}', \{-1\})$ is dense in the entire complex plane.[32] The

---

[31] **STOP PRESS!!!** Gordon Royle (private communication) has found a family of counterexamples to Conjectures 9.5 and 9.5$'$. His graphs are planar (but not series-parallel) and all vertices but one have degree 3 (hence $\Lambda = \Delta_2 = 3$). Included in his family are a 47-vertex graph with chromatic roots $q \approx 3.0129950712 \pm 0.8089628639\, i$, a 95-vertex graph with chromatic roots $q \approx 3.0536525915 \pm 0.7547530551\, i$, a 191-vertex graph with chromatic roots $q \approx 3.07174237056 \pm 0.7105232675\, i$, and a 383-vertex graph with chromatic roots $q \approx 3.0766232972 \pm 0.6746120243\, i$.

[32] The *join* of G and H, denoted $G + H$, is the graph obtained from the disjoint union of G and H by adding one edge connecting each pair of vertices $x \in V(G)$, $y \in V(H)$.

idea of the construction in [122] is to exploit the reduction formulae of Section 4.6: suitably chosen 2-rooted subgraphs (in this case single edges) are concatenated in order to create a larger 2-rooted subgraph that has some desired value of the effective coupling $v_{\rm eff}$. Such a construction is obviously inapplicable to 3-connected graphs, in which 2-rooted subgraphs of more than a single edge cannot occur. It is thus reasonable to ask: What is the closure of the set of chromatic roots of *3-connected* graphs? The answer, alas, is still the whole complex plane [71]. To see this, consider the graphs $\Theta^{(s,p)} + K_n$ for fixed n and varying s, p: they are $(n+2)$-connected, and their chromatic roots (taken together) are dense in the region $|q - (n+1)| \geq 1$. In particular, considering both n and $n+2$, the chromatic roots are dense in the whole complex plane. So even arbitrarily high connectedness does not, by itself, stop the chromatic roots from being dense in the whole complex plane.

On the other hand, the graphs $\Theta^{(s,p)} + K_n$ are non-planar whenever $n \geq 1$ and $p \geq 3$. This suggests posing a more restricted question:

Question 9.8 *What is the closure of the set of chromatic roots for 3-connected* planar *graphs?*

Here the answer may well be much smaller than $\mathbb{C}$ or $\mathbb{C} \setminus \{|q-1| < 1\}$. But it will not be a bounded set [71], since the bipyramids $C_n + \bar{K}_2$ are 4-connected plane triangulations, but their chromatic roots are unbounded [106, 117, 118].

9.4 A final remark

The ultimate goal of this research is to prove theorems of the type "such-and-such graph property (or conjunction of properties) implies such-and-such bound on the chromatic roots". But there is a virtually unlimited number of possible assertions of this type; the hard part is to figure out which ones are true!

There is thus, in my opinion, much room for numerical experiment as a means for obtaining intuition about which graph properties (e.g. maximum degree and its relatives, girth, planarity, connectivity, ...) affect which aspects of the chromatic roots (e.g. large positive real roots, large positive or negative real part, large imaginary part, etc.). Some information can be obtained by systematic calculation on all graphs of a specified type up to as many vertices as can be handled—or, for suitable classes (e.g. cubic graphs), by random generation of typical graphs with a rather larger number of vertices—in an attempt to correlate properties of the graph with properties of the chromatic roots. Guided by this information, one may then choose to focus on special classes of graphs that potentially exhibit some type of "extremal" behavior, and either pursue the numerical calculation to larger graphs or, better yet, find infinite families of graphs that exhibit such "extremal" behavior and for which one can compute exactly the asymptotic behavior of the chromatic roots.

It is important to keep in mind that many asymptotic properties of chromatic roots are achieved very slowly as the number of vertices grows: for instance, the roots of cubic graphs move quite slowly towards negative real part [105], and the roots of the generalized theta graphs $\Theta^{(s,p)}$ fill out the complex plane extremely slowly as $s, p \to \infty$ [18]. As a result, numerical experiments can be deceptive unless interpreted with extreme care. It is for this reason that exact computations with well-chosen infinite families of graphs are particularly valuable.

Acknowledgements

Many of the results discussed here were obtained in joint work with Jason Brown, Sergio Caracciolo, Young-Bin Choe, Carl Hickman, Bill Jackson, Jesper Jacobsen, James Oxley, Gordon Royle, Jesús Salas, Hubert Saleur, Alex Scott, Andrea Sportiello and Dave Wagner. I thank them all for the pleasure of doing mathematics together, as well as for permission to plagiarize here some of our collective prose. I also thank Abdelmalek Abdesselam, Norman Biggs, Graham Brightwell, David Brydges, Geoffrey Grimmett, John Imbrie, Marco Polin, Robert Shrock, Jan van den Heuvel, Dominic Welsh and Geoff Whittle for many helpful discussions; and Bill Jackson, Mark Jerrum, Gordon Royle, Dave Wagner and Geoff Whittle for helpful comments on a first draft of this paper.

The author's research was supported in part by U.S. National Science Foundation grants PHY–0099393 and PHY–0424082 and by U.K. Engineering and Physical Sciences Research Council grant GR/S26323/01.

References

[1] A. Abdesselam, Grassmann–Berezin calculus and theorems of the matrix-tree type, *Adv. Appl. Math.* **33** (2004), 51–70. Also math.CO/0306396 at arXiv.org.

[2] W.A. Adkins and S.H. Weintraub, *Algebra: An Approach via Module Theory*, Springer-Verlag, New York–Berlin–Heidelberg (1992).

[3] J. Ashkin and E. Teller, Statistics of two-dimensional lattices with four components, *Phys. Rev.* **64** (1943), 178–184.

[4] N. Balabanian and T.A. Bickart, *Electrical Network Theory*, Wiley, New York (1969).

[5] R.J. Baxter, *Exactly Solved Models in Statistical Mechanics*, Academic Press, London–New York (1982).

[6] N. Biggs, *Algebraic Graph Theory*, 2nd ed., Cambridge University Press, Cambridge–New York (1993).

[7] N.L. Biggs, R.M. Damerell and D.A. Sands, Recursive families of graphs, *J. Combin. Theory Ser. B* **12** (1972), 123–131.

[8] G.D. Birkhoff, A determinantal formula for the number of ways of coloring a map, *Ann. Math.* **14** (1912), 42–46.

[9] B. Bollobás and O. Riordan, A Tutte polynomial for coloured graphs, *Combin. Probab. Comput.* **8** (1999), 45–93.

[10] C. Borgs, R. Kotecký and S. Miracle-Solé, Finite-size scaling for Potts models, *J. Statist. Phys.* **62** (1991), 529–551.

[11] C. Borgs, *Expansion Methods in Combinatorics*, Conference Board of the Mathematical Sciences book series (to appear).

[12] A. Bovier and M. Zahradník, A simple inductive approach to the problem of convergence of cluster expansions of polymer models, *J. Statist. Phys.* **100** (2000), 765–778.

[13] A. Brandstädt, Le Van Bang and J.P. Spinrad, *Graph Classes: A Survey*, SIAM, Philadelphia (1999).

[14] F. Brenti, G.F. Royle and D.G. Wagner, Location of zeros of chromatic and related polynomials of graphs, *Canad. J. Math.* **46** (1994), 55–80.

[15] R.L. Brooks, C.A.B. Smith, A.H. Stone and W.T. Tutte, The dissection of rectangles into squares, *Duke Math. J.* **7** (1940), 312–340.

[16] J.I. Brown and C.J. Colbourn, Roots of the reliability polynomial, *SIAM J. Discrete Math.* **5** (1992), 571–585.

[17] J.I. Brown and C.A. Hickman, On chromatic roots of large subdivisions of graphs, *Discrete Math.* **242** (2002), 17–30.

[18] J.I. Brown, C.A. Hickman, A.D. Sokal and D.G. Wagner, On the chromatic roots of generalized theta graphs, *J. Combin. Theory Ser. B* **83** (2001), 272–297.

[19] W.C. Brown, *Matrices over Commutative Rings*, Dekker, New York (1993).

[20] S.G. Brush, History of the Lenz–Ising model, *Rev. Mod. Phys.* **39** (1967), 883–893.

[21] D.C. Brydges, A short course on cluster expansions, in *Phénomènes Critiques, Systèmes Aléatoires, Théories de Jauge / Critical Phenomena, Random Systems, Gauge Theories* (eds. K. Osterwalder and R. Stora), Les Houches summer school, Session XLIII, 1984, Elsevier/North-Holland, Amsterdam (1986), pp. 129–183.

[22] D.C. Brydges and J.C. Imbrie, Branched polymers and dimensional reduction, *Ann. Math.* **158** (2003), 1019–1039. Also math-ph/0107005 at arXiv.org.

[23] D.C. Brydges and J.C. Imbrie, Dimensional reduction formulas for branched polymer correlation functions, *J. Statist. Phys.* **110** (2003), 503–518. Also math-ph/0203055 at arXiv.org.

[24] D.C. Brydges and T. Kennedy, Mayer expansions and the Hamilton-Jacobi equation, *J. Statist. Phys.* **48** (1987), 19–49.

[25] D.C. Brydges and Ph.A. Martin, Coulomb systems at low density: A review, *J. Statist. Phys.* **96** (1999), 1163–1330. Also cond-mat/9904122 at arXiv.org.

[26] D.C. Brydges and J.D. Wright, Mayer expansions and the Hamilton-Jacobi equation. II. Fermions, dimensional reduction formulas, *J. Statist. Phys.* **51** (1988), 435–456. Erratum **97** (1999), 1027.

[27] T.H. Brylawski, A decomposition for combinatorial geometries, *Trans. Amer. Math. Soc.* **171** (1972), 235–282.

[28] T. Brylawski and J. Oxley, The Tutte polynomial and its applications, in *Matroid Applications* (ed. N. White), *Encyclopedia of Mathematics and its Applications*, 40, Cambridge University Press, Cambridge (1992), pp. 123–225.

[29] C. Cammarota, Decay of correlations for infinite range interactions in unbounded spin systems, *Commun. Math. Phys.* **85** (1982), 517–528.

[30] S. Caracciolo, J.L. Jacobsen, H. Saleur, A.D. Sokal and A. Sportiello, Fermionic field theory for trees and forests, *Phys. Rev. Lett.* **93** (2004), 080601. Also cond-mat/0403271 at arXiv.org.

[31] S. Caracciolo and A. Sportiello, General duality for abelian-group-valued statistical-mechanics models, *J. Phys. A: Math. Gen.* **37** (2004), 7407–7432. Also cond-mat/0308515 at arXiv.org.

[32] S. Caracciolo, A. Sportiello and A.D. Sokal, Generalized Potts models and the multivariate Tutte polynomial for matroids, in preparation.

[33] S. Chaiken, A combinatorial proof of the all minors matrix-tree theorem, *SIAM J. Alg. Disc. Meth.* **3** (1982), 319–329.

[34] S. Chaiken and D.J. Kleitman, Matrix tree theorems, *J. Combin. Theory Ser. A* **24** (1978), 377–381.

[35] W.-K. Chen, *Applied Graph Theory*, North-Holland, Amsterdam–London (1971).

[36] Y.-B. Choe, J.G. Oxley, A.D. Sokal and D.G. Wagner, Homogeneous multivariate polynomials with the half-plane property, *Adv. Appl. Math.* **32** (2004), 88–187. Also math.CO/0202034 at arXiv.org.

[37] Y.-B. Choe and D.G. Wagner, Rayleigh matroids, *Combin. Probab. Comput.*, in press. Also math.CO/0307096 at arXiv.org.

[38] C.J. Colbourn, *The Combinatorics of Network Reliability*, Oxford University Press, New York–Oxford (1987).

[39] R.M. Corless, G.H. Gonnet, D.E.G. Hare, D.J. Jeffrey and D.E. Knuth, On the Lambert W function, *Adv. Comput. Math.* **5** (1996), 329–359.

[40] H.H. Crapo, The Tutte polynomial, *Aequationes Math.* **3** (1969), 211–229.

[41] H.H. Crapo and G.-C. Rota, *On the Foundations of Combinatorial Theory: Combinatorial Geometries*, MIT Press, Cambridge MA (1970).

[42] C.A. Desoer and E.S. Kuh, *Basic Circuit Theory*, McGraw-Hill, New York (1969).

[43] R. Diestel, *Graph Theory*, Springer-Verlag, New York (1997).

[44] P. Di Francesco, P. Mathieu and D. Sénéchal, *Conformal Field Theory*, Springer-Verlag, New York (1997).

[45] R.L. Dobrushin, Estimates of semi-invariants for the Ising model at low temperatures, in *Topics in Statistical and Theoretical Physics*, American Mathematical Society Translations, Ser. 2, no. 177 (1996), pp. 59–81.

[46] R.L. Dobrushin, Perturbation methods of the theory of Gibbsian fields, in *Lectures on Probability Theory and Statistics* (ed. P. Bernard), Ecole d'Eté de Probabilités de Saint-Flour XXIV – 1994, *Lecture Notes in Mathematics*, 1648, Springer–Verlag, Berlin (1996), pp. 1–66.

[47] C. Domb, Configurational studies of the Potts models, *J. Phys. A* **7** (1974), 1335–1348.

[48] P.G. Doyle and J.L. Snell, *Random Walks and Electric Networks*, Mathematical Association of America, New York (1984).

[49] K. Drühl and H. Wagner, Algebraic formulation of duality transformations for abelian lattice models, *Ann. of Phys.* **141** (1982), 225–253.

[50] R.J. Duffin, Topology of series-parallel graphs, *J. Math. Anal. Appl.* **10** (1965), 303–318.

[51] B. Duplantier and F. David, Exact partition functions and correlation functions of multiple Hamiltonian walks on the Manhattan lattice, *J. Statist. Phys.* **51** (1988), 327–434.

[52] H. Edwards, R. Hierons and B. Jackson, The zero-free intervals for characteristic polynomials of matroids, *Combin. Probab. Comput.* **7** (1998), 153–165.

[53] R.G. Edwards and A.D. Sokal, Generalization of the Fortuin-Kasteleyn-Swendsen-Wang representation and Monte Carlo algorithm, *Phys. Rev. D* **38** (1988), 2009–2012.

[54] J.A. Ellis-Monaghan and L. Traldi, Parametrized Tutte polynomials of graphs and matroids, preprint, June 2004.

[55] P. Erdős and L. Lovász, Problems and results on 3-chromatic hypergraphs and some related questions, in *Infinite and Finite Sets, Vol. II* (eds. A. Hajnal, R. Rado and Vera T. Sós), *Colloq. Math. Soc. Janos Bolyai*, 10, North-Holland, Amsterdam (1975), pp. 609–627.

[56] P. Erdős and J. Spencer, Lopsided Lovász local lemma and Latin transversals, *Discrete Appl. Math.* **30** (1991), 151–154.

[57] T. Feder and M. Mihail, Balanced matroids, in *Proceedings of the Twenty-Fourth Annual ACM Symposium on the Theory of Computing*, Association for Computing Machinery, New York (1992), pp. 26–38.

[58] C.M. Fortuin and P.W. Kasteleyn, On the random-cluster model. I. Introduction and relation to other models, *Physica* **57** (1972), 536–564.

[59] J. Glimm and A. Jaffe, *Quantum Physics: A Functional Integral Point of View*, 2nd ed., Springer-Verlag, New York (1987).

[60] G. Grimmett, The rank polynomials of large random lattices, *J. London Math. Soc.* **18** (1978), 567–575.

[61] G. Grimmett, The stochastic random-cluster process and the uniqueness of random-cluster measures, *Ann. Probab.* **23** (1995), 1461–1510.

[62] G. Grimmett, *The Random-Cluster Model*, book in preparation.

[63] O. Häggström, Random-cluster measures and uniform spanning trees, *Stoch. Proc. Appl.* **59** (1995), 267–275.

[64] O.J. Heilmann and E.H. Lieb, Theory of monomer-dimer systems, *Commun. Math. Phys.* **25** (1972), 190–232.

[65] J.Z. Imbrie, Dimensional reduction and crossover to mean-field behavior for branched polymers, *Ann. Henri Poincaré* **4** (2003), S445–S458. Also math-ph/0303015 at arXiv.org.

[66] J.Z. Imbrie, Dimensional reduction for directed branched polymers, *J. Phys. A: Math. Gen.* **37** (2004), L137–L142. Also math-ph/0402074 at arXiv.org.

[67] E. Ising, Beitrag zur Theorie des Ferromagnetismus, *Z. Phys.* **31** (1925), 253–258.

[68] R.B. Israel, *Convexity in the Theory of Lattice Gases*, Princeton University Press, Princeton NJ (1979).

[69] B. Jackson, A zero-free interval for chromatic polynomials of graphs, *Combin. Probab. Comput.* **2** (1993), 325–336.

[70] B. Jackson, Zeros of chromatic and flow polynomials of graphs, *J. Geom.* **76** (2003), 95–109. Also math.CO/0205047 at arXiv.org.

[71] B. Jackson, private communication.

[72] B. Jackson and A.D. Sokal, Maxmaxflow and counting subgraphs, in preparation.

[73] B. Jackson and A.D. Sokal, Counting connected subgraphs with a specified number of vertices, in preparation.

[74] B. Jackson and A.D. Sokal, Zero-free intervals for multivariate Tutte polynomials (alias Potts-model partition functions) of graphs and matroids, in preparation.

[75] J.L. Jacobsen, J. Salas and A.D. Sokal, Spanning forests and the q-state Potts model in the limit $q \to 0$, preprint (2004). Also cond-mat/0401026 at arXiv.org.

[76] P.W. Kasteleyn and C.M. Fortuin, Phase transitions in lattice systems with random local properties, *J. Phys. Soc. Japan* **26** (Suppl.) (1969), 11–14.

[77] G. Kirchhoff, Über die Auflösung der Gleichungen, auf welche man bei der Untersuchung der linearen Verteilung galvanischer Ströme gefürht wird, *Ann. Phys. Chem.* **72** (1847), 497–508.

[78] S. Kobe, Ernst Ising – physicist and teacher, *J. Statist. Phys.* **88** (1997), 991–995.

[79] R. Kotecký, L. Laanait, A. Messager and J. Ruiz, The q-state Potts model in the standard Pirogov-Sinai theory: Surface tensions and Wilson loops, *J. Statist. Phys.* **58** (1990), 199–248.

[80] R. Kotecký and D. Preiss, Cluster expansion for abstract polymer models, *Commun. Math. Phys.* **103** (1986), 491–498.

[81] L. Laanait, A. Messager, S. Miracle-Solé, J. Ruiz and S. Shlosman, Interfaces in the Potts model I. Pirogov-Sinai theory of the Fortuin-Kasteleyn representation, *Commun. Math. Phys.* **140** (1991), 81–91.

[82] L. Laanait, A. Messager and J. Ruiz, Phases coexistence and surface tensions for the Potts model, *Commun. Math. Phys.* **105** (1986), 527–545.

[83] T.D. Lee and C.N. Yang, Statistical theory of equations of state and phase transitions. II. Lattice gas and Ising model, *Phys. Rev.* **87** (1952), 410–419.

[84] W. Lenz, Beitrag zum Verständnis der magnetischen Erscheinungen in festen Körpern, *Z. Phys.* **21** (1920), 613–615.

[85] E.H. Lieb and A.D. Sokal, A general Lee-Yang theorem for one-component and multicomponent ferromagnets, *Commun. Math. Phys.* **80** (1981), 153–179.

[86] C.J. Liu and Y. Chow, Enumeration of forests in a graph, *Proc. Amer. Math. Soc.* **83** (1981), 659–662.

[87] R. Lyons, A bird's-eye view of uniform spanning trees and forests, in *Microsurveys in Discrete Probability* (eds. D. Aldous and J. Propp), *DIMACS Series in Discrete Mathematics and Theoretical Computer Science*, 41, American Mathematical Society, Providence RI (1998), pp. 135–162.

[88] R. Lyons and Y. Peres, *Probability on Trees and Networks*, Cambridge University Press, Cambridge (to appear). Draft available at `http://mypage.iu.edu/~rdlyons/prbtree/prbtree.html`

[89] M. Marcus, *Introduction to Modern Algebra*, Dekker, New York (1978).

[90] D.H. Martirosian, Translation invariant Gibbs states in the q-state Potts model, *Commun. Math. Phys.* **105** (1986), 281–290.

[91] S. Miracle-Solé, On the convergence of cluster expansions, *Physica A* **279** (2000), 244–249.

[92] J.W. Moon, *Counting Labelled Trees*, Canadian Mathematical Congress, Montreal (1970).

[93] J.W. Moon, Some determinant expansions and the matrix-tree theorem, *Discrete Math.* **124** (1994), 163–171.

[94] A. Nerode and H. Shank, An algebraic proof of Kirchhoff's network theorem, *Amer. Math. Monthly* **68** (1961), 244–247.

[95] M. Newman, The Smith normal form, *Lin. Alg. Appl.* **254** (1987), 367–381.

[96] B. Nienhuis, Critical behavior of two-dimensional spin models and charge asymmetry in the Coulomb gas, *J. Statist. Phys.* **34** (1984), 731–761.

[97] J. Oxley, Graphs and series-parallel networks, in *Theory of Matroids* (ed. N. White), Cambridge University Press, Cambridge (1986), Chapter 6, pp. 97–126.

[98] J.G. Oxley, *Matroid Theory*, Oxford University Press, New York (1992).

[99] R. Pemantle, Uniform random spanning trees, in *Topics in Contemporary Probability and its Applications* (ed. J.L. Snell), CRC Press, Boca Raton FL (1995), pp. 1–54.

[100] P. Penfield, R. Spence and S. Duinker, *Tellegen's Theorem and Electrical Networks*, MIT Press, Cambridge MA (1970).

[101] O. Penrose, Convergence of fugacity expansions for classical systems, in *Statistical Mechanics: Foundations and Applications* (ed. T.A. Bak), Benjamin, New York–Amsterdam (1967), pp. 101–109.

[102] R.B. Potts, Some generalized order-disorder transformations, *Proc. Cambridge Philos. Soc.* **48** (1952), 106–109.

[103] A. Procacci, B.N.B. de Lima and B. Scoppola, A remark on high temperature polymer expansion for lattice systems with infinite range pair interactions, *Lett. Math. Phys.* **45** (1998), 303–322.

[104] A. Procacci and B. Scoppola, Polymer gas approach to N-body lattice systems, *J. Statist. Phys.* **96** (1999), 49–68.

[105] R.C. Read and G.F. Royle, Chromatic roots of families of graphs, in *Graph Theory, Combinatorics, and Applications*, vol. 2 (eds. Y. Alavi, G. Chartrand, O.R. Oellermann and A.J. Schwenk), Wiley, New York (1991), pp. 1009–1029.

[106] R.C. Read and W.T. Tutte, Chromatic polynomials, in *Selected Topics in Graph Theory 3* (eds. L.W. Beineke and R.J. Wilson), Academic Press, London (1988), pp. 15–42.

[107] V. Rivasseau, *From Perturbative to Constructive Renormalization*, Princeton University Press, Princeton NJ (1991).

[108] G. Royle and A.D. Sokal, The Brown–Colbourn conjecture on zeros of reliability polynomials is false, *J. Combin. Theory Ser. B* **91** (2004), 345–360. Also math.CO/0301199 at arXiv.org.

[109] G. Royle and A.D. Sokal, The Brown–Colbourn property for matroids, in preparation.

[110] D. Ruelle, Counting unbranched subgraphs, *J. Algebraic Combin.* **9** (1999), 157–160.

[111] D. Ruelle, Zeros of graph-counting polynomials, *Commun. Math. Phys.* **200** (1999), 43–56.

[112] J. Salas and A.D. Sokal, Chromatic roots of the complete bipartite graphs, in preparation.

[113] R. Savit, Duality in field theory and statistical systems, *Rev. Mod. Phys.* **52** (1980), 453–487.

[114] A.D. Scott and A.D. Sokal, The repulsive lattice gas, the independent-set polynomial, and the Lovász local lemma, *J. Statist. Phys.*, in press. Also cond-mat/0309352 at arXiv.org.

[115] E. Seiler, *Gauge Theories as a Problem of Constructive Quantum Field Theory and Statistical Mechanics*, *Lecture Notes in Physics*, 159, Springer-Verlag, Berlin–New York (1982).

[116] T. Seppäläinen, Entropy for translation-invariant random-cluster measures, *Ann. Probab.* **26** (1998), 1139–1178.

[117] R. Shrock and S.-H. Tsai, Asymptotic limits and zeros of chromatic polynomials and ground state entropy of Potts antiferromagnets, *Phys. Rev. E* **55** (1997), 5165–5178. Also cond-mat/9612249 at arXiv.org.

[118] R. Shrock and S.-H. Tsai, Families of graphs with $W_r(\{G\}, q)$ functions that are nonanalytic at $1/q = 0$, *Phys. Rev. E* **56** (1997), 3935–3943. Also cond-mat/9707096 at arXiv.org.

[119] R. Shrock and S.-H. Tsai, Ground state degeneracy of Potts antiferromagnets: Homeomorphic classes with noncompact W boundaries, *Physica A* **265** (1999), 186–223. Also cond-mat/9811410 at arXiv.org.

[120] B. Simon, *The Statistical Mechanics of Lattice Gases*, Princeton University Press, Princeton NJ (1993).

[121] A.D. Sokal, Bounds on the complex zeros of (di)chromatic polynomials and Potts-model partition functions, *Combin. Probab. Comput.* **10** (2001), 41–77. Also cond-mat/9904146 at arXiv.org.

[122] A.D. Sokal, Chromatic roots are dense in the whole complex plane, *Combin. Probab. Comput.* **13** (2004), 221–261. Also cond-mat/0012369 at arXiv.org.

[123] A.D. Sokal, More on the half-plane property for matroids, in preparation.

[124] J. Spencer, Ramsey's Theorem—a new lower bound, *J. Combin. Theory Ser. A* **18** (1975), 108–115.

[125] J. Spencer, Asymptotic lower bounds for Ramsey functions, *Discrete Math.* **20** (1977/78), 69–76.

[126] M.J. Stephen, Percolation problems and the Potts model, *Phys. Lett. A* **56** (1976), 149–150.

[127] C. Stutz and B. Williams, Ernst Ising [obituary], Phys. Today **52(3)** (1999), 106–108 (March).

[128] R.H. Swendsen and J.-S. Wang, Nonuniversal critical dynamics in Monte Carlo simulations, *Phys. Rev. Lett.* **58** (1987), 86–88.

[129] C. Thomassen, The zero-free intervals for chromatic polynomials of graphs, *Combin. Probab. Comput.* **6** (1997), 497–506.

[130] W.T. Tutte, A ring in graph theory, *Proc. Cambridge Philos. Soc.* **43** (1947), 26–40.

[131] W.T. Tutte, A contribution to the theory of chromatic polynomials, *Canad. J. Math.* **6** (1954), 80–91.

[132] G.E. Uhlenbeck and G.W. Ford, The theory of linear graphs with applications to the theory of the virial development of the properties of gases, in *Studies in Statistical Mechanics, vol. I* (eds. J. de Boer and G.E. Uhlenbeck), North-Holland, Amsterdam (1962), pp. 119–211.

[133] D.G. Wagner, Zeros of reliability polynomials and f-vectors of matroids, *Combin. Probab. Comput.* **9** (2000), 167–190.

[134] D.G. Wagner, Rank three matroids are Rayleigh, preprint, University of Waterloo, 2004. Also math.CO/0403216 at arXiv.org.

[135] D.J.A. Welsh, *Complexity: Knots, Colourings, and Counting*, *London Mathematical Society Lecture Note Series*, 186, Cambridge University Press, Cambridge–New York (1993).

[136] H. Whitney, A logical expansion in mathematics, *Bull. Amer. Math. Soc.* **38** (1932), 572–579.

[137] G. Whittle, On matroids representable over $GF(3)$ and other fields, *Trans. Amer. Math. Soc.* **349** (1997), 579–603.

[138] D.R. Woodall, Tutte polynomial expansions for 2-separable graphs, *Discrete Math.* **247** (2002), 201–213.

[139] F.Y. Wu, Number of spanning trees on a lattice, *J. Phys. A: Math. Gen.* **10** (1977), L113–L115.

[140] F.Y. Wu, The Potts model, *Rev. Mod. Phys.* **54** (1982), 235–268. Erratum **55** (1983), 315.

[141] F.Y. Wu, Potts model of magnetism (invited), *J. Appl. Phys.* **55** (1984), 2421–2425.

[142] F.Y. Wu and Y.K. Wang, Duality transformation in a many-component spin model, *J. Math. Phys.* **17** (1976), 439–440.

[143] C.N. Yang and T.D. Lee, Statistical theory of equations of state and phase transitions. I. Theory of condensation, *Phys. Rev.* **87** (1952), 404–409.

[144] T. Zaslavsky, The Möbius function and the characteristic polynomial, in *Combinatorial Geometries* (ed. N. White), *Encyclopedia of Mathematics and its Applications*, 29, Cambridge University Press, Cambridge (1987), pp. 114–138.

[145] T. Zaslavsky, Strong Tutte functions of matroids and graphs, *Trans. Amer. Math. Soc.* **334** (1992), 317–347.

[146] D. Zeilberger, A combinatorial approach to matrix algebra, *Discrete Math.* **56** (1985), 61–72.

[147] J. Zinn-Justin, *Quantum Field Theory and Critical Phenomena*, 3rd ed., Clarendon Press, Oxford (1996).

Department of Physics
New York University
4 Washington Place
New York, NY 10003
USA
sokal@nyu.edu

The sparse regularity lemma and its applications

Stefanie Gerke and Angelika Steger

Abstract

Szemerédi's regularity lemma is one of the most celebrated results in modern graph theory. However, in its original setting it is only helpful for studying large dense graphs, that is, graphs with n vertices and $\Theta(n^2)$ edges. The main reason for this is that the underlying concept of ε-regularity is not meaningful when dealing with sparse graphs, since for large enough n every graph with $o(n^2)$ edges is ε-regular. In 1997 Kohayakawa and Rödl independently introduced a modified definition of ε-regularity which is also useful for sparse graphs, and used it to prove an analogue of Szemerédi's regularity lemma for sparse graphs. However, some of the key tools for the application of the regularity lemma in the dense setting, the so-called embedding lemmas or, in their stronger forms, counting lemmas, are not known to be true in the sparse setting. In fact, counterexamples show that these lemmas do not always hold. However, Kohayakawa, Łuczak, and Rödl formulated a probabilistic embedding lemma that, if true, would solve several long-standing open problems in random graph theory. In this survey we give an introduction to Szemerédi's regularity lemma and its generalisation to the sparse setting, describe embedding lemmas and their applications, and discuss recent progress towards a proof of the probabilistic embedding lemma. In particular, we present various properties of ε-regular graphs in the sparse setting. We also show how to use these results to prove a weak version of the conjectured probabilistic embedding lemma.

1 Introduction

Szemerédi's regularity lemma is one of the most powerful tools in graph theory. Roughly speaking, it says that the vertex set of every sufficiently large graph can be partitioned into a *constant* number of roughly equal-sized parts such that the edges between most pairs of these parts are distributed regularly. In other words Szemerédi's regularity lemma says that any large graph contains a large subgraph with a certain regular structure. The regularity lemma is particularly useful when one is interested in finding subgraphs of large graphs, because then one can search for these subgraphs in the partition guaranteed by the lemma and employ its structure. Lemmas that guarantee substructures in the partition provided by Szemerédi's regularity lemma are usually called embedding or counting lemmas.

The main drawback of Szemerédi's regularity lemma is the fact that the regularity of the edges between the pairs of partition classes is only meaningful if there are many edges, that is, if there are $\Theta(n^2)$ edges between two vertex sets of size n. Independently, Kohayakawa [22] and Rödl (unpublished) introduced a more general concept of regularity that takes the density d of the graph into account. They showed that for a large class of graphs one can substitute the new regularity concept for the regularity in Szemerédi's regularity lemma. Unfortunately, there do not (yet) exist corresponding embedding lemmas. In this article we discuss recent progress towards proofs of embedding lemmas for sparse graphs.

The paper is organised as follows. In Section 2 we introduce the concept of regularity and present Szemerédi's regularity lemma and an embedding lemma. We then show how to use these results to derive a well known theorem of extremal graph theory concerning the maximal number of edges a graph on n vertices may have while avoiding a fixed graph H.

In Section 3 we present the more general version of Szemerédi's regularity lemma mentioned earlier that was introduced by Kohayakawa and Rödl, and restate a conjectured embedding lemma originally formulated by Kohayakawa, Łuczak and Rödl [22] together with some of its implications. In particular we show that if the conjectured embedding lemma is true, then we can derive an extremal result for random graphs that is similar to that mentioned above.

Section 4 is the central part of this survey where we investigate properties of regular graphs. By definition, regularity gives a priori only information on the number of edges between linear sized sets. However, it is well known that one can derive bounds on the degree of most vertices of regular graphs. We introduce corresponding results for sets, that is, we show that almost all subsets of a regular graph together with their neighbourhood induce graphs that have the expected structure. In particular, we shall see that regularity is hereditary, that is, almost all subsets of a regular graph contain a large subgraph that is again regular.

In Section 5 and Section 6 we show how to use the results of Section 4 to obtain results on almost all graphs and prove the conjectured embedding lemma for some special cases.

2 Szemerédi's regularity lemma

In this section we state Szemerédi's regularity lemma together with an embedding lemma. As mentioned in the introduction Szemerédi's lemma says roughly speaking that the vertex set of any large graph can be partitioned in such a way that most pairs of partition classes are distributed regularly. The next definition makes precise what is meant by "distributed regularly".

Definition 2.1 A bipartite graph $B = (V_1 \cup V_2, E)$ is called *ε-regular* if for all $V_1' \subseteq V_1$ and $V_2' \subseteq V_2$ with $|V_1'| \geq \varepsilon |V_1|$ and $|V_2'| \geq \varepsilon |V_2|$,

$$\left| \frac{|E(V_1', V_2')|}{|V_1'||V_2'|} - \frac{|E|}{|V_1||V_2|} \right| \leq \varepsilon \tag{2.1}$$

where $E(V_1', V_2')$ denotes the set of edges with one endpoint in V_1' and one endpoint in V_2'.

Thus a bipartite graph B is ε-regular if the density of any large induced subgraph is only ε away from the density of the graph B itself. Often a graph is called regular if the degrees of all vertices are the same. One easily checks that ε-regularity implies a rather tight bound on the degree of most vertices, that is, the graph is in some vague sense only ε away from being regular in the classical sense. Here, as in the remainder of the paper we denote by $\Gamma(v)$ the neighbourhood of a vertex v.

Proposition 2.2 *Let $0 < \varepsilon \leq 1/2$, and let $B = (V_1 \cup V_2, E)$ be an ε-regular graph of density $d = |E|/(|V_1||V_2|)$. If $V_2' \subseteq V_2$ satisfies $|V_2'| \geq \varepsilon |V_2|$, then there are at most $2\varepsilon |V_1|$ vertices $v_1 \in V_1$ that do not satisfy*

$$\lceil (d-\varepsilon)|V_2'| \rceil \leq |\Gamma(v_1) \cap V_2'| \leq \lfloor (d+\varepsilon)|V_2'| \rfloor . \tag{2.2}$$

Let us remark that most bipartite graphs are ε-regular as the following example shows.

Example 2.3 Consider two sets of vertices V_1 and V_2 and choose m edges between V_1 and V_2 at random. We want to show that with high probability the resulting graph is ε-regular if $m \geq C(|V_1| + |V_2|)$ for an appropriate constant C depending on ε. Using Hoeffding's inequality one can show that two fixed sets $V_1' \subseteq V_1$ and $V_2' \subseteq V_2$ with $|V_1'| \geq \varepsilon|V_1|$ and $|V_2'| \geq \varepsilon|V_2|$ satisfy Condition 2.1 with probability $1-o(2^{-(|V_1|+|V_2|)})$. It now follows that the graph is ε-regular with probability tending to 1 as $|V_1| + |V_2|$ tends to infinity since there are at most $2^{|V_1|+|V_2|}$ subsets of V_1 and V_2. For details, see Lemma 4.3 where we consider a more general concept than ε-regularity and show in particular that almost all subgraphs of ε-regular graphs are 2ε-regular.

It is easily checked that if in the above example $m \leq \varepsilon^3|V_1||V_2|$ then *every* graph on V_1 and V_2 with m edges is ε-regular and not just almost all such graphs. Thus when dealing with ε-regular graphs, one is usually interested in dense graphs (but not in too dense graphs either since all very dense graphs are also ε-regular).

One can also show [32, Theorem 1.1] that for every $0 < \varepsilon, d < 1$, every bipartite graph $B = (V_1 \cup V_2, E)$ with $|V_1| = |V_2| = n$ and density d contains an ε-regular subgraph $B' = (U_1 \cup U_2, E')$ with density $d' \geq (1 - \varepsilon/3)d$ and $|U_1| = |U_2| \geq (n/2)d^{50/\varepsilon^2}$. As we have seen before it is important to have the lower bound on the density of B' since trivially any bipartite graph contains the empty graph which is ε-regular.

Given an ε-regular graph, it is not hard to verify that every large induced subgraph is ε'-regular again, for a slightly larger ε'. For ease of reference, we state this observation as a proposition.

Proposition 2.4 *Let $0 < \varepsilon \leq 1/2$, and let $B = (V_1 \cup V_2, E)$ be an ε-regular graph of density d. If $\varepsilon \leq \alpha \leq 1/2$, then two subsets $V_1' \subseteq V_1$ and $V_2' \subseteq V_2$ with $|V_1'| \geq \alpha|V_1|$ and $|V_2'| \geq \alpha|V_2|$ induce an ε/α-regular graph with density at least $d - \varepsilon$.*

To state Szemerédi's regularity lemma we need one more definition.

Definition 2.5 A partition $(C_i)_{i=0}^k$ of the vertex set V is called an *equitable partition with exceptional class* C_0 if $|C_1| = |C_2| = \ldots = |C_k|$. An *$\varepsilon$-regular partition* of a graph G is an equitable partition $(C_i)_{i=0}^k$ of its vertex set such that the exceptional class C_0 has at most εn vertices and all but at most εk^2 of the pairs (C_i, C_j), $1 \leq i < j \leq k$, induce ε-regular graphs.

The celebrated lemma by Szemerédi states that all graphs that are sufficiently large can be decomposed into an ε-regular partition of constant size.

Theorem 2.6 (Szemerédi 1978 [34]) *For any $0 < \varepsilon < 1/2$ and $m_0 \geq 1$, there is a constant $M_0 = M_0(\varepsilon, m_0) \geq m_0$ such that any graph with at least m_0 vertices admits an ε-regular partition $(C_i)_{i=0}^k$ with exceptional class C_0 such that $m_0 \leq k \leq M_0$.*

The proof of Szemerédi's lemma can be found in [34] but also in some textbooks on graph theory; see for example [3], [4]. We shall not give a proof of this result here but instead we shall present a proof (which closely follows the original proof of Szemerédi's regularity lemma) of a generalisation later, see Theorem 3.4.

The bound on $M_0(\varepsilon, m_0)$ that will follow from the proof is a tower of twos of height proportional to ε^{-5} which is a huge constant for small ε. It was proved by Gowers [16] that one cannot hope for a small constant as there are graphs for which $M_0(\varepsilon, 2)$ is a tower of twos of height about $\varepsilon^{-1/16}$.

As was observed by several researchers, one needs to allow some irregular pairs in an ε-regular partition [1]. For example, there is no ε-regular partition without any irregular pairs of the vertex set of the bipartite graph with vertex classes $U = \{u_1, \ldots, u_n\}$ and $V = \{v_1, \ldots, v_n\}$ and edges $\{a_i, b_j\}$ for all $i \leq j$.

Usually one applies the lemma for sufficiently small ε and appropriate m_0. It is easily seen that for any graph G on n vertices the trivial partitions in either 1 or n partition classes are always ε-regular. Since for an ε-regular partition $(C_i)_{i=0}^k$ with exceptional class C_0 we have no control over the edges inside the partition classes nor over the edges that are incident to C_0, one wants to choose m_0 in such a way that the number of these edges is negligible; see the proof of Theorem 2.11 where these edges are counted carefully.

Finally, let us remark, that given a graph G, an integer $k \geq 1$, an $\varepsilon > 0$, and a partition of the vertex set into $k + 1$ parts, it is co-NP-complete to decide whether this partition is ε-regular [1]. Nevertheless, one can find an ε-regular partition as in Theorem 2.6 in polynomial time [1], see also [2, 9, 23, 27].

2.1 Embedding and counting lemmas

One of the most important aspects of Szemerédi's regularity is the fact that it can be used to find large structures of pairwise regular bipartite graphs, and that, once these regular structures are found, one can use these to prove certain properties of the underlying graph. Here so called *embedding lemmas* or, in a stronger version, *counting lemmas* turned out to be particularly useful.

We illustrate these lemmas in the case of finding a given fixed graph H. The set-up is as follows. We consider "blown-up" copies of H, where each vertex is replaced by a set of size n and each edge of H is replaced by a bipartite graph. Clearly, if we just allow any bipartite graph then there is no guarantee that the blown-up graph contains a copy of H. It is a routine exercise to construct a blown-up graph where all bipartite graphs have density, say $1/4$, but nevertheless the graph contains no copy of H.

The main idea of the embedding lemma is that if instead of requiring that the bipartite graphs are very dense we require that they are very regular, then we can always find a copy of H in the blown-up graph.

In order to make this statement precise, we first define what we mean by regular blown-up graphs.

Definition 2.7 For a graph H, let $\mathcal{P}(H, n, d, \varepsilon)$ be the class of graphs with vertex set $V = \bigcup_{x \in V(H)} V_x$, where the sets V_x are pairwise disjoint sets of vertices of size n, and with edge set $E = \bigcup_{\{x,y\} \in E(H)} E_{xy}$, where E_{xy} is the edge set of an ε-regular graph with density at least d between V_x and V_y.

The main assertion of the embedding lemma is that regardless how small d is, we can find an $\varepsilon > 0$ such that the ε-regularity implies that *every* graph in $\mathcal{P}(H, n, d, \varepsilon)$ contains a copy of H.

Lemma 2.8 *Let H be a fixed graph and let $d \in (0, 1]$. Then there exists an $\varepsilon_0 = \varepsilon_0(H, d)$ such that for all $0 < \varepsilon \leq \varepsilon_0$, any graph in $\mathcal{P}(H, n, d, \varepsilon)$ contains a copy of H as a subgraph.*

Proof We prove the lemma by induction on $v_H = |V(H)|$. For $v_H = 1$ the lemma is trivially true for $\varepsilon_0(H, d) = 1$, say. So assume $v_H > 1$. In order to define ε_0 we choose $x \in V(H)$ arbitrarily, and let

$$\varepsilon_0(H, d) = \min\left\{\varepsilon_0(H \setminus x, d/2), \frac{d}{2}, \frac{1}{2v_H}\right\}.$$

Now consider a graph G of $\mathcal{P}(H, n, d, \varepsilon)$ for some $0 < \varepsilon \leq \varepsilon_0(H, d)$. We need to show that G contains a copy of H. We will proceed as follows. We will find a vertex $v_x \in V_x$ and sets $V'_z \subseteq V_z$ of size $\lceil (d - \varepsilon)|V_z| \rceil$ such that $V'_z \subseteq \Gamma(v_x)$ for all $z \in \Gamma_H(x)$. It then follows by Lemma 2.4 that the sets V'_z induce a graph in $\mathcal{P}(H \setminus x, \lceil (d - \varepsilon)n \rceil, d - \varepsilon, \varepsilon/(d - \varepsilon))$. Now, by choice of $\varepsilon_0(H, d)$ and the induction hypothesis, the graph induced by the sets V'_z contains a copy of $H \setminus x$, which together with v_x forms a copy of H in G.

For a vertex $z \in V(H)$, that is not a neighbour of x, we arbitrarily choose a set V'_z of size $\lceil (d - \varepsilon)|V_z| \rceil$. For a neighbour y of x in H, the ε-regularity of the graph between V_x and V_y and Proposition 2.2 imply that all but $2\varepsilon n$ vertices in V_x are such that they have at least $\lceil (d - \varepsilon)|V_y| \rceil \geq 1$ neighbours in V_y. Hence, at least $n - 2(v_H - 1)\varepsilon n > 0$ vertices in V_x are such that we can find sets $V'_z \subset V_z$ of size $|V'_z| = \lceil (d - \varepsilon)|V_z| \rceil$ such that $V'_z \subseteq \Gamma(v_x) \cap V_z$ for all $z \in \Gamma_H(x)$. □

As the reader may have observed, the bound given on ε_0 in the proof of Lemma 2.8 is rather generous. While we could easily improve this, a different generalisation turns out to be more rewarding. In the definition of the class $\mathcal{P}(H, n, d, \varepsilon)$ we have blown-up *every* vertex of H. In fact, we can show that we can already find the graph H in a blown-up copy of the complete graph $K_{\chi(H)}$ on $\chi(H)$ vertices, where $\chi(H)$ denotes the chromatic number of H.

Lemma 2.9 *Let H be a fixed graph and let $d \in (0, 1]$. Then there exists an $\varepsilon_1 = \varepsilon_1(H, d)$ such that for all $0 < \varepsilon < \varepsilon_1$ and $n \geq |V(H)|$ any graph in $\mathcal{P}(K_{\chi(H)}, n, d, \varepsilon)$ contains a copy of H as a subgraph.*

Proof Let

$$\varepsilon_1(H, d) = \min\left\{\frac{\varepsilon_0(H, d/2)}{2|V(H)|}, \frac{d}{2}\right\},$$

where ε_0 is the function from Lemma 2.8, and let $0 < \varepsilon \leq \varepsilon_1$. By the definition of the chromatic number we can partition the vertex set of H into $\chi(H)$ independent sets of size at most $v_H := |V(H)|$. Given a graph G in $\mathcal{P}(K_{\chi(H)}, n, d, \varepsilon)$, we can therefore partition each vertex set into v_H parts of size $\lfloor n/v_H \rfloor$ (and at most $v_H - 1$ remaining vertices) such that, by Proposition 2.4, the resulting graph contains a graph from $\mathcal{P}(H, \lfloor n/v_H \rfloor, d - \varepsilon, \varepsilon \cdot 2v_H)$. (Here the factor 2 is used to generously take care of the Gauss brackets.) The claim of the lemma now follows from Lemma 2.8. □

An alert reader might have observed that in proving the above lemmas we were rather generous. Recall for example the proof of Lemma 2.8. We argued that at least $n-(|V(H)|-1)\varepsilon n$ vertices in V_x have $\lceil d-\varepsilon\rceil|V_z|$ neighbours in V_z for all $z\in\Gamma_H(x)$. But then we only used one such vertex to find the desired copy. Of course the same proof shows that we indeed find a copy of H for each such vertex. Using this idea inductively, one can show the following *counting version* of Lemma 2.9.

Lemma 2.10 *Let H be a fixed graph and let $d\in(0,1]$. Then there exist an $\alpha=\alpha(H,d)$ and an $\varepsilon_1=\varepsilon_1(H,d)$ such that for all $0<\varepsilon<\varepsilon_1$ and $n\geq|V(H)|$ any graph in $\mathcal{P}(K_{\chi(H)},n,d,\varepsilon)$ contains at least $\alpha n^{|V(H)|}$ copies of H as a subgraph.*

For the numerous applications of Szemerédi's regularity lemma, which use the embedding and counting lemmas, and also for the statement and applications of much more powerful embedding lemmas (as for example the blow-up lemma [28]), we refer the reader to the overview articles [30, 29]. Here we just present one application which we shall later generalise to the sparse case.

2.2 An application of Szemerédi's regularity lemma to extremal graph theory

A fundamental problem in extremal graph theory is the study of the parameter $\mathrm{ex}_n(H)$, which denotes the maximal number of edges a graph on n vertices may have without containing a copy of H. A celebrated theorem of Turán from 1941 [35] provides an exact value of this function for all complete graphs K_ℓ. In fact, Turán showed that the complete $(\ell-1)$-partite graph in which all class sizes differ by at most one is (up to isomorphisms) the only graph with $\mathrm{ex}_n(K_{\ell-1})$ edges. Note that this implies that

$$\mathrm{ex}_n(K_\ell)\leq\frac{1}{2}\left(1-\frac{1}{\ell-1}\right)n^2, \tag{2.3}$$

with equality if n is divisible by $\ell-1$.

In this section we want to use Szemerédi's regularity lemma and the embedding lemma introduced in the last section to show a theorem of Erdős and Stone [6], and Erdős and Simonovits [5] concerning the maximal number $\mathrm{ex}_n(H)$ of edges a graph on n vertices may have without containing a copy of H.

Theorem 2.11 ([6, 5]) *For each graph H with chromatic number $\chi(H)\geq2$, we have*

$$\mathrm{ex}_n(H)=\left(1-\frac{1}{\chi(H)-1}+o(1)\right)\frac{n^2}{2}.$$

Proof The lower bound follows immediately from the fact that a $(\chi(H)-1)$-partite graph on n vertices cannot contain H as a subgraph.

For the upper bound we first observe that it suffices to show that for every $0<\delta<1$ a graph G on n vertices with

$$|E(G)|\geq\left(1-\frac{1}{\chi(H)-1}+\delta\right)\frac{n^2}{2} \tag{2.4}$$

edges contains H as a subgraph, whenever n is sufficiently large. We intend to find such a copy by applying Szemerédi's regularity lemma.

Given H and δ we let $d = \delta/4$ and choose an $\varepsilon > 0$ such that $\varepsilon < \delta/10$ and $\varepsilon < \varepsilon_1(H, d)$, where ε_1 denotes the function from Lemma 2.9, and let $m_0 = \lceil \varepsilon^{-1} \rceil$. The regularity lemma now implies that there exists a constant $M_0 = M_0(\varepsilon, m_0)$ such that whenever $n \geq m_0$ a graph G on n vertices contains an ε-regular partition $(C_i)_{i=0}^k$ with $m_0 \leq k \leq M_0$. We may assume that $n \geq M_0|V(H)|/(1-\varepsilon) \geq m_0$ since it suffices to to find a copy of H in all large graphs.

Our strategy is now to show that the graph induced by $C_1, \ldots, C_k$ contains a graph in $\mathcal{P}(K_{\chi(H)}, |C_1|, d, \varepsilon)$ and thus by Lemma 2.9 contains H as $|C_1| \geq (1 - \varepsilon)n/M_0 \geq |V(H)|$. Let R be the graph consisting of k vertices corresponding to the partition classes $C_1, \ldots, C_k$, with an edge between two vertices if the corresponding partition classes form an ε-regular graph of density at least d. In order to apply Lemma 2.9 we need to show that R contains a clique of size $\chi(H)$. We show this in two steps. First we show that R contains many edges, and then we apply Turan's theorem.

In order to give a lower bound on the number of edges in R we first give an upper bound on the number of edges in G that are *not* involved in pairs (C_i, C_j), $0 \leq i \neq j \leq k$ that are ε-regular with density at least d. Clearly, the number of edges in R is at least the number of the remaining edges divided by $(n/k)^2$ since there are at most $(n/k)^2$ edges between any pair (C_i, C_j), $1 \leq i \neq j \leq k$.

There are at most

- εn^2 edges incident with a vertex in C_0 (as $|C_0| \leq \varepsilon n$ and each vertex in C_0 is incident to less than n edges),
- εn^2 edges between partition classes that do not form an ε-regular graph (as there are at most εk^2 such pairs and each such pair contains at most $(\frac{n}{k})^2$ edges),
- $\frac{1}{2} dn^2$ edges between partition classes that form a graph with density less than d (as each such pair can contain at most $d(\frac{n}{k})^2$ edges and there are at most $\binom{k}{2}$ such pairs),
- $\frac{1}{2}\varepsilon n^2$ edges that lie inside a partition class (as there are at most $\binom{\lfloor n/k \rfloor}{2}$ edges inside each of the k partition classes, and as we assumed that $k \geq m_0 \geq \varepsilon^{-1}$).

As each edge of R can represent at most $(n/k)^2$ edges of G we deduce that

$$\begin{aligned} |E(R)| &\geq \frac{|E(G)| - \frac{5}{2}\varepsilon n^2 - \frac{1}{2} dn^2}{(n/k)^2} \\ &\overset{(2.4)}{\geq} \left(1 - \frac{1}{\chi(H) - 1} + \delta - 5\varepsilon - d\right) \cdot \frac{k^2}{2}. \end{aligned}$$

Now recall that we have chosen $d = \delta/4$ and $\varepsilon < \delta/10$. Hence we get that

$$|E(R)| \geq \left(1 - \frac{1}{\chi(H) - 1} + \frac{1}{4}\delta\right) \cdot \frac{k^2}{2} \overset{(2.3)}{\geq} \operatorname{ex}_k(K_{\chi(H)}).$$

Hence the graph R contains a clique on $\chi(H)$ vertices, and the corresponding partition classes therefore induce a graph from $\mathcal{P}(K_{\chi(H)}, |C_1|, d, \varepsilon)$. By the assumption on n, we have that $|C_1| \geq |V(H)|$ and hence Lemma 2.9 implies that G contains the desired copy of H. $\square$

3 Szemerédi's regularity lemma for sparse graphs

For the remainder of this paper we are concerned with graphs that have $o(n^2)$ edges. As we have noted before any bipartite graph with n vertices in each partition class and with less than $\varepsilon^3 n^2$ edges is ε-regular. Hence nearly any partition of sparse graphs is ε-regular. In the following, we present a variant of ε-regularity that takes the density into account.

Definition 3.1 A bipartite graph $B = (V_1 \cup V_2, E)$ is called *(ε, p)-regular* if for all $V_1' \subseteq V_1$ and $V_2' \subseteq V_2$ with $|V_1'| \geq \varepsilon |V_1|$ and $|V_2'| \geq \varepsilon |V_2|$,

$$\left| \frac{|E(V_1', V_2')|}{|V_1'||V_2'|} - \frac{|E|}{|V_1||V_2|} \right| \leq \varepsilon p.$$

It is called *(ε)-regular* if p equals the density of B, that is, if $p = |E|/(|V_1||V_2|)$.

We use the notation (ε)-regular to distinguish it from the classical definition of ε-regularity which is $(\varepsilon, 1)$-regularity in this notation. Observe that (ε, p)-regularity implies (ε, p')-regularity for all $p' \geq p$, and thus ε-regularity is a weaker property than (ε)-regularity.

As in Example 2.3 one can show that the random bipartite graph with n vertices in each partition class and with $m \gg n$ edges is asymptotically almost surely (a.a.s., that is with probability tending to one as n tends to infinity) (ε)-regular. Moreover if the number of edges is less than $\varepsilon^3 pn^2$, then *every* bipartite graph with n vertices in each partition class is (ε, p)-regular.

As before with ε-regularity, we can define partitions where most pairs of partition classes form (ε)-regular graphs.

Definition 3.2 An *(ε, p)-regular partition* is an equitable partition $(C_i)_{i=0}^k$ such that the exceptional class C_0 has at most εn vertices and, with the exception of at most εk^2 pairs, the pairs (C_i, C_j) $1 \leq i \leq j \leq k$ are (ε, p)-regular.

It turns out that one can prove Szemerédi's lemma with ε-regularity replaced by (ε, p)-regularity for the following class of graphs that do not contain large dense parts.

Definition 3.3 Let $G = (V, E)$ be a graph. Let $0 < \eta \leq 1$, $0 < p \leq 1$ and $b \geq 1$. We say that G is *(η, b, p)-upper-uniform*, if for all disjoint sets V_1 and V_2 with $|V_1|, |V_2| \geq \eta |V|$,

$$\frac{|E(V_1, V_2)|}{|V_1||V_2|} \leq bp.$$

Note that every graph is $(\eta, 1, 1)$-upper-uniform. Usually one sets p equal to the density. Then (η, b, p)-upper-uniformity implies that $G = (V, E)$ does not contain a bipartite graph with partition classes bigger than $\eta |V|$ whose density exceeds the density of G by a factor of b. It is easily verified that for every (η, b, p)-uniform graph G, any set S of size at least $2\eta |V(G)|$, contains at most $(1/2)b|S|^2 p$ edges. For any $\eta > 0$, $b > 1$ and any $p \gg 1/n$, it is not hard to verify using Chernoff's inequality that the random graph $G_{n,p}$ is a.a.s. (η, b, p)-upper-uniform. Since any spanning subgraph of an (η, b, p)-upper-uniform graph is (η, b, p)-upper uniform, it follows that a.a.s. every spanning subgraph of the random graph $G_{n,p}$ is (η, b, p)-upper-uniform.

We can now state a variant of Szemerédi's regularity lemma which was first observed by Kohayakawa and independently by Rödl, see [20, 24].

Theorem 3.4 ([20]) *For any $0 < \varepsilon < 1/2$ and $b, m_0 \geq 1$, there are constants $\eta = \eta(\varepsilon, b, m_0) > 0$ and $M_0 = M_0(\varepsilon, m_0) \geq m_0$ such that for any $p > 0$, any (η, b, p)-upper-uniform graph with at least m_0 vertices admits an (ε, p)-regular partition $(C_i)_{i=0}^k$ with exceptional class C_0 such that $m_0 \leq k \leq M_0$.*

Note that this variant is more general than the original lemma since—as we have noted before—every graph is $(\eta, 1, 1)$-upper-uniform and (ε, p)-regularity implies ε-regularity. Hence most of the remarks after Szmerédi's regularity lemma also apply to this theorem.

It is not known whether the theorem remains true if the (η, b, p)-upper-uniformity condition on the graph is dropped. On the other hand it is known that there are graphs that only allow an (ε, p)-regular partition in which all (ε, p)-regular pairs are empty [32]. Of course this is not very helpful when one is interested in finding substructures. We shall see in the proof of Proposition 3.16 that the (η, b, p)-upper-uniformity of a graph insures that there are some (ε, p)-regular pairs (C_i, C_j) with density about p.

3.1 Proof of the sparse regularity lemma

This subsection is devoted to the proof of Theorem 3.4. On first reading the reader is invited to continue with the next section since the proof is somewhat technical. To prove the theorem, we need the following definition and three lemmas.

Definition 3.5 Let $G = (V, E)$ be a graph. For any two disjoint sets $U, W \subset V$, we define

$$d_p(U, W) = \frac{|E(U, W)|}{p|U||W|}.$$

The *p-index* of an equitable partition $\mathcal{P} = (C_i)_{i=0}^k$ with exceptional class C_0 is defined by

$$\text{index}_p(\mathcal{P}) = \frac{1}{k^2} \sum_{1 \leq i < j \leq k} d_p^2(C_i, C_j).$$

Observe that for any (η, b, p)-upper-uniform graph G and any equitable partition $\mathcal{P} = (C_i)_{i=0}^k$ with $|C_1| \geq \eta |V(G)|$, we have $\text{index}_p(\mathcal{P}) < b^2/2$. The main ingredient of the proof is the fact that given a partition that is not (ε, p)-regular one can refine it to obtain a partition whose index is larger than the original index by an additive constant. Since the index of any (η, b, p)-upper-uniform graph is bounded from above by $b^2/2$, it follows that one can refine the partition only a constant number of times.

The next lemma states that $d_p(X, Y)$ and $d_p(X^*, Y^*)$ do not differ much if the sets X^* and Y^* are large subsets of X and Y respectively, and we are concerned with (η, b, p)-upper-uniform graphs.

Lemma 3.6 *Let $0 \le \delta \le 1/2$. Suppose X and Y are disjoint sets of vertices of an (η, b, p)-upper-uniform graph G, and $X^* \subseteq X$ and $Y^* \subseteq Y$ are such that $|X^*| \ge (1-\delta)|X|$ and $|Y^*| \ge (1-\delta)|Y|$. If $\lfloor \delta|X| \rfloor, \lfloor \delta|Y| \rfloor \ge \eta|V(G)|$, then*

$$|d_p(X^*, Y^*) - d_p(X, Y)| \le 2b\delta \tag{3.1}$$

and

$$|d_p^2(X^*, Y^*) - d_p^2(X, Y)| \le 4b^2\delta. \tag{3.2}$$

Proof Clearly, $|E(X^*, Y^*)| \le |E(X, Y)|$ and hence

$$\frac{(1-\delta)^2|E(X^*, Y^*)|}{p\,|X^*||Y^*|} \le \frac{|E(X^*, Y^*)|}{p\,|X||Y|} \le \frac{|E(X, Y)|}{p\,|X||Y|}. \tag{3.3}$$

Since $|X^*| \ge (1-\delta)|X| \ge \delta|X| \ge \eta|V(G)|$, and similarly $|Y^*| \ge \eta|V(G)|$ it follows from the (η, b, p)-upper-uniformity of G, that $|E(X^*, Y^*)| \le bp|X^*||Y^*|$ and hence (3.3) implies

$$\frac{|E(X^*, Y^*)|}{p\,|X^*||Y^*|} - 2\delta b \le \frac{|E(X, Y)|}{p\,|X||Y|}. \tag{3.4}$$

Let $\tilde{X} \subseteq X^*$, and $\tilde{Y} \subseteq Y^*$, be such that $|\tilde{X}| = \lceil (1-\delta)|X| \rceil$ and $|\tilde{Y}| = \lceil (1-\delta)|Y| \rceil$. Then

$$|E(X, Y)| \le |E(X^*, Y^*)| + |E(X \setminus \tilde{X}, Y)| + |E(X, Y \setminus \tilde{Y})|,$$

and since $|X \setminus \tilde{X}| \ge \eta|V(G)|$ and $|Y \setminus \tilde{Y}| \ge \eta|V(G)|$,

$$\frac{|E(X \setminus \tilde{X}, Y)|}{p\,|X||Y|} \le \frac{bp\,|X \setminus \tilde{X}||Y|}{p\,|X||Y|} \le \delta b,$$

and similarly

$$\frac{|E(X, Y \setminus \tilde{Y})|}{p\,|X||Y|} \le \frac{bp\,|X||Y \setminus \tilde{Y}|}{p\,|X||Y|} \le \delta b.$$

Hence

$$\frac{|E(X, Y)|}{p\,|X||Y|} \le \frac{|E(X^*, Y^*)|}{p\,|X||Y|} + 2\delta b \le \frac{|E(X^*, Y^*)|}{p\,|X^*||Y^*|} + 2\delta b,$$

and (3.1) now follows with (3.4).

Now (3.2) follows, since

$$\begin{aligned} &|d_p^2(X^*, Y^*) - d_p^2(X, Y)| \\ &\quad = |d_p(X^*, Y^*) + d_p(X, Y)|\,|d_p(X^*, Y^*) - d_p(X, Y)| \\ &\quad \le 4b^2\delta. \end{aligned}$$

□

The proof of the next lemma, the defect form of the Cauchy-Schwarz inequality, can be found for example in [3].

Lemma 3.7 *Let $(d_i)_{i=1}^s \subseteq \mathbf{R}$, $1 \le t < s$, $D = \sum_{i=1}^s d_i/s$, and $d = \sum_{i=1}^t d_i/t$. Then*

$$\frac{1}{s}\sum_{i=1}^s d_i^2 \ge D^2 + \frac{t}{s-t}(D-d)^2 \ge D^2 + \frac{t}{s}(D-d)^2.$$

In particular, if $t \geq \gamma s$ and $|D - d| \geq \delta > 0$, then

$$\frac{1}{s}\sum_{i=1}^{s} d_i^2 \geq D^2 + \gamma\delta^2.$$

The next lemma states that given a partition that is not (ε, p)-regular we can find a new partition with an index that is an additive constant larger.

Lemma 3.8 *Let $0 < \varepsilon < 1/2$, and $2^{-k} \leq \varepsilon^5/(8b^2)$. Let G be an (η, b, p)-bounded graph with an equitable partition $\mathcal{P} = (C_i)_{i=0}^k$ with exceptional class C_0 and*

$$|C_1| = |C_2| = \ldots = |C_k| = c \geq 2^{3k+1}$$

such that $\lfloor \varepsilon^5/(8b^2)|C_1| \rfloor \geq \eta|V|$. If the partition is not (ε, p)-regular, then there exists an equitable partition $\mathcal{P}' = (C_i')_{i=1}^l$ with $l = k(4^k - 2^{k-1})$ and exceptional class C_0' such that

$$|C_0'| \leq |C_0| + \frac{n}{2^k}$$

and

$$\text{index}_p(\mathcal{P}') \geq \text{index}_p(\mathcal{P}) + \frac{\varepsilon^5}{2}.$$

Proof For any pair (C_i, C_j) that is not (ε, p)-regular, let $C_{ij} \subseteq C_i$ and $C_{ji} \subseteq C_j$ be certificates that (C_i, C_j) is not (ε, p)-regular, that is, $|C_{ij}| \geq \varepsilon|C_i|$, $|C_{ji}| \geq \varepsilon|C_j|$ and $|d_p(C_{ij}, C_{ji}) - d_p(C_i, C_j)| > \varepsilon$. For an (ε, p)-regular pair (C_i, C_j), we set $C_{ij} = C_{ji} = \emptyset$.

The main idea of the proof is as follows. Let C_i and C_j be partition classes that do not form an (ε, p)-regular pair. Assume we can partition C_i and C_j into sets $D_{ih} \subseteq C_i$ and $D_{jh} \subseteq C_j$ of size d each such that C_{ij} and C_{ji} are the union of some of these sets. Then by Lemma 3.7 with $s = |C_i||C_j|/d^2$ and $t = |C_{ij}||C_{ji}|/d^2 \geq \varepsilon^2 s$, we have

$$\begin{aligned}\frac{d}{|C_i|}\frac{d}{|C_j|}\sum_{h=1}^{|C_i|/d}\sum_{k=1}^{|C_i|/d} d_p^2(D_{ih}, D_{jk}) &\geq \left(\sum_{h=1}^{|C_i|/d}\sum_{k=1}^{|C_i|/d} \frac{d_p(D_{ih}, D_{jk})d^2}{|C_i||C_j|}\right)^2 + \varepsilon^4 \qquad (3.5)\\ &= \left(\sum_{h=1}^{|C_i|/d}\sum_{k=1}^{|C_i|/d} \frac{E(D_{ih}, D_{jk})}{p\,|C_i||C_j|}\right)^2 + \varepsilon^4 = d_p^2(C_i, C_j) + \varepsilon^4,\end{aligned}$$

that is, the index has increased. We want to choose the sets D_{ih} in such a way that they are either contained in C_{ij} for all $j \neq i$ or the intersection is empty. The price we have to pay for it is that we might only partition a large fraction of C_i and not necessarily the entire set. All the vertices that are not covered are moved to the set C_0. By Lemma 3.6 it also does not matter too much if we only consider large subsets and not the original sets.

Let two vertices $u, v \in C_i$ be equivalent if for all $1 \leq j \neq i \leq k$, $u \in C_{ij}$ if and only if $v \in C_{ij}$. Note that there are at most 2^{k-1} equivalence classes. Set $H = 4^k - 2^{k-1}$, and set $d = \lfloor c/4^k \rfloor$, so that $d \geq 2^{k+1}$ and $4^k d \leq c \leq 4^k(d+1) - 1$. Let $D_{ih} \subseteq C_i$, $1 \leq h \leq H$ be pairwise disjoint sets of size d such that each D_{ih} is contained in some equivalent class. Observe that such sets D_{ih} exist since all but at most $d-1$ elements of each equivalent class can be partitioned into sets of size d, and

$$Hd + 2^{k-1}(d-1) = 4^k d - 2^{k-1} < 4^k d \le c.$$

For $1 \le i \ne j \le k$, set $\tilde{C}_i = \bigcup_{i=1}^{H} D_{ih}$ and $\tilde{C}_{ij} = \bigcup\{D_{ih} : D_{ih} \subseteq C_{ij}\}$. We will rename the sets D_{ih} as C'_j, but first, we want to show that $d_p(C_i, C_j)$ and $d_p(\tilde{C}_i, \tilde{C}_j)$ are close and that the same is true for $d_p(C_{ij}, C_{ji})$ and $d_p(\tilde{C}_{ij}, \tilde{C}_{ji})$. Now,

$$\frac{|C_i \setminus \tilde{C}_i|}{|C_i|} \le 1 - \frac{(4^k - 2^{k-1})d}{4^k(d+1)} \le \frac{1}{d} + \frac{1}{2^{k+1}} \le 2^{-k} \le \frac{\varepsilon^5}{8b^2}. \tag{3.6}$$

Since η is sufficiently small it follows from Lemma 3.6 that

$$|d_p(\tilde{C}_i, \tilde{C}_j) - d_p(C_i, C_j)| \le \frac{\varepsilon^5}{4b} \le \frac{\varepsilon^5}{4} \tag{3.7}$$

and

$$|d_p^2(\tilde{C}_i, \tilde{C}_j) - d_p^2(C_i, C_j)| \le \frac{\varepsilon^5}{2}. \tag{3.8}$$

Hence

$$\frac{1}{k^2} \sum_{1 \le i < j \le k} d_p^2(\tilde{C}_i, \tilde{C}_j) \ge \frac{1}{k^2} \sum_{1 \le i < j \le k} d_p^2(C_i, C_j) - \frac{\varepsilon^5}{4} \tag{3.9}$$

Suppose (C_i, C_j) are not (ε, p)-regular. Then by (3.6) and since $|C_{ij}| \ge \varepsilon |C_i|$,

$$\frac{|C_{ij} \setminus \tilde{C}_{ij}|}{|C_{ij}|} \le \frac{|C_i \setminus \tilde{C}_i|}{|C_{ij}|} \le \frac{|C_i \setminus \tilde{C}_i|}{|C_i|} \frac{|C_i|}{|C_{ij}|} \le \frac{\varepsilon^4}{8b^2} \tag{3.10}$$

and since $\varepsilon \le 1/2$,

$$|\tilde{C}_{ij}| = |C_{ij}| - |C_{ij} \setminus \tilde{C}_{ij}| \ge (1 - \varepsilon^4/(8b^2))|C_{ij}| \ge (1 - 2^{-7})\varepsilon |C_i|. \tag{3.11}$$

It follows from (3.10) and Lemma 3.6 that

$$|d_p(\tilde{C}_{ij}, \tilde{C}_{ji}) - d_p(C_{ij}, C_{ji})| \le \frac{\varepsilon^4}{4b} \le \frac{\varepsilon^4}{4}. \tag{3.12}$$

Since C_{ij} and C_{ji} are a certificate for (C_i, C_j) not being (ε, p)-regular and by (3.7), (3.12), it follows that

$$\begin{aligned} |d_p(\tilde{C}_{ij}, \tilde{C}_{ji}) - d_p(\tilde{C}_i, \tilde{C}_j)| &\ge |d_p(C_{ij}, C_{ji}) - d_p(C_i, C_j)| \\ &- \ |d_p(\tilde{C}_{ij}, \tilde{C}_{ji}) - d_p(C_{ij}, C_{ji})| - |d_p(\tilde{C}_i, \tilde{C}_j) - d_p(C_i, C_j)| \qquad (3.13) \\ &> \ \varepsilon - \frac{\varepsilon^4}{4} - \frac{\varepsilon^5}{4} \ge \frac{15}{16}\varepsilon. \qquad (3.14) \end{aligned}$$

Hence if (C_i, C_j) are not (ε, p)-regular, than we can proceed as in (3.5) and use Lemma 3.7 to obtain

$$\begin{aligned} \frac{1}{H^2} \sum_{u=1}^{H} \sum_{v=1}^{H} d_p^2(D_{iu}, D_{jv}) &\ge d_p^2(\tilde{C}_i, \tilde{C}_j) + \frac{|\tilde{C}_{ij}||\tilde{C}_{ji}|}{|\tilde{C}_i||\tilde{C}_j|} \left(\frac{15}{16}\varepsilon\right)^2 \qquad (3.15) \\ &\stackrel{(3.11)}{\ge} d_p^2(\tilde{C}_i, \tilde{C}_j) + \frac{3}{4}\varepsilon^4. \end{aligned}$$

Also, for every pair (C_i, C_j) we have by the Cauchy-Schwarz-inequality that

$$\frac{1}{H^2}\sum_{u=1}^{H}\sum_{v=1}^{H} d_p^2(D_{iu}, D_{jv}) \geq d_p^2(\tilde{C}_i, \tilde{C}_j).$$

As mentioned before we rename the sets D_{ih} as C'_j. It remains to show that the obtained partition has the required properties. Thus let $\{C'_1, \ldots, C'_l\} = \{D_{ih} : 1 \leq i \leq k, 1 \leq h \leq H\}$, and $C'_0 = V \setminus \bigcup_{j=1}^{l} C'_j$. Then by (3.6)

$$|C'_0 \setminus C_0| = \sum_{i=1}^{k} |C_i \setminus \tilde{C}_i| \leq \frac{n}{2^k},$$

and finally by (3.9),(3.15), and the fact that there are at least εk^2 pairs (C_i, C_j) that are not (ε, p)-regular

$$\begin{aligned}\text{index}_p(\mathcal{P}') &= \frac{1}{l^2}\sum_{1\leq i<j\leq l} d_p^2(C'_i, C'_j) \geq \frac{1}{k^2}\sum_{1\leq i<j\leq k} d_p^2(\tilde{C}_i, \tilde{C}_j) + \frac{3}{4}\varepsilon^4\frac{\varepsilon k^2}{k^2}\\ &\geq \frac{1}{k^2}\sum_{1\leq i<j\leq k} d_p^2(C_i, C_j) - \frac{\varepsilon^5}{4} + \frac{3}{4}\varepsilon^5 \geq \text{index}_p(\mathcal{P}) + \frac{\varepsilon^5}{2}.\end{aligned}$$ □

Finally, we can prove Theorem 3.4. We have seen that an irregular partition gives rise to a partition with higher index. Since the index of any partition of an (η, b, p)-upper uniform graph is bounded from above by $b^2/2$, it follows that we can only refine the partition a constant number of times before we reach a regular partition.

Proof of Theorem 3.4 Set $t = \lfloor b^2\varepsilon^{-5}\rfloor$ and define $k_0, \ldots, k_{t+1}$ by letting k_0 be the minimal integer satisfying $k_0 \geq m_0$ and $2^{-k_0} \leq \varepsilon^5/8b^2$, and setting $k_{i+1} = k_i(4^{k_i} - 2^{k_i - 1})$. We claim that $M_0 = k_t 2^{3k_t+2}$ and $\eta = \varepsilon^5/(16b^2 k_{t+1})$ will do. Let G be a graph of order n. We may assume that $n \geq M_0$ since otherwise the trivial partition into singletons will do. Let $\mathcal{P}_0 = (C_i^{(0)})_{i=0}^{k_0}$ be an equitable partition of the vertex set of G with exceptional class C_0 such that $|C_1^{(0)}| = \ldots = |C_{k_0}^{(0)}| = \lfloor n/k_0\rfloor$ and $0 \leq |C_0^{(0)}| \leq n - k_0\lfloor n/k_0\rfloor \leq k_0 \leq \varepsilon n/2$. Either $\mathcal{P}_0$ is (ε, p)-regular and we are done, or we can use Lemma 3.8 to obtain a partition $\mathcal{P}_1 = (C_i^{(1)})_{i=0}^{k_1}$. Continuing like this we obtain partitions $\mathcal{P}_2, \ldots, \mathcal{P}_{t+1}$. Note that for all $0 \leq j \leq t$, we have

$$|C_0^{(j)}| \leq |C_0^{(0)}| + n(2^{-k_0} + 2^{-k_1} + \ldots + 2^{-k_j}) \leq \varepsilon n$$

and

$$\frac{\varepsilon^5}{8b^2}|C_1^{(j)}| \geq \frac{\varepsilon^5}{8b^2}\frac{(1-\varepsilon)n}{k_j} \geq \frac{\varepsilon^5}{8b^2}\frac{n}{2k_{t+1}} \geq \eta n$$

and hence Lemma 3.8 is applicable. Continuing this way, we obtain an (ε, p)-regular partition $\mathcal{P}_j = (C_i^{(j)})_{i=1}^{k_j}$ for some j with $0 \leq j \leq t$ or we end up with the partition $\mathcal{P}_{t+1}(C_i^{(t+1)})_{i=1}^{k_{t+1}}$ with exceptional set $C_0^{(t+1)}$ of size $|C_0^{(t+1)}| \leq \varepsilon n$ and with index

$$\text{index}_p(\mathcal{P}_{t+1}) \geq (t+1)\frac{\varepsilon^5}{2} \geq \frac{b}{2}$$

which is impossible since $\text{index}_p(\mathcal{P}) < b^2/2$ for every (η, b, p)-bounded graph G and every equitable partition $\mathcal{P} = (C_i)_{i=1}^{k}$ of G with $|C_1| \geq \eta n$. □

3.2 Conjectured embedding and counting lemmas

As mentioned before an important aspect of Szemerédi's regularity lemma is the fact that there are corresponding embedding and counting lemmas. To state an embedding lemma for sparse graphs, we first need to adapt the concept of a "blown-up" graph, that is, we need a class of graphs corresponding to $\mathcal{P}(H, n, d, \varepsilon)$.

Definition 3.9 For a graph H, let $\mathcal{G}(H, n, m, \varepsilon)$ be the class of graphs on vertex set $V = \bigcup_{x \in V(H)} V_x$, where the sets V_x are pairwise disjoint sets of vertices of size n, and edge set $E = \bigcup_{\{x,y\} \in E(H)} E_{xy}$, where E_{xy} is the edge set of an (ε)-regular graph with m edges between V_x and V_y.

The straight-forward generalisation of Lemma 2.8 would state that a copy of H is contained in any graph of $\mathcal{G}(H, n, m, \varepsilon)$ if ε is sufficiently small. Unfortunately this is not true as the following example of Łuczak shows; see [25] where Łuczak is quoted.

Example 3.10 Let $\ell > 3$ be integral. Consider ℓ sets of vertices of size n, and assume that each possible edge between two vertices of different sets is present with $p = \tilde{C}/n$ for an appropriate constant $\tilde{C}$. For sufficiently large n one can use the FKG inequality by Fortuin, Kasteleyn and Ginibre [7] (or see [19]), to show that the probability that this graph does not contain a complete graph on l vertices is at least $1/2e^{-\tilde{C}^l}$. Moreover, it can be shown as in Example 2.3 that a.a.s. the bipartite graph between any two partition classes is (ε)-regular and has a density d satisfying $(1-\varepsilon)p \leq d \leq (1+\varepsilon)p$. We shall see later (cf. Lemma 4.3) that it follows that there exists a graph G in $\mathcal{G}(K_\ell, n, (1-\varepsilon)\tilde{C}n, 2\varepsilon)$ that does not contain a copy of K_ℓ. Now, one can replace each vertex of G by N vertices and each edge of G by a complete bipartite graph. One can check that this graph is again (ε)-regular with density $d = (1-\varepsilon)C/n$ and clearly this graph does not contain a copy of K_ℓ. Note that d depends on n and ε but not on N. Therefore we can find for each $\varepsilon > 0$, a constant $0 < d < 1$ such that there exists an K_ℓ-free graph in $\mathcal{G}(K_\ell, n, dn, \varepsilon)$ for sufficiently large n.

As we have just seen, we cannot hope that all graphs in $\mathcal{G}(H, n, m, \varepsilon)$ contain a copy of H, but one can hope that only a tiny fraction of graphs in $\mathcal{G}(H, n, m, \varepsilon)$ contain no copy of H at least if m is sufficiently large and ε is sufficiently small. This is what the next conjecture states which was first stated by Kohayakawa, Łuczak and Rödl in [22].

Conjecture 3.11 ([22]) *Let H be a fixed graph, and let*

$$\mathcal{F}(H, n, m, \varepsilon) = \{G \in \mathcal{G}(H, n, m, \varepsilon) : H \textit{ is not a subgraph of } G\}.$$

For any $\beta > 0$, there exist constants $\varepsilon_0 > 0$, $C > 0$, $n_0 > 0$ such that

$$|\mathcal{F}(H, n, m, \varepsilon)| \leq \beta^m \binom{n^2}{m}^{|E(H)|}$$

for all $m \geq Cn^{2-1/d_2(H)}$, $n \geq n_0$, and $0 < \varepsilon \leq \varepsilon_0$. Here

$$d_2(H) = \max\left\{\frac{|E(F)|-1}{|V(F)|-2} \mid F \subseteq H, |V(F)| \geq 3\right\}.$$

We shall present proofs of Conjecture 3.11 for some special graphs H in Section 6. Note that Conjecture 3.11 essentially resembles Lemma 2.8 for the sparse case. Having Conjecture 3.11 at hand, we then could also easily derive a version which resembles Lemma 2.9. This brings up the question whether perhaps also some stronger version of Conjecture 3.11 could be true that would generalise Lemma 2.10 to the sparse case. As far as we know such a conjecture has not yet been formulated precisely. A rough calculation yields that one can hope that for $m \geq Cn^{2-1/d_2(H)}$ "almost all" graphs in $\mathcal{G}(H,n,m,\varepsilon)$ are such that they contain $\Theta(n^{|V(H)|} \cdot (m/n^2)^{|E(H)|})$ copies of H, but we restrain ourselves from making such a speculation precise.

The statement of Conjecture 3.11 is not true if $m \ll n^{2-1/d_2(H)}$. Also, it is essential to consider graphs in $\mathcal{G}(H,n,m;\varepsilon)$ and not the class of graphs that has a bipartite (not necessarily (ε)-regular) graph with m edges between any two partition classes that represent adjacent vertices of H. It is for example easy to show that if $H = K_3$ the number of such graphs that do not contain a triangle is bigger than $c^m \binom{n^2}{m}^3$ for a constant $c \geq 1/(4e^2)$ which is obviously not arbitrarily close to nought.

Since Conjecture 3.11 gives information on almost all graphs in $\mathcal{G}(H,n,m,\varepsilon)$ one can use it to derive results on almost all graphs. We will show in the next section how to use Conjecture 3.11 to prove a result on the maximal number of edges a subgraph of a random graph can have while avoiding a fixed graph H, but first let us state interesting implications of the conjecture that were proved by Łuczak [31].

Theorem 3.12 ([31]) *Let H be a bipartite graph for which Conjecture 3.11 holds. Then for every $\alpha > 0$ there exists $c = c(\alpha, H)$ and n_0 such that for $n \geq n_0$ and $m \geq cn^{2-1/d_2(H)}$ the number of H-free labelled graphs on n vertices and m edges is bounded from above by $\alpha^m \binom{\binom{n}{2}}{m}$.*

Theorem 3.13 ([31]) *Let H be a graph with $\chi(H) = h \geq 3$ for which Conjecture 3.11 holds. Then for every $\delta > 0$ there exists $c = c(\delta, H)$ such that the probability that a graph chosen uniformly at random from the family of all H-free labelled graphs on n vertices and $m \geq cn^{2-1/d_2(H)}$ edges can be made $(h-1)$-partite by removing δm edges tends to one as n tends to infinity.*

Theorem 3.14 ([31]) *Let H be a graph with $\chi(H) = h \geq 3$ for which Conjecture 3.11 holds. Then for every $\varepsilon > 0$, there exist $c = c(\varepsilon, H)$ and $n_0 = n_0(\varepsilon, H)$ such that for $n \geq n_0$ and $cn^{2-1/d_2(H)} \leq m \leq n^2/c$, a graph $G_{n,m}$ drawn uniformly at random from all labelled graphs on n vertices and m edges satisfies*

$$\left(\frac{h-2}{h-1} - \varepsilon\right)^m \leq \mathbb{P}\left(G(n,m) \text{ does not contain } H\right) \leq \left(\frac{h-2}{h-1} + \varepsilon\right)^m .$$

3.3 An application of the sparse regularity lemma to random extremal graph theory

In this section we are concerned with the maximal number $\mathrm{ex}(G,H)$ of edges a subgraph of G may have without containing the graph H. We shall consider the case when $G = G_{n,p}$, that is, the graph on n vertices where each edge is chosen with probability p independent of the presence or absence of all other edges. The question about $\mathrm{ex}(G_{n,p},H)$ has attracted considerable attention during the last decades (see [19] and the references therein), and has been formulated as the following conjecture.

Conjecture 3.15 ([22]) *Let H be a non-empty graph with $|V(H)| \geq 3$, and let $0 < p = p(n) < 1$ be such that $pn^{1/d_2(H)} \to \infty$ as $n \to \infty$. Then a.a.s.*

$$\operatorname{ex}(G_{n,p}, H) \leq \left(1 - \frac{1}{\chi(H) - 1} + o(1)\right) |E(G_{n,p})|.$$

It is known [19] that if Conjecture 3.15 is true for $G_{n,p}$ then it is also true for $G_{n,p'}$ with $p' \geq p$. Moreover, one cannot replace $pn^{1/d_2(H)} \to \infty$ by $pn^{1/d_2(H)} \geq C$ for a constant C.

Conjecture 3.15 has been verified for various special cases in a series of papers. It is not hard to see that the conjecture is true when H is a tree, and it is also known to hold when H is a cycle of arbitrary length [8, 10, 17, 18, 21], $H = K_4$[22], $H = K_5$ [15] and $H = K_6$ [11]. If denser random graphs are considered, where p is about the square root of the conjectured value, then the result is also known to be true for any complete graph H [26, 33], and more generally, for so-called d-generate graphs [26].

We shall show that Conjecture 3.11 implies Conjecture 3.15, a fact that was first noted without proof when Conjecture 3.11 was stated in [22].

Theorem 3.16 *If Conjecture 3.11 is true then Conjecture 3.15 is true.*

Proof Note that it is sufficient to show that for any $\alpha > 0$ there exists a $C = C(\alpha) > 0$ such that for all $p \geq Cn^{-1/d_2(H)}$ we have that the random graph $G_{n,p}$ a.a.s. satisfies

$$\operatorname{ex}(G_{n,p}, H) \leq \left(1 - \frac{1}{\chi(H) - 1} + 20\alpha\right) |E(G_{n,p})|.$$

So in the following let $0 < \alpha \leq \frac{1}{100}$ be arbitrary, but fixed.

First we collect some properties of random graphs. Using Chernoff's inequality it is not hard to verify that a.a.s. $G_{n,p}$ has more than $(1-\alpha)\frac{1}{2}n^2p$ edges and is $(\mu, (1+\alpha), p)$-uniform, for all fixed $\mu > 0$.

Next we show that there exists an $\varepsilon_\alpha > 0$ such that for all $\mu > 0$ there exists a $C_\mu > 0$ such that for all $p \geq C_\mu n^{-1/d_2(H)}$ we have that $G_{n,p}$ a.a.s. does not contain a graph from $\mathcal{F}(H, \tilde{n}, \alpha\tilde{n}^2 p, \varepsilon_\alpha)$[1] for all $\tilde{n} \geq \mu n$. We will prove this claim by computing the expected number $\mathbb{E}[X]$ of such copies. By Markov's inequality it suffices to show that this expectation is $o(1)$. Clearly,

$$\mathbb{E}[X] \leq \sum_{\tilde{n} \geq \mu n} n^{|V(H)|\tilde{n}} |\mathcal{F}(H, \tilde{n}, \alpha\tilde{n}^2 p, \varepsilon_\alpha)| p^{|E(H)| \cdot \alpha\tilde{n}^2 p}.$$

In order to give a bound on this value we apply Conjecture 3.11 for $\beta := (\alpha/e^2)^{|E(H)|}$. Let $n_\alpha = n_0(\beta, H)$, $\varepsilon_\alpha = \varepsilon_0(\beta, H)$ and $C_\alpha = C(\beta, H)$ denote the constants guaranteed by Conjecture 3.11. Clearly, for n sufficiently large we have that $\mu n \geq n_\alpha$. Furthermore for $C_\mu = C_\alpha/(\alpha\mu^2)$ we have that

$$\alpha\tilde{n}^2 p \geq \alpha(\mu n)^2 p \geq \alpha\mu^2 C_\mu n^{2-1/d_2(H)} \geq C_\alpha n^{2-1/d_2(H)}.$$

[1] To be precise we should write $\lfloor \alpha\tilde{n}^2 p \rfloor$ instead of just $\alpha\tilde{n}^2 p$ since the number of edges is integral. We will omit floors and ceilings in the remainder of this paper for easier exposition.

Hence, we can apply the bound from Conjecture 3.11 for $|\mathcal{F}(H,\tilde{n},\alpha\tilde{n}^2p,\varepsilon_\alpha)|$ for all $\tilde{n} \geq \mu n$ and therefore get that for all n sufficiently large

$$\mathbb{E}[X] \;\leq\; \sum_{\tilde{n}\geq\mu n} n^{|V(H)|\tilde{n}} \left(\frac{\alpha}{e^2}\right)^{|E(H)|\cdot\alpha\tilde{n}^2p} \binom{\tilde{n}^2}{\alpha\tilde{n}^2p}^{|E(H)|} p^{|E(H)|\cdot\alpha\tilde{n}^2p}.$$

Using that $\binom{n}{k} \leq (en/k)^k$ and that $d_2(H) > 1$, this implies

$$\begin{aligned}\mathbb{E}[X] \;&\leq\; \sum_{\tilde{n}\geq\mu n} n^{|V(H)|\tilde{n}} \left(\frac{\alpha}{e^2}\cdot\frac{e\tilde{n}^2}{\alpha\tilde{n}^2p}\cdot p\right)^{|E(H)|\cdot\alpha\tilde{n}^2p}\\ &\leq\; n\cdot n^{|V(H)|n}\cdot e^{-|E(H)|\alpha\mu^2n^2p}\\ &\leq\; n^{\theta(n)}\cdot e^{-\theta(n^{2-1/d_2(H)})} = o(1),\end{aligned}$$

as desired.

Let us summarise what we have shown so far. There exists a constant ε_α such that for any fixed $\mu > 0$ there exists a constants C_μ such that for $p \geq C_\mu n^{-1/d_2(H)}$ the random graph $G_{n,p}$ a.a.s has more than $(1-\alpha)\frac{1}{2}n^2p$ edges, is $(\mu,(1+\alpha),p)$-uniform, and contains no graph from $\mathcal{F}(H,\tilde{n},\alpha\tilde{n}^2p,\varepsilon_\alpha)$, where $\tilde{n} \geq \mu n$. Now we can proceed in a similar way as in the proof of Theorem 2.11. That is, we will apply Theorem 3.4 for $\varepsilon := \min\{\varepsilon_\alpha/2,\alpha\}$ and $m_0 \geq 1/\varepsilon$ such that

$$\left(1-\frac{1}{\chi(H)-1}+\alpha\right)\cdot\frac{k^2}{2} \geq \mathrm{ex}_k(H) \qquad \text{for all } k \geq m_0. \tag{3.16}$$

(Note that Theorem 2.11 guarantees that such an m_0 exists.) Theorem 3.4 defines constants $\eta = \eta(\varepsilon, 1+\alpha, m_0)$ and a constant $M_0 = M_0(\varepsilon, m_0)$.

Let $\mu = \min\{\eta, (1-\varepsilon)/(2M_0)\}$ and consider a graph J with at least $(1-\alpha)\frac{1}{2}n^2p$ edges that is $(\mu,(1+\alpha),p)$-upper-uniform (and hence in particular $(\eta,(1+\alpha),p)$-upper-uniform), and that does not contain a graph from $\mathcal{F}(H,\tilde{n},\alpha\tilde{n}^2p,\varepsilon_\alpha)$ for any $\tilde{n} \geq \mu n$. (Recall that a.a.s. $G_{n,p}$ satisfies all these properties.) Consider an arbitrary subgraph G of J with at least

$$\left(1-\frac{1}{\chi(H)-1}+6\alpha\right)|E(J)| \geq \left(1-\frac{1}{\chi(H)-1}+6\alpha\right)\cdot(1-\alpha)\frac{1}{2}n^2p \tag{3.17}$$

edges. We will show that G contains a copy of H, which then completes the proof of the theorem.

We proceed similarly as in the proof of Theorem 2.11. Namely, we first apply Theorem 3.4 to G to obtain an (ε)-regular partition $(C_i)_{i=0}^k$ with $m_0 \leq k \leq M_0$ and then show that $C_1,\ldots,C_k$ contains a graph $G' \in \mathcal{G}(H,|C_1|,\alpha|C_1|^2p,\varepsilon_\alpha)$. As $|C_1| \geq (1-\varepsilon)n/M_0 \geq \mu n$ the fact that J does not contain a graph from $\mathcal{F}(H,\tilde{n},\alpha\tilde{n}^2p,\varepsilon_\alpha)$ for $\tilde{n} \geq \mu n$ implies that G' does contain the desired copy of H.

So let us see how we can find G'. As in the proof of Theorem 2.11 let R be the graph consisting of k vertices corresponding to the partition classes $C_1,\ldots,C_k$, and an edge between two vertices whenever the corresponding partition classes form an (ε,p)-regular graph with at least $\alpha|C_i|^2p$ edges.

In order to show that R contains a K_k we first determine an upper bound on the number of edges of G that are not involved in edges of R. There are at most

- $\varepsilon(1+\alpha)n^2p$ edges incident to vertices in C_0. (As we have observed already $|C_i| \geq \mu n$ for all $1 \leq i \leq k$. Since J is $(\mu, (1+\alpha), p)$-upper-uniform, it follows that for all $1 \leq i \leq k$, and all sets $S \subseteq V \setminus C_i$ of size $\varepsilon n \geq \mu n$ containing C_0, there are at most $(1+\alpha)p|S||C_i| \leq (1+\alpha)p\varepsilon n^2/k$ edges between S and C_i and thus between C_0 and C_i.)

- $\varepsilon(1+\alpha)n^2p$ edges between partition classes that do not form an (ε, p)-regular graph (as there are at most εk^2 such pairs and by the upper-uniformity there are at most $(1+\alpha)\left(\frac{n}{k}\right)^2 p$ edges contained in each such pair).

- $\frac{1}{2}\alpha n^2p$ edges between partition classes that form a graph with less than $\alpha\left(\frac{n}{k}\right)^2 p$ edges (as each such pair contains at most $\alpha\left(\frac{n}{k}\right)^2 p$ edges and there are at most $\binom{k}{2}$ such pairs).

- $\varepsilon(1+\alpha)n^2p$ edges that lie inside partition classes (as the graph is $(\eta, (1+\alpha), p)$-upper-uniform and thus there are less than $(1+\alpha)\left(\frac{n}{k}\right)^2 p$ edges inside each of the k partition classes, and we have assumed that $k \geq m_0 \geq \varepsilon^{-1}$).

As each edge of R can represent at most $(1+\alpha)(n/k)^2p \leq (n/k)^2p/(1-\alpha)$ edges of G (as G is $(\eta, 1+\alpha, p)$-upper-uniform), we get

$$\begin{aligned}
|E(R)| &\geq \frac{|E(G)| - 3(1+\alpha)\varepsilon n^2p - \frac{1}{2}\alpha n^2 p}{(n/k)^2p/(1-\alpha)} \\
&\overset{(3.17),\, \varepsilon\leq\alpha}{\geq} \frac{(1 - \frac{1}{\chi(H)-1} + 20\alpha)(1-\alpha)\frac{1}{2}n^2p - 7\alpha n^2 p}{(n/k)^2p/(1-\alpha)} \\
&\geq \left(1 - \frac{1}{\chi(H)-1} + 20\alpha\right)(1-2\alpha)\frac{k^2}{2} - 7\alpha k^2 \\
&\geq \left(1 - \frac{1}{\chi(H)-1} + \alpha\right)\frac{k^2}{2}.
\end{aligned}$$

Hence the definition of m_0 in (3.16) implies that R contains a copy of H. Recall that the definition of R implies that each edge of this H corresponds to an (ε, p)-regular pair (C_i, C_j) with at least $\alpha|C_1|^2p$ edges. It remains to find an (ε_α, p)-regular graph of density αp in each of the (ε, p)-regular graphs of density at least αp. That this is possible we shall show in Lemma 4.3 (recall that $\varepsilon \leq \varepsilon_\alpha/2$) and therefore we find the desired graph $G' \in \mathcal{G}(H, |C_1|, \alpha|C_1|^2p, \varepsilon_\alpha)$. □

Note that the proof of the previous theorem implies something a bit stronger than claimed: If Conjecture 3.11 is true for $m \geq f(n)$ for some function $f(n)$, then Conjecture 3.15 is true for $p \gg f(n)/n^2$. Moreover, in order to prove Conjecture 3.15 it would be sufficient to prove Conjecture 3.11 with some further conditions on the graph class $\mathcal{F}(H, n, m, \varepsilon)$ as long as these properties are a.a.s. satisfied for $G_{n,p}$ for the appropriate values of p. In [25, 24] properties of such (ε, p)-regular graphs with additional assumptions were studied and used to prove some special cases of Conjecture 3.15.

4 Properties of regular graphs

In this section we want to present some properties of (ε)-regular graphs. We are mainly interested in graphs that are induced by small subsets of the original (ε)-regular graph. It will turn out that most sets together with their neighbourhood induce a graph that has the structure that one expects. In the case of very small sets (and in particular vertices) this means that the size of the neighbourhood is as expected, and in case of larger sets it means that the edges are distributed as one expects and in particular contain a large (ε')-regular graph.

4.1 Vertices

Even though the definition of (ε)-regular graphs concerns only sets of linear size it is well-known that it is possible to derive bounds on the degree of nearly all vertices as we will see in the next lemma. In fact we have already seen in Proposition 2.2 that in the case of (dense) ε-regular graphs we can find a bound on the degree of most vertices.

Lemma 4.1 *Let $G = (V_1 \cup V_2, E)$ be an (ε)-regular graph with density d, and let $V_2' \subseteq V_2$ satisfy $|V_2'| \geq \varepsilon|V_2|$. Then at most $\lfloor \varepsilon|V_1| \rfloor$ vertices $v \in V_1$ do not satisfy*

$$|\Gamma(v) \cap V_2'| \geq (1-\varepsilon)d|V_2'|, \tag{4.1}$$

and at most $\lfloor \varepsilon|V_1| \rfloor$ vertices $u \in V_1$ do not satisfy

$$|\Gamma(u) \cap V_2'| \leq (1+\varepsilon)d|V_2'|.$$

Proof Let V_1' consist of all vertices v for which equation (4.1) does not hold. Assume that $|V_1'| \geq \lfloor \varepsilon|V_1| \rfloor + 1 > \varepsilon|V_1|$, then

$$|E(V_1', V_2')| < (1-\varepsilon)d|V_1'||V_2'|$$

which contradicts the assumption that G is (ε)-regular. The second statement follows analogously. □

4.2 Large subgraphs of regular graphs

In this subsection we want to show that large subgraphs of an (ε)-regular graph are again (ε')-regular for appropriate ε'. The first result concerns induced graphs of large vertex sets and resembles Proposition 2.4. The second result deals with spanning subgraphs and generalises Example 2.3.

Proposition 4.2 *Let $G = (V_1 \cup V_2, E)$ be an (ε)-regular graph of density d. Then V_2 together with any subset V_1' of V_1 of size at least $\alpha|V_1|$ with $\alpha > \varepsilon > 0$ induces an (α')-regular graph with $\alpha' = \max\{\varepsilon/\alpha, 2\varepsilon/(1-\varepsilon)\}$ and density $d' \geq (1-\varepsilon)d$.*

Proof Since $\alpha \geq \varepsilon$, it follows from the definition of (ε)-regularity that

$$d' = \frac{|E(V_1', V_2)|}{|V_1'||V_2|} \geq (1-\varepsilon)d.$$

Let $W \subseteq V_1'$ and $Z \subseteq V_2$ be such that $|W| \geq \alpha'|V_1'|$ and $|Z| \geq \alpha'|V_2|$. It follows that $|W| \geq \varepsilon|V_1|$ and $|Z| \geq \varepsilon|V_2|$ and hence

$$\begin{aligned}\left|\frac{|E(W,Z)|}{|W||Z|} - \frac{|E(V_1',V_2)|}{|V_1'||V_2|}\right| &\leq \left|\frac{|E(W,Z)|}{|W||Z|} - d\right| + \left|d - \frac{|E(V_1',V_2)|}{|V_1'||V_2|}\right| \\ &\leq 2\varepsilon d \leq \frac{2\varepsilon}{1-\varepsilon}\frac{|E(V_1',V_2)|}{|V_1'||V_2|} \leq \alpha' d'.\end{aligned}$$

□

The next lemma concerns spanning subgraphs of (ε)-regular graphs.

Lemma 4.3 *For all $0 < \varepsilon \leq 1/6$, there exists a constant $C = C(\varepsilon)$ such that any (ε)-regular graph $B = (V_1 \cup V_2, E)$ contains a (2ε)-regular subgraph $B = (V_1 \cup V_2, E')$ with $|E'| = m$ edges for all m satisfying $C|V(B)| \leq m \leq |E(B)|$.*

Proof We choose a set $E' \subseteq E$ of size m uniformly at random. Let d be the density of B, that is $d = |E|/(|V_1||V_2|)$ and let $d' = |E'|/(|V_1||V_2|)$ be the density of $B' = (V_1 \cup V_2, E')$. Let $V_1' \subseteq V_1$ and $V_2' \subseteq V_2$ be such that $|V_1'| \geq 2\varepsilon|V_1|$ and $|V_2'| \geq 2\varepsilon|V_2|$. We want to apply Hoeffding's inequality to show that with very high probability our random set E' satisfies the (2ε)-regularity condition with respect to V_1' and V_2', that is, E' intersects $E(V_1', V_2')$ in at least $(1-2\varepsilon)d'|V_1'||V_2'|$ and at most $(1+2\varepsilon)d'|V_1'||V_2'|$ edges.

Hoeffding's inequality states that if there is a set of size N and a subset E of size n, then a random variable X that counts $|E \cap E'|$ for a random set E' of m elements satisfies for all $0 < \lambda \leq 3/2$,

$$\mathbb{P}(|X - mn/N| \geq \lambda mn/N) \leq 2e^{-\frac{\lambda^2}{3}\frac{mn}{N}};$$

see for example Theorem 2.10 of [19]. We want to consider $N = |E| = |E(V_1, V_2)|$, $n = |E(V_1', V_2')|$ and $m = |E'|$. First, we want to derive bounds on mn/N.

Since B is (ε)-regular, the number of edges $|E(V_1', V_2')|$ in B between V_1' and V_2' satisfies

$$(1-\varepsilon)|V_1'||V_2'|d \leq |E(V_1', V_2')| \leq (1+\varepsilon)|V_1'||V_2'|d.$$

Hence

$$\frac{mn}{N} \geq (1-\varepsilon)|V_1'||V_2'|d\frac{|E'|}{|E|} = (1-\varepsilon)|E'|\frac{|V_1'||V_2'|}{|V_1||V_2|} \geq (1-\varepsilon)4\varepsilon^2|E'|$$

and

$$\frac{mn}{N} \leq (1+\varepsilon)|V_1'||V_2'|d\frac{|E'|}{|E|} = (1+\varepsilon)d'|V_1'||V_2'|.$$

By the above, it follows that

$$\begin{aligned}&\mathbb{P}\left(\big|\,|E'(V_1', V_2')| - d'|V_1'||V_2'|\,\big| \geq (\lambda(1+\varepsilon)+\varepsilon)d'|V_1'||V_2'|\right) \\ &\leq \mathbb{P}\left(\left|\,|E'(V_1', V_2')| - \frac{mn}{N}\right| \geq \lambda\frac{mn}{N}\right) \leq 2e^{-(1-\varepsilon)4\varepsilon^2|E'|\frac{\lambda^2}{3}}.\end{aligned}$$

With $\lambda = \varepsilon/2$,

$$\left|\frac{|E'(V_1', V_2')|}{|V_1'||V_2'|} - d'\right| \geq 2\varepsilon d',$$

implies

$$\big|\,|E'(V_1', V_2')| - d'|V_1'||V_2'|\,\big| \geq (\lambda(1+\varepsilon)+\varepsilon)d'|V_1'||V_2'|.$$

Thus the probability that the condition for (2ε)-regularity is not satisfied for two fixed sets V_1' and V_2' with $|V_1'| \geq 2\varepsilon|V_1|$ and $|V_2'| \geq 2\varepsilon|V_2|$ is at most $2e^{-(1-\varepsilon)\varepsilon^4|E'|/3}$.

Since there are at most $2^{|V_1|+|V_2|}$ choices for V_1' and V_2', we obtain

$$\mathbb{P}(G' \text{ not } (2\varepsilon)\text{-regular}) \leq 2^{|V_1|+|V_2|}2e^{-(1-\varepsilon)\varepsilon^4|E'|/3} < 1$$

if $|E'| > (3\ln 2)(|V_1|+|V_2|+1)/((1-\varepsilon)\varepsilon^4)$. Hence in this case there exists a subset of E of size m' that forms an (2ε)-regular graph. □

Observe that the previous proof implies that the probability that m edges chosen uniformly at random form a (2ε)-regular graph tends to one as $|V(B)|$ tends to infinity. It also implies that for $\alpha > 0$, one can obtain from an (ε)-regular graph B a $((1+\alpha)\varepsilon)$-regular spanning subgraph B' with m' edges, where $C(\alpha,\varepsilon)|V(B)| \leq m' \leq |E(B)|$.

4.3 Small subgraphs of regular graphs

As we have seen in Proposition 4.1, most vertices of an (ε)-regular graph $B = (V_1 \cup V_2, E)$ with $|V_1| = |V_2| = n$ and $|E| = m$ satisfy that their degree is at least $(1-\varepsilon)m/n$. If $m/n \ll n$ and we consider pairs of vertices then we expect that most pairs of vertices satisfy that the union of their neighbourhoods is only a little bit less than $2m/n$. More generally if we consider a set S with $|S|m/n \ll n$, then the neighbourhoods of S should have size about $|S|m/n$. Lemma 4.4 states that in fact most small sets S of size at most about n^2/m have the property that their neighbourhood is roughly m/n times their size.

Lemma 4.4 ([12]) *For all $\beta, \nu > 0$, there exists $\varepsilon_0 = \varepsilon_0(\beta,\nu) > 0$ such that for all $\varepsilon \leq \varepsilon_0$, every (ε)-regular graph $G(V_1 \cup V_2, E)$ of density d satisfies that the number of sets C of size $c \geq \tilde{c}$ with $\tilde{c} \leq \nu/(3d)$ that satisfy $|\Gamma(C)| \geq (1-\nu)\tilde{c}d|V_2|$ is at least*

$$(1-\beta^c)\binom{|V_1|}{c}.$$

The proof of this lemma (and of the following lemmas in this subsection) is deferred to the next subsection. When considering slightly bigger sets, that is, sets S of size a little bit bigger than n^2/m it cannot be true that the size of the neighbourhood of S is about $|S|m/n$, since otherwise the neighbourhood would be bigger than n. Instead we expect that a set $S \subseteq V_1$ of size Dn^2/m is such that most vertices of V_2 have about D neighbours in S. To make this precise we need one definition before we can state the lemma.

Definition 4.5 Let $G = (V_1 \cup V_2, E)$ be a bipartite graph. A set $C \subseteq V_1$ is called a *(ν, D)-cover* of V_2 if at least $(1-\nu)|V_2|$ vertices of V_2 have degree at least $(1-\nu)D$ into C.

Lemma 4.6 ([12]) *For all $\beta, \nu > 0$, there exist $D = D(\nu)$ and $\varepsilon_0 = \varepsilon_0(\nu, \beta) > 0$ such that, for all $0 < \varepsilon \le \varepsilon_0$, and all (ε)-regular graphs $G = (V_1 \cup V_2, E)$ of density d, the number of sets $C \subseteq V_1$ of size $c = \lceil D/d \rceil$ that are (ν, cd)-covers of V_2 is at least*

$$(1-\beta^c)\binom{|V_1|}{c}.$$

Most sets of size bigger than n^2/m are such that not only most vertices have the correct degree, but also that the edges are distributed regularly, that is, we can show that these sets contain a large regular subgraph with about the expected density.

Theorem 4.7 ([12]) *For all $0 < \beta, \varepsilon' < 1$, there exist $\varepsilon_0 = \varepsilon_0(\beta, \varepsilon') > 0$ and $C = C(\varepsilon')$ such that, for all $0 < \varepsilon \le \varepsilon_0$, all (ε)-regular graphs $G = (V_1 \cup V_2, E)$ of density d, and all $q \ge Cd^{-1}$, there are at least*

$$(1-\beta^q)\binom{|V_1|}{q}$$

sets $Q \subset V_1$ of size q whose subset $\{v \in Q : (1-\varepsilon)d|V_2| \le |\Gamma(v)| \le (1+\varepsilon)d|V_2|\}$ is of size $(1-\varepsilon')q$ and with V_2 induces an (ε')-regular graph.

It is easily verified that for sufficiently small ε, if one adds to the vertex set V_1 of an (ε)-regular graph $G = (V_1 \cup V_2, E)$ of density d less than $\varepsilon/(1-\varepsilon)|V_1|$ isolated vertices to obtain V_1', then the graph G' induced by V_1' and V_2 is $(\sqrt{\varepsilon})$-regular and of density at least $(1-\varepsilon)d$. Thus together with Lemma 4.3 we obtain the following corollary of Theorem 4.7.

Corollary 4.8 *For $0 < \beta, \varepsilon' < 1$, there exist $\varepsilon_0 = \varepsilon_0(\beta, \varepsilon') > 0$ and $C = C(\varepsilon')$ such that, for all $0 < \varepsilon \le \varepsilon_0$, every (ε)-regular graph $G = (V_1 \cup V_2, E)$ of density d satisfies that the number of sets $Q \subset V_1$ of size $q = |Q| \ge Cd^{-1}$ that together with V_2 induce a graph that contains a spanning (ε')-regular graph of density $(1-\varepsilon')d$ is at least*

$$(1-\beta^q)\binom{|V_1|}{q}.$$

Of course we can apply Theorem 4.7 repeatedly if we have more than one (ε)-regular graph. Recall, that $\mathcal{G}(H, n, m, \varepsilon)$ consists of graphs with $|V(H)|$ disjoint vertex sets of size n and there is an (ε)-regular graph between two such vertex sets whenever the corresponding vertices are adjacent in H.

Corollary 4.9 *Let H be a fixed graph. For all $\varepsilon', \beta > 0$, there exist an $\varepsilon_0 = \varepsilon_0(H, \beta, \varepsilon') > 0$ and $C = C(H, \varepsilon')$ such that, for all $0 < \varepsilon \le \varepsilon_0$, every graph in $\mathcal{G}(H, n, m, \varepsilon)$ satisfies the following: For all $q \ge Cn^2/m$, there are at least $(1-\beta^q)\binom{n}{q}^{|V(H)|}$ $|V(H)|$-tuples $Q_i \subseteq V_i$ of size q that contain a graph from $\mathcal{G}(H, q, m', \varepsilon')$ with $m' = \lfloor (1-\varepsilon')q^2m/n^2 \rfloor$.*

Let us remark that it would be desirable if in Theorem 4.7 the entire set Q induced an (ε')-regular and does not only contain a spanning subgraph that is regular. This is not true in general since one can show [12] that there are (ε)-regular graphs $G = (V_1 \cup V_2, E)$ with $|V_1| = |V_2| = n$ such that the probability that a subset Q of V_1 of sufficient size contains a subset Q' of, say, half the size with $|E(Q', V_2)| \leq |E(Q, V_2)| / \log n$ is at least $(1/(2e))^{|Q|}$. This shows that we have to take subsets in Theorem 4.7 since the set Q' does not satisfy the regularity constraint for any ε' if n is sufficiently large.

4.4 Proofs of Lemma 4.4, Lemma 4.6, and Theorem 4.7

In this subsection we give the missing proofs of the results of the last subsection. For easier reading we have repeated the statements.

Lemma 4.4 *For all $\beta, \nu > 0$, there exists $\varepsilon_0 = \varepsilon_0(\beta, \nu) > 0$ such that for all $\varepsilon \leq \varepsilon_0$, every (ε)-regular graph $G(V_1 \cup V_2, E)$ of density d satisfies that the number of sets C of size $c \geq \tilde{c}$ with $\tilde{c} \leq \nu/(3d)$ that satisfy $|\Gamma(C)| \geq (1-\nu)\tilde{c}d|V_2|$ is at least*

$$(1 - \beta^c)\binom{|V_1|}{c}.$$

Proof Let C be a set with $|\Gamma(C)| < (1-\nu)\tilde{c}d|V_2|$. Consider a subset $C' \subseteq C$ of maximal cardinality that satisfies

$$|\Gamma(C')| \geq \left(1 - \frac{\nu}{2}\right)|C'|d|V_2|.$$

Clearly $|C'| \leq (1 - \nu/2)\tilde{c}$ since otherwise

$$|\Gamma(C)| \geq |\Gamma(C')| \geq \left(1 - \frac{\nu}{2}\right)\left(1 - \frac{\nu}{2}\right)\tilde{c}d|V_2| \geq (1-\nu)\tilde{c}d|V_2|.$$

By the choice of C and since $\tilde{c} \leq \nu/(3d)$, we have

$$|\Gamma(C')| \leq |\Gamma(C)| < (1-\nu)\frac{\nu}{3d}d|V_2| \leq \frac{\nu}{3}|V_2|$$

and therefore

$$|V_2 \setminus \Gamma(C')| \geq \left(1 - \frac{\nu}{3}\right)|V_2|.$$

Let $\varepsilon_0 < \nu/6$. By the maximality of C' all vertices $v \in C \setminus C'$ must satisfy

$$|\Gamma(v) \setminus \Gamma(C')| \leq \left(1 - \frac{\nu}{2}\right)d|V_2| \leq (1-\varepsilon)\left(1 - \frac{\nu}{3}\right)d|V_2| \leq (1-\varepsilon)d|V_2 \setminus \Gamma(C')|,$$

but since $|V_2 \setminus \Gamma(C')| \geq \varepsilon|V_2|$, there are at most $\lfloor \varepsilon|V_1| \rfloor$ such vertices v in V_1 by Proposition 4.1. Hence there are at most

$$\begin{aligned}\sum_{c' \leq (1-\nu/2)\tilde{c}} \binom{|V_1|}{c'}\binom{\lfloor \varepsilon|V_1| \rfloor}{c-c'} &\overset{(4.2)}{\leq} \sum_{c' \leq (1-\nu/2)\tilde{c}} \binom{|V_1|}{c'}\varepsilon^{c-c'}\binom{|V_1|}{c-c'} \\ &\overset{(4.3)}{\leq} \left(1 - \frac{\nu}{2}\right)c \cdot \varepsilon^{\frac{\nu c}{2}} 4^c \binom{|V_1|}{c} \leq \beta^c \binom{|V_1|}{c}\end{aligned}$$

sets of size $c \geq \tilde{c}$ with $|\Gamma(C)| < (1-\nu)\tilde{c}d|V_2|$ for sufficiently small ε. In the last calculation we used that for $0 \leq x \leq 1$,

$$\binom{xa}{b} \leq x^b \binom{a}{b} \tag{4.2}$$

and for all a, b, c with $a \geq c > b$,

$$\binom{a}{b}\binom{a}{c-b} \leq 4^c \binom{a}{c}. \tag{4.3}$$

□

Lemma 4.6 *For all $\beta, \nu > 0$, there exist $D = D(\nu)$ and $\varepsilon_0 = \varepsilon_0(\nu, \beta) > 0$ such that, for all $0 < \varepsilon \leq \varepsilon_0$, and all (ε)-regular graphs $G = (V_1 \cup V_2, E)$ of density d, the number of sets $C \subseteq V_1$ of size $c = \lceil D/d \rceil$ that are (ν, cd)-covers of V_2 is at least*

$$(1 - \beta^c)\binom{|V_1|}{c}.$$

Proof Without loss of generality we may assume that $\nu \leq 1/3$. Let ε_0 and D satisfy $\varepsilon_0 \leq \nu^3/2$, $D \geq 40/\nu^4$, $4D\varepsilon_0 \leq \nu$ and (4.4) below.

Let $G = (V_1 \cup V_2, E)$ be an (ε)-regular graph for some $0 < \varepsilon \leq \varepsilon_0$. If $c \geq |V_1|/2$ then by Proposition 4.1 at least $(1-\varepsilon)|V_2| \geq (1-\nu)|V_2|$ vertices $v \in V_2$ satisfy

$$(1-\nu)cd \leq (1-\varepsilon)cd \leq |\Gamma(v) \cap C|.$$

So assume $c < |V_1|/2$. We generate a set C of size $c = \lceil D/d \rceil$ by randomly picking its elements one at a time. At each step t we call some vertices *useful*. If a useful vertex is picked, we call it *good* and declare some of its edges *relevant*. For $t = 1, 2, 3, \ldots, c$ and $i = 0, \ldots, D$, let $G(t,i) \subseteq V_2$ be the set of all vertices that are incident to i relevant edges after t vertices have been selected, and let $g(t,i) = |G(t,i)|$. At time $t+1$ all vertices in V_1 are useful that have at least $(1-\varepsilon)dg(t,i)$ neighbours into each $G(t,i)$ with $g(t,i) \geq \varepsilon|V_2|$. If a useful vertex is selected, we arbitrarily choose $(1-\varepsilon)dg(t,i)$ of its edges into $G(t,i)$ for each $G(t,i)$ with $g(t,i) \geq \varepsilon|V_2|$ and declare them relevant. Observe that at the beginning all vertices of V_1 with degree at least $(1-\varepsilon)d|V_2|$ are useful.

At any time t, we only consider degrees into $D+1$ sets to determine the set of useful vertices. By Proposition 4.1 for any $G(t,i)$ with $g(t,i) \geq \varepsilon|V_2|$, there are at most $\varepsilon|V_1|$ vertices in V_1 that do not have at least $(1-\varepsilon)dg(t,i)$ neighbours in $G(t,i)$. Thus at each time step there are at most $(D+1)\varepsilon|V_1|$ vertices that are not useful. Since we select the vertices from a set of size at least $|V_1| - t \geq |V_1|/2$, the probability that C contains at least $\nu c/2$ vertices that are not good is at most

$$\binom{c}{\lceil \nu c/2 \rceil}\left(\frac{(D+1)\varepsilon|V_1|}{|V_1| - c}\right)^{\lceil \frac{\nu c}{2} \rceil} \leq 2^c\left((D+1)2\varepsilon_0\right)^{\lceil \frac{\nu}{2} c \rceil} \leq \beta^c. \tag{4.4}$$

It remains to show that if we select $\lfloor (1-\nu/2)c \rfloor$ good vertices, these vertices form a (ν, D)-cover. This then implies that an (unordered) set only fails to be a (ν, D)-cover if all $c!$ orders contain less than $\lfloor (1-\nu/2)c \rfloor$ good vertices. Since there are at most $\beta^c n!/(n-c)!$ such orders there are at most $\beta^c \binom{n}{c}$ sets that are not (ν, D)-covers.

Since selected vertices that are not good do not affect the sets $G(t,i)$ for any t or i, we may ignore these time steps and assume that we have selected t good vertices at time t. For a good vertex $v \in C$, let $\tilde{\Gamma}(v) \subseteq \Gamma(v)$ denote the set of all vertices $u \in V_2$ that are connected to v by a relevant edge. Let $|V_2| = n$. Consider the sets $G(t,i)$ for $t = 0, 1, 2, \ldots, c$, $i = 0, \ldots, D$. At time $t \geq 1$ we add a good vertex v, so in particular if $g(t-1,i) \geq \varepsilon n$, then the vertex v satisfies

$$(1-\varepsilon)dg(t-1,i) = |\tilde{\Gamma}(v) \cap G(t-1,i)|,$$

and if $g(t-1,i) < \varepsilon n$, then v satisfies

$$0 = |\tilde{\Gamma}(v) \cap G(t-1,i)|.$$

Hence the following inequality is always satisfied:

$$dg(t-1,i) - \varepsilon dn \leq |\tilde{\Gamma}(v) \cap G(t-1,i)| \leq dg(t-1,i). \tag{4.5}$$

At step $t = 0$ there are no good vertices and hence all vertices in V_2 have 0 good neighbours, i.e., $G(0,0) = V_2$, $g(0,0) = n$. At step $t \geq 1$ the (good) vertex v is added, and we have

$$G(t,0) = G(t-1,0) \setminus \tilde{\Gamma}(v)$$

and hence by (4.5)

$$g(t,0) = g(t-1,0)(1-d) + f(t,0)$$

where $|f(t,0)| \leq \varepsilon dn$. Now consider $i = 1, \ldots, D$. If $t < i$, we clearly have $g(t,i) = 0$, so let $t \geq i$. Observe that when we add the good vertex v at time t, we have

$$G(t,i) = (G(t-1,i) \setminus \tilde{\Gamma}(v)) \cup (G(t-1,i-1) \cap \tilde{\Gamma}(v)).$$

Hence by (4.5),

$$g(t,i) = g(t-1,i-1)d + g(t-1,i)(1-d) + f(t,i)$$

where $|f(t,i)| \leq 2\varepsilon dn$. Summing up, we have to solve the following recursion:

$$\begin{aligned}
g(0,0) &= n \\
g(t,0) &= g(t-1,0)(1-d) + f(t,0) \\
g(t,i) &= 0 && t < i \\
g(t,i) &= g(t-1,i-1)d + g(t-1,i)(1-d) + f(t,i) && t \geq i \geq 1
\end{aligned}$$

where $f(t,i)$ satisfies $|f(t,i)| \leq 2\varepsilon dn$ for all $i = 0, \ldots, D$, $t \geq i$. It is not hard to check that for $t \geq i$, we have

$$\left| g(t,i) - n\binom{t}{i} d^i (1-d)^{t-i} \right| \leq 2t\varepsilon dn, \tag{4.6}$$

that is $g(t,i)$ has roughly binomial distribution.

Let $t_0 = \lfloor (1-\nu/2)c \rfloor$. Using Chernoff's inequality one can verify that that

$$\sum_{i=0}^{(1-\nu)cd} g(t_0,i) \leq \nu |V_2|$$

which implies that C is a (ν, cd)-cover since adding edges (like non-relevant ones, or edges incident to vertices that are not good) does not affect a (ν, cd)-cover. □

Theorem 4.7 *For all $0 < \beta, \varepsilon' < 1$, there exist $\varepsilon_0 = \varepsilon_0(\beta, \varepsilon') > 0$ and $C = C(\varepsilon')$ such that, for all $0 < \varepsilon \leq \varepsilon_0$, all (ε)-regular graphs $G = (V_1 \cup V_2, E)$ of density d, and all $q \geq Cd^{-1}$, there are at least*

$$(1 - \beta^q)\binom{|V_1|}{q}$$

sets $Q \subset V_1$ of size q whose subset $\{v \in Q : (1-\varepsilon)d|V_2| \leq |\Gamma(v)| \leq (1+\varepsilon)d|V_2|\}$ is of size $(1-\varepsilon')q$ and with V_2 induces an (ε')-regular graph.

Proof We first want to show that most sets of size $\lceil C/d \rceil$ have the required properties. By Lemma 4.1 all but at most $\varepsilon|V_1|$ vertices of V_1 have a degree between $(1-\varepsilon)d|V_2|$ and $(1+\varepsilon)d|V_2|$. Using this property it is not hard to verify that most sets Q of size q have at least $(1-\varepsilon')q$ vertices with a degree between $(1-\varepsilon)d|V_2|$ and $(1+\varepsilon)d|V_2|$. Furthermore, for sufficiently small ε, a bipartite graph $B = (Q \cup V_2, E)$ with density $\tilde{d}$, $(1-\varepsilon)d \leq \tilde{d} \leq (1+\varepsilon)d$ that satisfies for all sets $Q' \subset Q$ with $|Q'| \geq \varepsilon|Q|$ and $V_2' \subset V_2$ with $|V_2'| \geq \varepsilon|V_2|$, $|E(V_1', V_2')| \geq (1-\varepsilon)d|Q'||V_2'|$ is (ε')-regular. Hence it suffices to consider the lower bound on the number of edges.

Assume that we are given a (ν, C) cover W of V_2 of size $\lceil C/d \rceil$. For any set V_2' of size $\varepsilon'|V_2|$ there are at least $(\varepsilon' - \nu)|V_2|(1-\nu)C = (1-\nu/\varepsilon')(1-\nu)d|V_2'||W|$ edges between V_2' and W. Thus if ν is sufficiently small the lower bound on the edges of the regularity condition is satisfied for all $V_2' \subset V_2$ with $|V_2'| \geq \varepsilon'|V_2|$ and W. It remains to find a structure that allows us to take large subsets of W and not only the entire set W to ensure that there are many edges between the sets.

We say that a set $S \subseteq V_1$ is a (ν, C, c)-supercover of the set V_2 if every subset $S' \subseteq S$ of size $|S'| = c$ is a (ν, C)-cover. We first claim that for all $\beta, \nu > 0$ there exist $C = C(\nu)$ and $\varepsilon_0 = \varepsilon_0(\nu, \beta) > 0$ such that for any $0 < \varepsilon \leq \varepsilon_0$, every (ε)-regular graph $G = (V_1 \cup V_2, E)$ of density $d > 0$ satisfies that the number of (ν, C, c)-supercovers $S \subseteq V_1$ of V_2 of size $s \leq \nu^{-1}c$ with $c = \lceil C/d \rceil$ is at least

$$(1 - \beta^s)\binom{|V_1|}{s}.$$

Let $S \subset V_1$ be a set of size s that is not a (ν, C, c)-supercover of V_2. By definition it must contain a set of size c that is not a (ν, C)-cover. By Lemma 4.6 applied to ν and $\beta \to (\beta/4)^{1/\nu}$ there exist at most

$$\left(\frac{\beta}{4}\right)^{\frac{c}{\nu}}\binom{|V_1|}{c}$$

such sets for appropriate values of $C = C(\nu)$ and $\varepsilon_0(C, \nu, \beta)$. Hence the number of sets that are not (ν, C, c)-supercovers can be bounded from above by

$$\left(\frac{\beta}{4}\right)^{\frac{c}{\nu}}\binom{|V_1|}{c}\binom{|V_1|-c}{s-c} \overset{(4.2)(4.3)}{\leq} \left(\frac{\beta}{4}\right)^{s} 4^s \binom{|V_1|}{s}$$

and the claim follows.

As at the beginning of the proof it is easily verified that a (ν, C, c)-supercover of V_2 of size $\nu^{-1}c$ is (ε')-regular with V_2 if ν is sufficiently small (for example $\nu \leq \varepsilon'/12$

will certainly do). To prove that most larger sets are (ε')-regular it suffices to show that most such sets consist of many supercovers and have the correct density, for details see [12]. □

5 Properties of almost all (ε)-regular graphs

In this section we want to show how to use the results of the previous section to prove properties of almost all graphs in $\mathcal{G}(H, n, m, \varepsilon)$. We will concentrate on the case when H is the complete graph K_ℓ on ℓ vertices. Consider a graph in $G \in \mathcal{G}(K_\ell, n, m, \varepsilon)$ with vertex set $V_1 \cup \ldots \cup V_l$. By Lemma 4.1 most vertices of V_1 see about m/n vertices in all the other partition classes. On the other hand by Corollary 4.9 there are only very few subgraphs of this size in G that are not in $\mathcal{G}(K_{\ell-1}, \approx m/n, \approx m^3/n^4, \varepsilon')$. Thus one expects that most vertices of $v \in V_1$ of G contain a graph in $\mathcal{G}(K_{\ell-1}, \approx m/n, \approx m^3/n^4, \varepsilon')$ in their neighbourhood. Our first result states that this is the case for almost all graphs in $\mathcal{G}(K_\ell, m, n, \varepsilon)$.

The second result is a bit more general. Instead of just requiring that a graph induced by the neighbourhood of most vertices of V_1 is regular and has approximately the expected number of edges, we want to be able to deduce that it also has the properties of a "typical" such regular structure. By "typical", we mean that only a tiny fraction of all structures does not satisfy the property.

Lemma 5.1 *Let $\ell \geq 3$ be any integer. Then there exist for all $\alpha > 0$, $\beta > 0$ and $\varepsilon' > 0$, constants $\varepsilon_0 = \varepsilon_0(\alpha, \beta, \varepsilon) > 0$ and $C = C(\varepsilon') > 0$ such that for all $0 < \varepsilon \leq \varepsilon_0$ and for all sufficiently large n and $m \geq Cn^{3/2}$ all but at most $\beta^m \binom{n^2}{m}^{\binom{\ell}{2}}$ graphs in $\mathcal{G}(K_\ell, n, m, \varepsilon)$ satisfy that all but αn vertices v in V_1 have neighbourhoods that contain a graph in $\mathcal{G}(K_{\ell-1}, \tilde{n}, \tilde{m}, \varepsilon')$ with $\tilde{n} = \lfloor (1-\varepsilon)m/n \rfloor$ and $\tilde{m} = \lfloor (1-\varepsilon')\tilde{n}^2 m/n^2 \rfloor$.*

Proof Let $\hat{\beta}$ satisfy $32^l \hat{\beta}^{((1-\varepsilon')\alpha)/2} \leq \beta$. We apply Corollary 4.9 with $H = K_{\ell-1}$, $\beta \to \hat{\beta}$ and $\varepsilon' \to \varepsilon'/2$ to obtain constants ε_{cor} and C_{cor}. We prove the lemma for $\varepsilon_0 = \min\{\varepsilon_{\text{cor}}, \alpha/(2\ell), \varepsilon'/2\}$ and $C = \max\{C_{\text{cor}}/(1-\varepsilon), 1\}$.

We want to construct all graphs that do not satisfy the conditions of the lemma and show that there are only few of them. Firstly, we select all the edges between V_i, V_j for all $2 \leq i < j \leq \ell$. There are at most $\binom{n^2}{m}^{\binom{\ell-1}{2}}$ possibilities. Secondly, we select the degrees for all the vertices of V_1 into V_j for $j \geq 2$. There are at most $n^{\ell n} \leq 2^m$ possibilities for sufficiently large n. By Proposition 4.1 and since we are constructing an (ε)-regular graph between V_1 and V_j we have to choose for at least $(1 - \varepsilon\ell)n$ vertices of V_1, degrees into all the sets V_j that are at least $(1-\varepsilon)m/n$. Now we choose a set of at least αn vertices that have neighbourhoods that do not contain a graph in $\mathcal{G}(K_{\ell-1}, \tilde{n}, \tilde{m}, \varepsilon')$. There are at most $2^n \leq 2^m$ possibilities to choose these vertices. We denote by A the set of all such vertices that have a degree at least $\tilde{n} = \lfloor (1-\varepsilon)m/n \rfloor$. Note that $|A| \geq (\alpha - \ell\varepsilon)n \geq (\alpha/2)n$. Now we select the neighbourhoods for the vertices in $V_1 \setminus A$. There are at most $\binom{n}{d_j(v)}$ possibilities for each vertex v to choose its neighbourhood in V_j where $d_j(v)$ is the chosen size of the neighbourhood of v in V_j. For the vertices of A, we first choose a set of size $\tilde{n}$ in each partition class $V_2, \ldots, V_\ell$. We require that these sets induce a graph that does not contain a graph in $\mathcal{G}(K_{\ell-1}, \tilde{n}, \tilde{m}, \varepsilon')$. Since

$$(1-\varepsilon)\frac{m}{n} \geq C_{\mathrm{cor}}\frac{n^2}{m},$$

it follows from Corollary 4.9 that there are at most $\hat{\beta}^{\tilde{n}}\binom{n}{\tilde{n}}^{l-1}$ ways to choose such sets. Now we have to choose the remaining neighbours for every vertex $v \in A$. There are at most $\prod_{j=2}^{\ell}\binom{n-\tilde{n}}{d_j(v)-\tilde{n}}$ ways to do this. There are at most

$$\prod_{v\in A}\left(\hat{\beta}^{\tilde{n}}\binom{n}{\tilde{n}}^{\ell-1}\prod_{j=2}^{\ell}\binom{n-\tilde{n}}{d_j(v)-\tilde{n}}\right) \overset{(4.3)}{\leq} \hat{\beta}^{\tilde{n}\frac{\alpha}{2}n}\left(\prod_{v\in A}\prod_{j=2}^{\ell}4^{d_j(v)}\binom{n}{d_j(v)}\right)$$

ways to select the neighbourhoods of vertices in A, and thus since $d_j(v) \leq (1+\varepsilon)m/n \leq 2m/n$ for all $v \in A$, there are at most

$$\hat{\beta}^{\tilde{n}\frac{\alpha}{2}n}4^{2\ell m}\left(\prod_{v\in V_1}\prod_{j=2}^{\ell}\binom{n}{d_j(v)}\right) \leq \left(\frac{\beta}{4}\right)^m\binom{n^2}{m}^{\ell-1}$$

ways to select the neighbourhoods of the vertices in V_1. Summarising, there are at most

$$4^m\left(\frac{\beta}{4}\right)^m\binom{n^2}{m}^{\ell-1}\binom{n^2}{m}^{\binom{\ell-1}{2}} \leq \beta^m\binom{n^2}{m}^{\binom{\ell}{2}}$$

ways to construct the graphs that do not satisfy the conditions of the lemma. □

To state the second result of this section, we need some more notation. We denote by $\mathcal{G}(K_\ell, n, m)$ the set of graphs that consists of ℓ disjoint vertex sets of size n, and there is a bipartite graph with m edges between each pair of vertex sets. We say that a family $\mathcal{B}(K_\ell, n, m) \subseteq \mathcal{G}(K_\ell, n, m)$ is small with respect to a function $m_0(n)$ if there exist for all $\beta > 0$ a constant $n_\beta \in \mathbb{N}$, a $C_\beta > 0$, and an $\varepsilon_\beta > 0$ such that for all $n \geq n_\beta$, $m \geq C_\beta m_0(n)$,

$$|\mathcal{B}(K_\ell, n, m) \cap \mathcal{G}(K_\ell, n, m, \varepsilon_\beta)| \leq \beta^m\binom{n^2}{m}^{\binom{\ell}{2}} \tag{5.1}$$

Note that the Conjecture 3.11 states that the K_ℓ-free graphs in $\mathcal{G}(K_\ell, n, m)$ form a small family with respect to $m = n^{2-1/d_2(K_\ell)}$.

We can now state the main result of this section formally.

Lemma 5.2 ([13]) *Let $m_0(x) \geq x$ be a monotone increasing function and let $\mathcal{B}(K_{\ell-1}, x, y)$ be small with respect to m_0. For all $\beta, \varepsilon' > 0$, there exist constants $\varepsilon_0 = \varepsilon_0(\beta, \varepsilon') > 0$ and $C = C(\beta, \varepsilon') > 0$ such that for all*

$$0 < \varepsilon \leq \varepsilon_0, \quad x = (1-\varepsilon')\frac{m}{n}, \quad m \gg n^{3/2}\sqrt{\log n},$$

$$Cy_0(x) \leq \frac{m^3}{n^4}, \quad \varepsilon'\frac{m^3}{n^4} \leq y \leq (1-\varepsilon')^3\frac{m^3}{n^4},$$

and n sufficiently large, all but at most $\beta^m \binom{n^2}{m}^{\binom{\ell}{2}}$ *graphs* $G \in \mathcal{G}(\ell, n, m, \varepsilon)$ *satisfy the following property: there at least* $(1-\varepsilon')n$ *vertices in* V_1 *that contain a member of the family*

$$\mathcal{G}(\ell-1, x, y, \varepsilon') \setminus \mathcal{B}(K_{\ell-1}, x, y)$$

in their neighborhood.

6 Proof of the embedding lemma (Conjecture 3.11) for some special cases

In this section we present proofs of Conjecture 3.11 and thus Conjecture 3.15 for some graphs H. The smallest graph for which the conjectures are interesting are triangles, that is, $H = K_3$. But in this case Conjecture 3.11 follows immediately from Lemma 5.1, since any vertex in V_1 that has neighbourhoods in V_2 and V_3 that form an (ε)-regular graph of positive density is involved in lots of triangles. Thus we have shown the following theorem that was already shown in [31] using different methods.

Theorem 6.1 ([31]) *Conjecture 3.11 is true for* $H = K_3$.

The next theorem states that a slightly weaker statement than Conjecture 3.11 is true for the complete graph K_ℓ on ℓ vertices when we are only interested in graphs in $\mathcal{G}(K_\ell, n, m, \varepsilon)$ when m is slightly larger than stated in the conjecture.

Theorem 6.2 *For any* $\beta, \ell > 0$, *there exist constants* $\varepsilon_0 > 0$, $C > 0$, $n_0 > 0$ *such that the set* $\mathcal{F}(K_\ell, n, m, \varepsilon) = \{G \in \mathcal{G}(K_\ell, n, m, \varepsilon) : K_\ell \text{ is not a subgraph of } G\}$ *satisfies*

$$|\mathcal{F}(K_\ell, n, m, \varepsilon)| \leq \beta^m \binom{n^2}{m}^{\binom{\ell}{2}}$$

for all $m \geq Cn^{2-1/(\ell-1)}$, $n \geq n_0$, *and* $0 < \varepsilon \leq \varepsilon_0$.

Proof We prove the theorem by induction on ℓ. For $\ell = 3$, the theorem is true by Theorem 6.1. So assume the theorem is true for $\ell - 1 \geq 3$. Note that this means that the class $\mathcal{B}(n, m) \subseteq \mathcal{G}(K_{\ell-1}, n, m)$ of $K_{\ell-1}$-free graphs is small with respect to $m_0 = n^{2-1/(\ell-2)}$. Now, we can apply Lemma 5.2 with $\beta \to \beta$ and $\varepsilon' \to 1/10$ to obtain constants C_0 and ε_0. Let $C = \max\{C_0, 1\}$, and let $m \geq Cn^{2-1/(\ell-1)}$. We have to show that $C_0 m_0((9/10)m/n) \leq m^3/n^4$, since then for all $\varepsilon \leq \varepsilon_0$ and n sufficiently large, all but at most

$$\beta^m \binom{n^2}{m}^{\binom{\ell}{2}}$$

graphs in $\mathcal{G}(K_\ell, m, n, \varepsilon)$ are such that they have more than $n/10$ vertices in V_1 that contain a graph in

$$\mathcal{G}\left(K_{\ell-1}, (1-\varepsilon')m/n, m^3(10n^4), \varepsilon'\right) \setminus \mathcal{B}\left((1-\varepsilon')m/n, m^3/(10n^4)\right)$$

in their neighbourhood. Clearly, each of these vertices are part of a K_ℓ. Now,

$$\begin{aligned} C_0 m_0((9/10)m/n) &= C_0 \left(\frac{9}{10}\frac{m}{n}\right)^{2-\frac{1}{\ell-2}} \\ &\leq C_0 \frac{m^3}{n^4} \frac{n^{2+\frac{1}{\ell-2}}}{m^{1+\frac{1}{\ell-2}}} \\ &\leq C_0 \frac{m^3}{n^4} \frac{n^{2+\frac{1}{\ell-2}}}{C_0(n^{2-\frac{1}{\ell-1}})^{1+\frac{1}{\ell-2}}} = \frac{m^3}{n^4} \end{aligned}$$

and the result follows. □

Let us remark that Conjecture 3.15 is also known to be true, if H is a cycle [12], if $H = K_4$ [14] and if $H = K_5$ [15]. If the last result is used for the induction hypothesis in the previous theorem, it follows that Conjecture 3.15 is true for the complete graph K_ℓ if $m \geq Cn^{2-1/(\ell-2)}$ for an appropriate constant C.

Acknowledgements

We would like to thank the referee for the prompt reply and the many helpful comments.

References

[1] N. Alon, R.A. Duke, H. Lefmann, V. Rödl and R. Yuster, The algorithmic aspects of the regularity lemma, *J. Algorithms* **16 (1)** (1994), 80–109.

[2] N. Alon and A. Assaf, Approximating the cut-norm via Grothendieck's inequality, in *Proceedings of the thirty-sixth annual ACM Symposium on Theory of Computing* (2004), pp. 72–80.

[3] B. Bollobás, *Modern Graph Theory*, Springer, New York (1998).

[4] R. Diestel, *Graph Theory*, Springer, New York (1997).

[5] P. Erdős and M. Simonovits, A limit theorem in graph theory, *Studia Sci. Math. Hungar.* **1** (1966), 51–57.

[6] P. Erdős and A.H. Stone, On the structure of linear graphs, *Bull. Amer. Math. Soc.* **52** (1946), 1087–1091.

[7] C.M. Fortuin, P.W. Kasteleyn and J. Ginibre, Correlation inequalities for some partially ordered sets, *Comm. Math. Phys.* **22** (1971), 89–103.

[8] P. Frankl and V. Rödl, Large triangle-free subgraphs in graphs without K_4, *Graphs Combin.* **2** (1986), 135–144.

[9] A. Frieze and R. Kannan, A simple algorithm for constructing Szemerédi's regularity partition, *Electr. J. Comb.* **6** (1999),

[10] Z. Füredi, Random Ramsey graphs for the four-cycle, *Discrete Math.* **126** (1994), 407–410.

[11] S. Gerke, Random graphs with constraints, Habilitationsschrift, TU München, 2004.

[12] S. Gerke, Y. Kohayakawa, V. Rödl and A. Steger, Small subsets inherit sparse ε-regularity, submitted, 2004.

[13] S. Gerke, M. Marciniszyn and A. Steger, A probabilistic counting lemma for complete graphs, submitted, 2005.

[14] S. Gerke, H.J. Prömel, T. Schickinger, A. Steger and A. Taraz, K_4-free subgraphs of random graphs revisited, submitted, 2002.

[15] S. Gerke, T. Schickinger, and A. Steger, K_5-free subgraphs of random graphs, *Random Structures Algorithms* **24 (2)** (2004), 194–232.

[16] W.T. Gowers, Lower bounds of tower type for Szemerédi's uniformity lemma, *Geom. Funct. Anal.* **7 (2)** (1997), 322–337.

[17] P.E. Haxell, Y. Kohayakawa and T. Łuczak, Turán's extremal problem in random graphs: forbidding even cycles, *J. Combin. Theory Ser. B* **64** (1995), 273–287.

[18] P.E. Haxell, Y. Kohayakawa and T. Łuczak, Turán's extremal problem in random graphs: forbidding odd cycles, *Combinatorica* **16(1)** (1996), 107–122.

[19] S. Janson, T. Łuczak and A. Rucinski, *Random Graphs*, John Wiley & Sons, New York (2000).

[20] Y. Kohayakawa, Szemerédi's regularity lemma for sparse graphs, in *Foundations of Computational Mathematics* (eds. F. Cucker and M. Shub), Springer-Verlag, Berlin, Heidelberg (1997), pp. 216–230.

[21] Y. Kohayakawa and B. Kreuter, Threshold functions for asymmetric Ramsey properties involving cycles, *Random Structures Algorithms* **11** (1997), 245–276.

[22] Y. Kohayakawa, T. Łuczak and V. Rödl, On K^4-free subgraphs of random graphs, *Combinatorica* **17(2)** (1997), 173–213.

[23] Y. Kohayakawa and V. Rödl, Algorithmic Aspects of Regularity, in *Proceedings of the 4th Latin American Symposium on Theoretical Informatics* (eds. G. Gonnet, D. Panario, and V. Viola), *Lecture Notes in Computer Science*, (2000), Springer-Verlag, Berlin pp. 1–17.

[24] Y. Kohayakawa and V. Rödl, Szemerédi's regularity lemma and quasi-randomness, in *Recent advances in algorithms and combinatorics* (eds. C. Linhares-Sales and B. Reed), *CMS Books Math./Ouvrages Math. SMC*, 11, (2003), Springer, New York pp. 289–351.

[25] Y. Kohayakawa and V. Rödl, Regular pairs in sparse random graphs (I), *Random Structures Algorithms* **22 (4)** (2003), 359–434.

[26] Y. Kohayakawa, V. Rödl and M. Schacht, The Turán theorem for random graphs, *Combin. Probab. Comput.* **13 (1)** (2004), 61–91.

[27] Y. Kohayakawa, V. Rödl and L. Thoma, An optimal algorithm for checking regularity, *SIAM J. Comput.* **32 (5)** (2003), 1210–1235 (electronic).

[28] J. Komlós, G. Sárközy and E. Szemerédi, Blow-up lemma, *Combinatorica* **17 (1)** (1997), 109–123.

[29] J. Komlós, A. Shokoufandeh, M. Simonovits and E. Szemerédi, The regularity lemma and its applications in graph theory, in *Theoretical Aspects of Computer science (Tehran, 2000) Lecture Notes in Comput. Sci.*, 2292, (2002), Springer, Berlin pp. 84–112.

[30] J. Komlós and M. Simonovits, Szeméredi's regularity lemma and its application in graph theory, in *Combinatorics, Paul Erdős is eighty, Vol. 2 (Keszthely, 1993)* (eds. D. Miklós, V.T. Sós and T. Szőnyi), *Bolyai Soc. Math. Stud.*, 2, János Bolyai Math. Soc., Budapest (1996), pp. 295–352.

[31] T. Łuczak, On triangle-free random graphs, *Random Structures Algorithms* **16** (2000), 260-276.

[32] Y. Peng, V. Rödl and A. Ruciński, Holes in graphs, *Electron. J. Combin.* **9 (1)** (2002), 1–18 (electronic).

[33] T. Szabó and V.H. Vu, Turán's Theorem in sparse random graphs, *Random Structures Algorithms* **23 (3)** (2003), 225–234.

[34] E. Szeméredi, Regular partitions of graphs, in *Problèmes combinatoires et théorie des graphes Colloques Internationaux CNRS*, Vol.260, (1978), pp. 399–401.

[35] P. Turán, Eine Extremalaufgabe aus der Graphentheorie, *Mat. Fiz. Lapok* **48** (1941), 436–452.

Stefanie Gerke, Angelika Steger
ETH Zürich
Institute of Theoretical
Computer Science
ETH Zentrum
Haldeneggsteig 4
CH-8092 Zürich
`sgerke@inf.ethz.ch`
`steger@inf.ethz.ch`